Geophysical Case Study of the Woodlawn Orebody, New South Wales, Australia

RELATED PERGAMON TITLES OF INTEREST

Books

GARLAND: The Earth's Shape and Gravity

GRIFFITHS & KING: Applied Geophysics for Geologists and Engineers, 2nd Edition

INSTITUTE OF APPLIED GEOLOGY: Evolution and Mineralization of the Arabian - Nubian Shield (in 4 volumes)

KELLER & FRISCHKNECHT: Electrical Methods in Geophysical Prospecting

MAURER: Novel Drilling Techniques

PATERSON: The Physics of Glaciers, 2nd Edition

RAMDOHR: The Ore Minerals and Their Intergrowths, 2nd Edition

RIDGE: Annotated Bibliographies of Mineral Deposits in Africa, Asia (Exclusive of the USSR) and Australasia

UDINTSEV: Geological/Geophysical Atlas of the Indian Ocean

Journals*

Computers and Geosciences
Geochimica et Cosmochimica Acta
Journal of Structural Geology
Organic Geochemistry

*Free specimen copy of any Pergamon journal available on request

Geophysical Case Study of the Woodlawn Orebody New South Wales, Australia

The first publication of methods and techniques tested over a base metal orebody of the type which yields the highest rate of return on mining investment with modest capital requirements

Editor
ROBERT J. WHITELEY
School of Applied Geology, The University of New South Wales, Australia

PERGAMON PRESS

OXFORD · NEW YORK · TORONTO · SYDNEY · PARIS · FRANKFURT

U.K.	Pergamon Press Ltd., Headington Hill Hall, Oxford OX3 0BW, England
U.S.A.	Pergamon Press Inc., Maxwell House, Fairview Park, Elmsford, New York 10523, U.S.A.
CANADA	Pergamon of Canada, Suite 104, 150 Consumers Road, Willowdale, Ontario M2J 1P9, Canada
AUSTRALIA	Pergamon Press (Aust.) Pty. Ltd., P.O. Box 544, Potts Point, N.S.W. 2011, Australia
FRANCE	Pergamon Press SARL, 24 rue des Ecoles, 75240 Paris, Cedex 05, France
FEDERAL REPUBLIC OF GERMANY	Pergamon Press GmbH, 6242 Kronberg-Taunus, Hammerweg 6, Federal Republic of Germany

First edition **1981**

British Library Cataloguing in Publication Data

Geophysical case study of the Woodlawn orebody, New South Wales, Australia.
1. Copper ores - Australia - Woodlawn
2. Prospecting - Australia - Woodlawn - Geophysical methods
I. Whiteley, Robert J
622'.18'43 TN271.C6 79-42637

ISBN 0-08-023996-X

In order to make this volume available as economically and as rapidly as possible the author's typescript has been reproduced in its original form. This method has its typographical limitations but it is hoped that they in no way distract the reader.

Printed in Great Britain by A. Wheaton & Co. Ltd., Exeter

EDITORIAL COMMITTEE

PARTICIPATING ORGANISATIONS

Bureau of Mineral Resources
Crone Geophysics Ltd
CSIRO (Mineral Research Laboratories)
Dighem Ltd, Toronto
Esso Australia Ltd
Geoex Pty Ltd
Geological Survey of New South Wales
Geoterrex Ltd
Jododex Aust Pty Ltd
Macquarie University
Newmont Pty Ltd
Phelps Dodge Exploration Corporation
St Joe Minerals Corporation
University of New South Wales
Woodlawn Mines

FOREWORD

Mineral exploration geophysics has advanced significantly over the last two decades with the incorporation of advanced electronic technology into practical exploration systems. This has led to the development of new geophysical methods, particularly electrical methods and the revitalization of old methods, particularly electromagnetic methods.

Ability to interpret geophysical data in geological terms has not kept pace with this rapid instrumental development even though computer modelling is commonplace. The major reason for the interpretative lag is the lack of freely available, comprehensive case studies over known, economic mineral deposits.

This geophysical case study of the Woodlawn orebody presents and discusses the detailed geophysical responses of an economic base metal deposit to a large number of geophysical techniques, including the latest airborne, ground and downhole methods. Results of computer and scale modelling, physical property measurements on drillcore, and detailed geology and soil geochemical surveys are also included together with a subordinate geophysical study of an uneconomic black shale, a common source of unwanted geophysical anomalies.

The Woodlawn orebody is located in a well populated, pastoral area of southeastern New South Wales, Australia. It consists of a shallow, massive sulphide deposit of complex shape with reserves of approximately 6.3 million tonnes plus a deeper, associated zone of copper mineralization with reserves of 3.7 million tonnes. The Woodlawn orebody is best geologically classified as a stratiform volcanogenic copper-lead-zinc deposit occurring within acid volcanic host rocks.

Woodlawn was discovered in late 1969 by Jododex Australia Proprietary Limited as a result of systematic application of geological, geochemical and geophysical methods. This discovery was of considerable importance since it was the first major find in the region for over seventy years and was twice the size of the largest known deposit. The Woodlawn discovery rejuvenated base metal exploration in southeastern Australia and sustained it through a period of low activity in the early to middle 1970's. Part of the reason for the importance of Woodlawn is that, under prevailing economic conditions, such deposits are highly prized since they are capable of providing high returns on mining capital at modest investment levels. They are, therefore, the object of considerable worldwide exploration effort.

The path from discovery of the Woodlawn deposit to the mining and processing of the ore has not been easy. Hampered by legal and environmental problems and by metallurgical problems due to the fine grained banded nature of the sulphide ore,it was 9 years from discovery to the commencement of mining in 1978. This extended delay had beneficial side effects, however, as it allowed this comprehensive geophysical case study to be completed over the period from 1974 until 1979 after mining had begun. Earlier surveys associated with discovery of the orebody and exploration of the surrounding region have also been incorporated in this case study.

Foreword

Although the Woodlawn orebody was not an initial geophysical discovery, it produces classical responses to and interpretative problems for many of the well established and recently developed geophysical exploration methods.

This volume reports the results of these geophysical studies and consists of 34 papers by geologists and geophysicists from exploration, government, research and contracting organizations. The papers are grouped into eleven chapters.

The geophysical response characteristics of the Woodlawn orebody are summarized in Chapter 1. Exploration history, geology, regional geophysics and mineralization are discussed in Chapters 2 and 3. The results of prediscovery induced polarization and magnetic surveys are reviewed in Chapter 4. Chapter 5 describes the physical property measurements on Woodlawn drillcore samples. Data from airborne electromagnetic surveys are reviewed in Chapter 6. Chapters 7 to 10 present results from extensive ground electromagnetic, electrical, potential field and seismic surveys completed over the Woodlawn orebody. The final chapter, Chapter 11, discusses downhole and inhole electrical, electromagnetic, gamma ray and applied potential surveys.

Individual papers in this case study vary somewhat in style and content reflecting the differing approaches of the various authors to the application of geophysical methods and the interpretation of geophysical data. Since this was felt to be a reflection of the real exploration situation, editing was kept to a minimum and aimed at reduction of redundant material and the inclusion of comprehensive cross references to other relevant papers in the volume.

The initial geophysical surveys were completed in Imperial units. However, SI units have been used throughout the case study with reference to Imperial units where appropriate.

Many dedicated people have contributed to this, the most complete geophysical case study of a single economic ore deposit ever undertaken for publication. The enlightened approach of the board of directors and management of the successful explorer, Jododex Australia Proprietary Limited, and of its parent companies, the St. Joe Minerals Corporation and the Phelps Dodge Corporation, cannot be praised highly enough. Their realization of the importance, to successful mineral exploration, of comprehensive case studies on economic targets and their open door approach made the Woodlawn geophysical case study possible. Hopefully this study will help convince others to embark on similar projects over a wide range of exploration targets. Only in this way can exploration science really progress.

Original listings of field data and base maps at an enlarged scale can be made available if sufficient interest is forthcoming. These should enable students and professional geoscientists to carry out their own studies in the reduction, modelling and detailed interpretation of the geophysical data. Further information can be obtained by contacting me directly.

Robert J. Whiteley

Editor

School of Applied Geology,

University of New South Wales,

P.O. Box 1,

Kensington, Sydney, N.S.W. 2033,

Australia.

3rd November 1979

PREFACE

Discovery of the Woodlawn deposit was important from both an economic and scientific viewpoint. Economically the Woodlawn deposit represented the first major new base metal deposit discovered in New South Wales for many years, although significant mineralization had been outlined in the Cobar Field in the 1950's associated with previously worked orebodies. It is, therefore, not surprising that the discovery gave great impetus to the search for new mineral deposits in this, the most populous State of Australia.

In the two to three years following the discovery hole, drilled in 1970, development drilling at Woodlawn outlined some ten million tonnes of both zinc-lead-copper and copper ore of mineable grade. At the same time geological investigations revealed that the deposit was of the stratiform massive sulphide type with features similar to a number of other volcanogenic deposits in the eastern part of the Lachlan Fold Belt e.g. Captains Flat, Lewis Ponds, Wisemans Creek and Mt. Bulga.

During this period the successful explorer, Jododex Australia Proprietary Limited, recognized that the Woodlawn discovery offered an unrivalled opportunity to complete detailed scientific studies on the deposit, not only with a view to better understanding its geological setting and mode of formation, but also to provide exploration criteria that could be used elsewhere in the Lachlan Fold Belt. From this, two separate projects arose: one dealing with the geology, mineralization and geochemistry of the deposit, part of which has recently been published by the Geological Society of Australia; the second, which is the subject of this case study volume, focussing on the detailed geophysical responses of the Woodlawn orebody. The most pleasing feature of these studies is that they represented co-operative research by industry, government and university.

Each organization has contributed its own specialist knowledge and expertise to produce a geophysical case study of the Woodlawn orebody which is undoubtedly the most complete geophysical study of an economic base metal deposit ever undertaken for publication. The variety of geophysical techniques which have been employed is impressive. Readers of this volume are brought into contact with both the well known and newer geophysical techniques and should be able to readily assess those which are most suited for use in the search for orebodies of the Woodlawn type.

Robert J. Whiteley, Editor, the editorial committee and authors of individual papers are to be congratulated on the contributions they have made during the preparation of this volume. The management of Jododex Australia deserves special recognition for foresight in encouraging a scientific study of the Woodlawn orebody and instilling a spirit of co-operation between the contributing organizations.

The efforts of all those who have been involved with this case study volume will be worthwhile if its publication leads to a better appreciation of modern geophysical techniques, their role in mineral exploration and ultimately to more effective search and discovery of further orebodies of the Woodlawn type.

N.L. Markham
Director
Geological Survey of New South Wales

September, 1979.

ACKNOWLEDGEMENTS

The following organizations have participated in the production of this case study volume and their assistance is gratefully acknowledged: Jododex Australia Proprietary Limited (Pty. Ltd.); The University of New South Wales; Geological Survey of New South Wales; Bureau of Mineral Resources; C.S.I.R.O., Division of Mineral Physics; Crone Geophysics Aust. Pty. Ltd.; Esso Minerals Aust. Pty. Ltd.; Geoex Pty. Ltd.; Geoterrex Pty. Ltd.; Macquarie University; Newmont Pty. Ltd.; Scintrex Pty. Ltd.

Many dedicated individuals, too many to acknowledge here, have contributed to this volume. In particular we are indebted to F. Olgers, T. Nicholas, F.G. Cucchi, W.J. McKay, J.G. Jung, R.S. Cook, L.F. Chaffer,W.J. Brydon, J.J. Ferguson, C.L. Phillips, M. Horvath, M. Clarke, G. Small, M. Vallentine, D. Wehr, B. O'Neill, J.P. Pik, R. Drinkrow, R. J. Searls, J. Haigh, E. Burnside, J. Chisholm, J. Ringis, D. Stewart, S. Collins, C. Smith

CONTENTS

CHAPTER 1

INTRODUCTION AND SUMMARY

1.1. Brief Description of the Woodlawn Orebody and Summary of Geophysical Responses

E. J. MALONE, R. J. WHITELEY, E. D. TYNE AND L. V. HAWKINS

Contents

The Woodlawn Deposit

The deposit is situated in southeastern New South Wales at Lat.35°04'S, Long.149°34'E. (Fig.1) The first phase of testing proved up a body of polymetallic massive sulphides and an adjacent zone of dominantly copper ore. The massive sulphides body consisted of about 6.3 million tonnes of copper, lead and zinc ore (1.7% Cu: 5.5% Pb; 14.4% Zn) and was the main target for geophysical investigation. The copper ore zone contains about 3.7 million tonnes averaging about 1.9% Cu, 0.1%Pb, 0.5% Zn in a number of ore shoots both along strike from and below the massive sulphides. Subsequent drilling indicated several concealed bodies of massive sulphides, separate from the first discovered body but these are at depths of more than 200m and hence too deep to have influenced the geophysical investigations.

The top of the massive sulphide body is close to the surface, being overlain in turn by a generally thin layer of supergene sulphides, then by a 5m to 15m thick ferruginous clay gossan zone, and finally by a few metres of colluvium and soil.

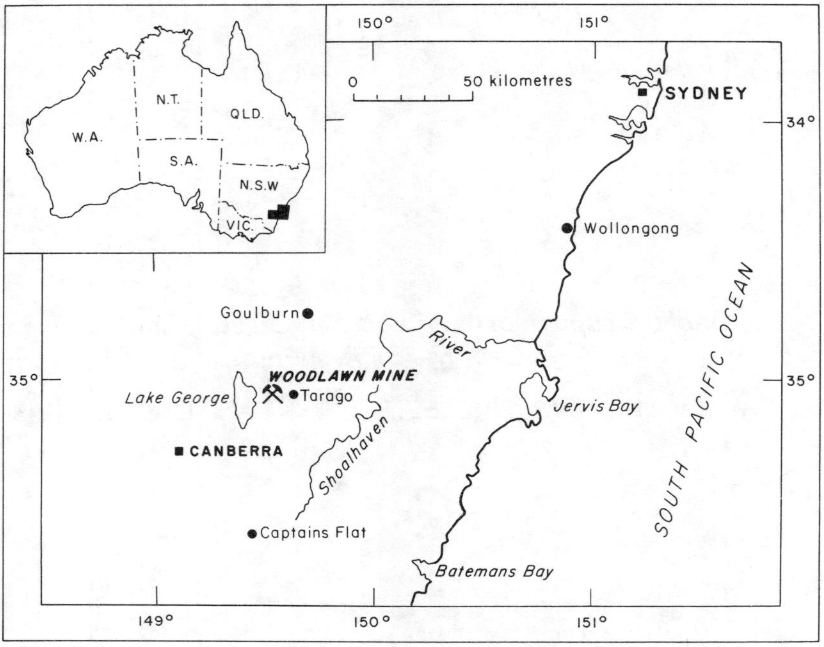

Figure 1. Locality map.

The massive sulphide body is geometrically complex. The southern part strikes 030°M and dips west at about 40°; the northern part strikes 305°M and is near vertical. In the hinge zone between the two parts the body is locally flat lying at depth. The body has an average thickness of about 25m and consists of about 75% sulphides, mainly pyrite, sphalerite, galena and chalcopyrite in decreasing order of abundance with quartz, volcanic detritus, chlorite and talc as the main gangue minerals. The most striking features of the massive sulphides are the fine grain size and the regular banding throughout most of the body. The near surface copper ore consists of veinlets and lenticles of pyrite and chalcopyrite in chloritic and siliceous volcanics.

The host rocks to the mineralization include felsic tuffs, fragmentals and flows, volcanic derived sediments, shale, chert and local chlorite rock. The sequence is intruded by tholeiitic metadolerite sills, both above and below the massive sulphides body.

The Woodlawn Test Grid

In the initial stages of exploration at Woodlawn, Jododex used an existing, near north to south rabbit-proof boundary fence as a base line. For inconvenience, the origin for the local grid was established at the northern end of the fence. This grid was used for initial grid geochemical sampling and for the prediscovery resistivity and induced polarization (IP) survey (Smith, this volume, p 79).

As exploration activity in the Woodlawn area increased, a base line in the middle of the prospect became inconvenient and the origin was shifted 10,000 feet grid north and 10,000 feet grid east. This placed the area of interest in the southwest quadrant. The fence, which was the original base line, was physically removed as it was a well grounded wire mesh fence which interfered considerably with electrical geophysical surveys.

The second grid was in use when the Woodlawn Test Grid was established. This grid was originally established by Jododex as Gradient Array Grid 2 after considerable geophysical work had outlined the area of maximum geophysical response. It was later decided to use this grid for the majority of the geophysical research carried out at Woodlawn. The Test Grid consists of 21 east to west lines, at intervals of 100ft (30.5m) apart and denoted by letters A to U from north to south. Stations on each line are also 100ft apart and are denoted by numbers 1 to 20 increasing from east to west. Relevant parts of the Woodlawn Test Grid are shown on many of the diagrams in this geophysical case study volume so that data from different techniques may be compared.

Subsequent to the establishment of the Woodlawn Test Grid, Australia adopted the metric system and a metric grid was established at Woodlawn. The origin for the metric grid was placed at 13,000m grid west and 13,000m grid south from the origin of the second Imperial System grid, and consequently the area of interest lies in the northeast quadrant of this grid.

Figure 2 shows the Woodlawn Test Grid, the gossan capping the massive sulphides and the location of the grounded fence used as the original baseline. The potential of the fence to interfere with the geophysical response of the orebody is obvious and this interference is illustrated in the paper on Early Geophysical Exploration of the Woodlawn Orebody (Smith, this volume, p 79).

Physical Properties of the Woodlawn Orebody

Table 1 summarizes the physical properties of the Woodlawn orebody which are of major significance in relation to the measured geophysical responses. Electrical resistivity, the IP parameter, percent frequency effect (PFE) density and seismic velocity were the most significant.

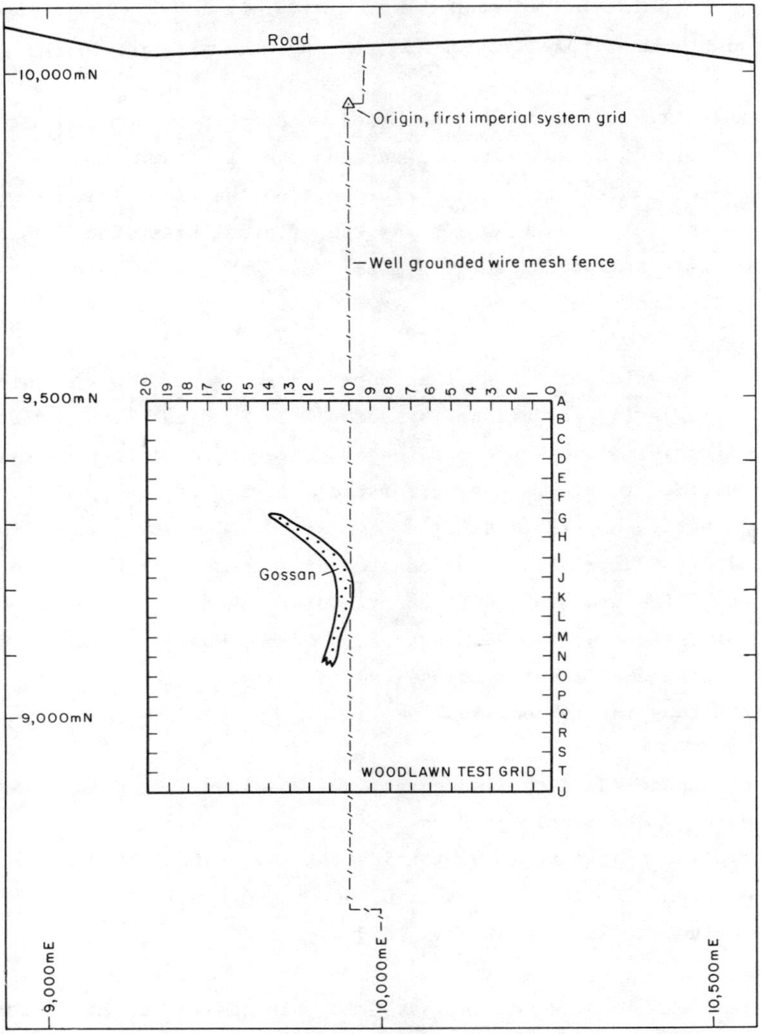

Figure 2. Woodlawn Test Grid Location

The largest contrast with the country rocks was in resistivity. This is typical of many massive sulphide orebodies and accounts for the widespread use of electromagnetic and electrical methods in exploration for such deposits. Meaningful PFE values for the massive sulphide body as a whole could not be determined from core samples or downhole measurements. Computer modelling of IP results indicated a value of around 5% for the massive orebody which was significantly greater than the PFE of some country rocks but was less than that of other country rocks.

Table 1.

Summary of Petrophysical Properties of the Woodlawn Massive Sulphide Orebody

Physical Property	Value	Contrast with Country Rocks
Resistivity	0.1 ohm m	1:1,000 - 1:10,000
PFE	∿ 5%	Variable
Density	3,900 kg/m^3	> 1,000 kg/m^3
Seismic velocity	∿ 1400 m/s	1:3
Magnetic susceptibility	0	Variable

Seismic compressional (P) wave velocities in the upper part of the massive sulphide body, obtained during a refraction survey, were much less than velocities in the host rocks.

The magnetic susceptibility of the orebody was practically zero, however, the contrast with the country rocks was variable as significant magnetic anomalies were obtained over some rock types in the Woodlawn region.

Character of the Geophysical Responses, Woodlawn Orebody

Table 2 is a subjective classification of the geophysical responses over the Woodlawn orebody.

The magnitude of response relative to background response for each particular method has been graded as weak, moderate or strong and intermediate between these levels. The airborne and ground magnetic methods gave no anomaly associated with the Woodlawn orebody.

The geophysical responses of the orebody were further classified as either simple, complex or intermediate. In the case of Woodlawn the sources of complexity were generally either associated with the proximity of other sources of response or with the complex geometry of the orebody itself.

Table 2.

Character of the geophysical responses of the Woodlawn orebody.

GEOPHYSICAL METHOD	CHARACTER OF GEOPHYSICAL RESPONSE				
	WEAK	MODERATE	STRONG	SIMPLE	COMPLEX
AIRBORNE					
Magnetics			No Anoma	ly	
HEM (Scintrex 701)	###########				###########
INPUT	######	######			###########
Dighem^{11}		#######	######	######	######
GROUND					
Electromagnetic (continuous wave)					
VLF Geonics (EM16)			###########	######	######
AFMAG (McPhar)		###########		######	######
Audio MT (Geotronics)			###########	######	######
Dip Angle (McPhar VHEM)					
: Broadside	###########				###########
: Fixed Transmitter			###########	######	######
Slingram (ABEM)			###########	######	######
Turam (ABEM)					
: Large Loop			###########	######	######
: Small Loop			###########	######	######
Electromagnetic (pulse transient)					
Crone PEM			###########	######	######
Newmont EMP			###########	######	######
MPPO 1			###########	######	######
SIROTEM			###########	######	######
Electrical					
SP			###########	###########	
Resistivity			###########	###########	
IP		###########			###########
MIP	######	######			###########
Potential Field					
Magnetics			No Anoma	ly	
Gravity		###########		###########	
Seismic Refraction			###########		###########
DOWNHOLE					
SP			###########	###########	
Gamma Ray		######	#####	######	######
EM (Scintrex DHP4)	##########				###########
Resistivity			###########	###########	
Applied Potential (Mise-à-la-Masse)			###########	###########	

Table 2 shows that, of the airborne electromagnetic (EM) methods, the latest Dighem^{II} helicopter system showed improved responses mainly due to the large number of coil geometries used and possibly because the measurements were made over the open pit. The Scintrex HEM701 and INPUT systems produced complex responses because of the complicated geometry and dip of the orebody. Irregular surrounding terrain (Fig.1), the short strike of the orebody and a large number of EM sources in the Woodlawn region also posed problems for these methods.

The wide variety of ground EM methods used at Woodlawn produced strong responses of intermediate complexity showing the Woodlawn orebody to be an excellent target for ground EM methods. This would be expected because of the high conductivity (Table 1) and shallowness of the massive orebody. Exceptions are the AFMAG method which suffered equipment problems and the broadside, dip-angle method which was used in a fairly limited survey, although the failure of the latter method was not unexpected under Australian conditions.

The large loop and newer pulse transient EM methods did not appear to show any particular advantages over, for example, the Slingram method. This was possibly due to the shallowness of the highly conductive orebody, so that large responses from the top part of the orebody were detected by all ground EM techniques. The sophisticated Newmont EMP and SIROTEM systems which are capable of measuring to long delay times would be expected to show advantages over the other techniques for somewhat deeper targets.

Of the electrical methods, the self potential (SP) and resistivity methods showed simple strong responses which closely outlined the shallow region of the massive orebody. The induced polarization (IP) and magnetic IP methods produced complex responses due to the proximity of polarizable, resistive sources in the footwall believed to be pyritic black shales. Computer modelling indicated the magnitude of the IP response of the orebody was of a moderate level and certainly no more than that of the polarizable footwall rocks.

The gravity method gave a simple response of moderate magnitude which defined the orebody accurately despite the presence of a strong regional gradient.

The seismic refraction method gave a strong response, with the upper part of the orebody appearing as a seismic low velocity zone. This response was complex due to the diffractions of seismic wave around the margins of the orebody.

Of the downhole methods, the SP, resistivity and applied potential methods gave strong, simple responses again reflecting the high conductivity and electrical continuity of the massive sulphide orebody.

Table 3.

Character of the geophysical responses of the Black Shale.

GEOPHYSICAL METHOD	CHARACTER OF GEOPHYSICAL RESPONSE				
	WEAK	MODERATE	STRONG	SIMPLE	COMPLEX
GROUND					
Electromagnetic (continuous wave)					
VLF (EMI6)			#########	#########	
Audio MT (Geotronics)			#########		#########
Dip angle: broadside (McPhar VHEM)		No	Anomaly		
Slingram (ABEM)	#########			#########	
Turam: Small Loop (ABEM)	#########			#########	
Electromagnetic (pulse transient)					
SIROTEM	#########			#########	
Electrical					
Self Potential			#########	#########	
Resistivity			#########	#########	
IP			#########	#########	#########
MIP		#########	#########	#########	
Potential Field					
Magnetics		No	Anomaly		
Gravity	#########			#########	
Seismic Refraction	#########			#########	

Natural gamma ray logging gave a moderate to strong response detecting low radioactivity associated with the massive sulphides. The response was complicated by features within the sulphides particularly near their upper margin.

The Scintrex DHP4 downhole to surface EM system gave weak and complex responses and was not considered to be very useful in this environment.

The Black Shale

Local zones of the several black shale units occuring near the Woodlawn orebody exhibit SP and other electrical responses similar to those of the massive sulphide body. One such black shale, located about 2.5km south of the Woodlawn deposit, was chosen for geophysical research in order to compare the results with those over the massive sulphide body. The black shale unit as a whole had a strike length of about 3km and an average width of about 40m and was generally interbedded with sandstone, shale and minor tuff. The part of the black shale unit, where the test work was done, was referred to as the Black Shale. It was found to be much more extensive at depth in the one drill hole than was apparent in outcrop, and was overlain by tuff, sandstone and shale intruded by small bodies of quartz feldspar phorphyry. Contact metamorphic effects associated with the intrusions may have contributed to the geophysical responses of the shale unit in this locality.

Geophysical Responses of the Black Shale

The geophysical responses of the Black Shale are classified in Table 3 in a similar manner to those of the Woodlawn orebody. This table shows that the electrical SP and resistivity methods and a number of EM methods produce strong responses. These reflect the high conductivity of this region of the Shale.

The responses produced by the Shale were generally simple because of the simpler geometry and larger width of the Shale relative to the Woodlawn orebody.

In order to distinguish the Black Shale, a common source of unwanted electrical and EM responses as can be seen in Table 3, a combination of the EM methods which produced weak simple responses (i.e. Slingram, Turam and SIROTEM) and gravity with possible seismic refraction would appear to be most appropriate.

CHAPTER 2

EXPLORATION HISTORY

2.1. Exploration History of the Woodlawn Orebody: Discovery to Development

D. F. O'DRISCOLL

Contents

Exploration History

Woodlawn was a geological/geochemical discovery, but from its earliest days geophysics have played a significant part in exploring and testing the orebody and its extensions.

In February, 1967, from information contained largely in published geological maps, the Goulburn-Tarago area was selected by Jododex geologists as prospective for base metal deposits of the Captains Flat type - i.e. polymetallic sulphide bodies in acid volcanic and associated rocks. Application was then made for an Exploration Licence to cover about 900 square miles (2,300 square kilometres) of the mapped Ordovician/Silurian belt extending northward from Captains Flat to Goulburn and beyond.

Jododex Australia Pty. Ltd. was a newly formed exploration company backed by two US principals, and staffed by experienced Australian geologists. In the summer of 1967/68, long before the exploration licence was granted, they carried out extensive orientation surveys in the Captains Flat area to establish the most appropriate geochemical methods and then began a systematic reconnaissance sampling programme in the selected area, using available road and track access to sample B or C soil horizons at sites adjacent to drainage channels.

During the early months of the programme, samples were collected from the roadside downstream of Woodlawn. One sample gave anomalously high results for Cu, Pb and Zn against generally low background values and above background values for Ag and Sn. When the geologist concerned, the late R.W. Haynes, returned to check the sample he noted the ferruginous outcrops half a mile to the south-west, and the several old shallow pits and sampled them. These chip samples also gave geochemically anomalous results for Cu, Pb, Zn, Ag and Sn, though the level of anomaly was much higher than for the soil sample.

By May, 1968, the reconnaissance sampling of the selected area was completed and only the Woodlawn area appeared to be sufficiently anomalous to justify further work. Grid geochemical sampling of the locality subsequently outlined a well defined and extensive geochemical anomaly. An induced polarization survey over the area in July, 1969, revealed an anomalous polarizable zone partly coincident with the geochemical anomaly, and drilling targets were selected.

Drilling Results and Metallurgy

A four hole drilling programme was commenced in December, 1969. Following an indecisive first hole, the second hole intersected 24.4m of massive sulphides of average grade 1.45% copper, 6.95% lead, 18.0% zinc and 72 gm/tonne silver. This established the presence of significant mineralization and the programme changed from one designed to locate mineralization to one designed to outline an orebody which was now sought with some degree of confidence. A total of 25 holes, of which 14 penetrated the orebody, were drilled in the next twelve months. They indicated a high grade orebody of about 5 million to 10 million tonnes size.

A secondary drilling programme of 85 holes, mainly on a 30m grid, was then undertaken from March, 1971 to May, 1972. This intensive drilling was necessary to define the limits of an irregular-shaped orebody and to obtain representative samples for metallurgical study. It was followed by the digging of a large trench some 14 metres deep, and by three shafts, two with cross cuts, to collect bulk samples for metallurgical testing. The ore was shown to be fine-grained, refractory and susceptible to oxidation and ultimately more than five years of intensive metallurgical research in laboratories in Australia and Canada, with supplementary studies in the

United States and Japan were undertaken to develop a satisfactory process for recovering the metals from the ore. This metallurgical difficulty delayed a planned development until mid 1975.

Delay while overcoming metallurgical problems is not unexpected in base metal development. However, the Woodlawn project also had to overcome a potentially more serious problem in a challenge to its title in the law courts.

Legal and Environmental Aspects

One of the partners of the syndicate which owned the Woodlawn grazing property, on becoming aware of Jododex's concentration of exploration effort in one part of the property, withdrew an earlier permission for Jododex to work there and then lodged applications for exploration titles in his own name over the whole of the area of interest. These applications clearly postdated the Jododex application. However, when they came up for hearing at Warden's Court, the applicant succeeded on several occasions in having the matter adjourned without decision. Thus he effectively prevented any start on the mine development programme because as long as his applications remained before the Court they created a bar which made it impossible for Jododex to be legally granted the necessary mining leases. In the meantime, Jododex had purchased the property from the syndicate, and with encouraging results coming from its metallurgical research, was ready to go to the feasibility study stage.

To resolve the stalemate, Jododex, acting on legal advice, sought and gained a declaratory injunction from the New South Wales Supreme Court requiring the Warden to reject the applications. Their opponent opposed the granting of this injunction and asserted that Jododex had been late in lodging an application for renewal of the licence. This argument was lost when it was proved that the renewal application had been lodged on the last day for lodgement under the procedures established and followed by the Mines Department.

The opponent then appealed to the High Court of Australia on the ground that the last day for lodgement of renewal of an Exploration Licence was not in law the last day as defined in the long-standing practice of the Mines Department, but the day before that day. It was argued that when counting back from a date of expiry, the counting should begin one day earlier thus setting the last day for application one day earlier also. This appeal was rejected by the High Court Bench in a majority 3-2 decision, but as the point was a novel one not previously argued before the Court, it granted leave of appeal to the Privy Council. In London some months later the Privy Council was quick to dismiss the case, thereby upholding the decisions of both the High Court and the Supreme Court of New South Wales, and ensuring the validity of possibly many other mining titles.

In the meantime new legislation had been introduced in Australia requiring environmental impact statements to be submitted and accepted before mining leases could be granted. Accordingly, a feasibility study and an Environmental Impact Statement (EIS) were prepared to support an application for a mining lease. The EIS was accepted by the State authorities and Special Mining Lease No.20 was granted on November 16, 1973. However, development did not commence immediately. Increasing inflation rates and falling metal prices meant that all financial projections had to be recast to meet the changed conditions and an improvement in some aspects of metal recoveries became desirable before it was felt wise to proceed. Metallurgical work therefore continued and a new feasibility study was carried out in 1975.

Emphasis on environmental protection had increased during these years, and when Commonwealth Government consent was sought for the introduction of funds to meet development costs, the Foreign Investment Review Board made it a condition that a second environmental impact statement should be prepared to meet the requireme of the Commonwealth Department of the Environment. This statement was completed and placed on public display from March 9 to April 9, 1976. When it finally received approval it was sufficient for all necessary consents from the various authorities in the three levels of government. The way was then clear to arrange the financing of the project.

Ownership and Development

Woodlawn was discovered before Australian ownership of resources had become an announced policy of Australian governments. However, the American parent companies had always made public their intention to invite some Australian equity in the project. No specific level of such equity had been set by any governme and the proposal crystallised into one which would seek a third equal joint-venture partner for the project. After several wholly Australian companies had examined the proposal without further interest, the CRA subsidiary, New Broken Hill Consolidated Ltd., signified its wish to participate and governmental approval was obtained for it to become a one-third participant.

The corporate ownership of the Woodlawn orebody was then reorganised into a joint venture between St. Joseph International Explorations Ltd., and Phelps Dodge

Exploration Corporation, (subsidiaries of the two parents of Jododex), and New Broken Hill Consolidated Ltd., as from January 1, 1976.

The Woodlawn joint venture is controlled and directed by a management committee made up of one representative appointed by each of the participants and the non-voting General Manager of Woodlawn Mines who reports to the Committee. The feasibility study completed in 1975 was updated during 1976 and on this basis final design and construction commenced in 1976. The overall project programme was based on carrying out concurrent activities in design, procurement, construction and commissioning, thus minimizing the total project duration.

The total project cost about $78 million to complete. Some major items in the project included the removal of 1.6 million m^3 of pit overburden, the placement of 700,000m^3 of tailings dam earthworks, of 19,000m^3 of concrete and of 2,800 tonnes of structural steel. At the peak of construction the work force totalled 520, of whom about 350 were accommodated in an on-site construction camp. Precommissioning testing commenced in February, 1978 followed by commissioning of sections of the plant which commenced in May 1978, for the copper circuit and September 1978 for the complex ore circuit.

The mine and mill were officially opened on December 15, 1978, more than 10 years after the first geochemical indications of the orebody were recorded and almost 9 years after the drilling of W2, the discovery drill hole.

Geological Hypotheses

The comprehensive papers presented in this volume are based on a geological interpretation of the Woodlawn orebody as one of a number of lenticular bodies in a wide zone of acid volcanics, pyroclastics and sediments which dip west. These rocks are considered to be transgressively invaded by thick lenses of dolerite which are conformable in places but break across the stratigraphy in other places.

However, a possible interpretation of a different nature was offered by a consultant geologist, Alan Blatchford, of Sydney, who undertook a commission at Woodlawn in 1978. He saw the data as being equally amenable to interpretation as a tightly folded overturned synclinal structure plunging north in which the ore horizons and the dolerites are conformably folded.

As in most mining fields alternative geological hypotheses are almost certain to develop until sufficient data have accumulated to swing the pendulum one way or the other. At Woodlawn the interpretation is of more than academic interest because i may affect the potential ore resources and can help to guide the path of future drilling. The next three or four years of mine development should therefore be of particular interest as fresh evidence is uncovered and the geological interpretation assumes a firmer shape.

The alternative geological interpretation does not affect interpretation of th geophysical data presented here as these data mainly come from investigation of the near surface massive sulphides body. However, it would require serious consideration in interpreting results of geophysical techniques attempting to investigate at depths of some hundreds of metres below the surface.

The accompanying plates illustrate the title of this paper.

Plate 1 shows Woodlawn in January, 1970, with the drilling rig in position, drilling the discovery hole, W2. Prior to that time, the geochemical anomaly had been outlined and a partly coincident I.P. anomaly located but the presence of economic mineralization remained to be proved. Hole W2 dramatically provided that proof though many more holes had to be drilled to prove the size and distribution of the ore body.

In Plate 1, Woodlawn is virtually unaffected by exploration.

Plate 2 is an aerial view of Woodlawn in mid 1979 showing the development which has taken place since then.

Plate 1

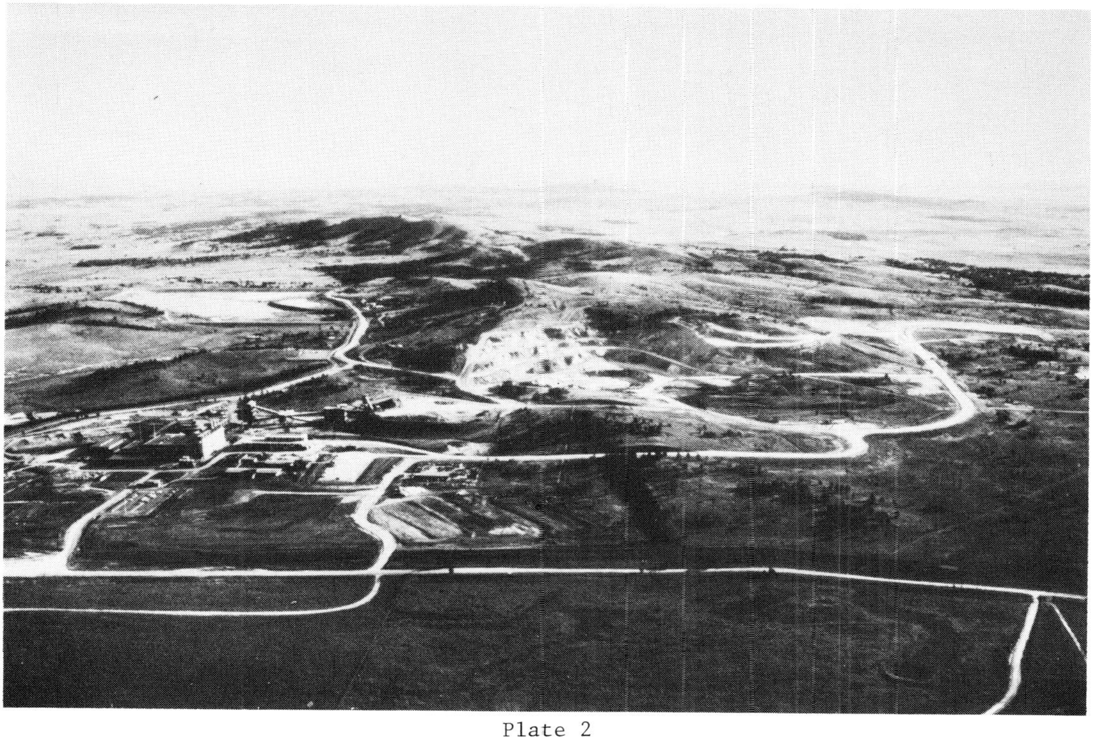

Plate 2

CHAPTER 3

REGIONAL AND LOCAL SETTING

3.1. Regional Mineralization and Geophysics of the Eastern Lachlan Fold Belt

E. J. MALONE AND R. J. WHITELEY

Contents

Abstract

Regional geophysical and geological data have been examined in relation to volcanogenic mineralization in the mid-late Silurian rocks of the eastern Lachlan Fold Belt. This part of the Belt contains a great number of deposits which exhibit many features common to volcanogenic base metal deposits of other parts of the world and other ages of deposition. The

Woodlawn orebody is the largest yet known in the belt and is a fully developed and well preserved volcanogenic stratiform base metal deposit.

Regional Bouguer gravity trends correlate in general with the major structural elements of the Belt which control the gross distribution of mineralization.

A more detailed examination of the regional gravity and aeromagnetic data of the southern part of the Captains Flat-Goulburn Synclinorial Zone revealed an apparent association of gravity highs with the major volcanogenic sulphide deposits.

Complex aeromagnetic anomalies are commonly associated with basic igneous material in the Woodlawn-Captains Flat region. Both the Woodlawn and Captains Flat deposits occur in broad magnetic lows near narrow magnetic highs and bounded on either side by large magnetic highs.

Many of the extensive granite bodies, mainly occupying anticlinorial zones, display substantial magnetic character due mainly to their magnetite content, particularly near their contacts with sediments and volcanics.

Introduction

Considerable regional information on the Lachlan Fold Belt is readily available. Large scale geological and geophysical maps of the Belt have been published by the Australian Bureau of Mineral Resources and the Geological Survey of New South Wales. These maps include the Geology of New South Wales at a scale of 1:1,000,000 and individual map sheets at scales of 1:250,000 and 1:100,000.

The published geophysical maps consist of regional Bouguer gravity together with incomplete aeromagnetic coverage.

Descriptions of the tectonics, regional geology and mineralization have been published by Packham (1969), Scheibner (1973), Markham and Basden (1974) and are incorporated within explanatory notes provided with the geological map sheets. The genesis of ore deposits in the Lachlan Fold Belt has also recently been the object of a specialist conference (McPowell and Johnson,1976)

In this paper, the nature and regional distribution of volcanogenic base metal mineralization within the eastern Lachlan Fold Belt are briefly reviewed and aspects of the regional geophysical data are examined in relation to the distribution of this mineralization. A much more detailed and comprehensive review of mineralization in the Lachlan Fold Belt is contained in Markham and Basden (1974).

Nature and Distribution of the Mineralization

A considerable number of volcanogenic, stratiform base metal deposits occur in the middle to late Silurian rocks of the eastern Lachlan Fold Belt. These are generally referred to as "Captains Flat type" (McClatchie, 1970) after the mine at Captains Flat which produced 4 million tonnes of copper, lead, zinc, gold and silver ore before it was mined out in 1962. The Woodlawn orebody is the largest such deposit yet discovered and is clearly a stratiform, volcanogenic massive sulphide deposit.

The distribution of most known deposits associated with the middle to late Silurian sequence in the eastern Lachlan Fold Belt is shown on Fig. 1.* The deposits are named in Table 1. Most of these deposits are sub-economic, generally

*The folded Figs. 1, 2 and 4 can be found in the envelope accompanying the book

because of small size. The list is not exhaustive but those omitted are very small occurrences of mineralization.

TABLE 1

Base Metal Occurrences, Eastern Lachlan Fold Belt

Key to location on figures

1.	Leadville	Zn, Pb, Cu
2.	Belara, Native Bee	Cu, Au, Ag
3.	Commonwealth, Stringers	Cu, Au, Ag
4.	Galwadgere	Cu, Au
5.	Mt. Lindsay	Cu, Pb, Zn
6.	Mt. Bulga	Cu, Pb, Zn
7.	Lewis Ponds	Cu, Pb, Zn
8.	Ophir	Cu
9.	Boyd	Cu, Pb, Ag
10.	Sunny Corner	Cu, Pb, Zn, Ag
11.	Cow Flat	Cu, Zn
12.	Apsley	Cu, Pb, Zn
13.	Wisemans Creek	Cu, Pb, Zn
14.	Kempfield, Colo Creek	Cu, Pb, Zn
15.	Burraga	Cu, Au, Ag
16.	Mt. Costigan	Cu, Pb, Zn
17.	Hurn's	Cu, Pb, Zn
18.	Cordillera	Cu, Pb, Zn
19.	Peelwood	Cu, Pb, Zn
20.	Elsinora	Au, Ag, Pb, Zn
21.	Black Springs	Pb, Zn, Ba
22.	Frogmore	Cu
23.	Rye Park	Cu, Pb, Zn, Ag
24.	Kangiara	Cu, Pb, Zn, Ag
25.	Breadalbane	Cu, Fe
26.	Clare Vale	Ba
27.	Currawang	Cu, Pb, Zn
28.	Wyelba, Sapling Point	Cu, Pb, Zn, Ag
29.	Woodlawn	Cu, Pb, Zn, Ag
30.	Boro	Cu, Pb, Zn, Ag
31.	Mulloon	Cu, Au, Ag
32.	Belconon	Pb, Ag, Au
33.	The Briars	Cu, Pb, Ag
34.	Valley Creek	Cu, Pb, Zn
35.	Little Bombay	Ba, Pb
36.	Captains Flat	Cu, Pb, Zn, Ag, Au
37.	Krawaree	Pb, Zn
38.	Michelago	Cu, Zn
39.	Ryries	Au
40.	Colinton	Cu, Pb, Zn, Ag
41.	Baalgammon	Cu, Pb, Ag

The deposits of Captains Flat type, such as Woodlawn, Captains Flat, Peelwood and Mt. Bulga, are fairly reliably identified as volcanogenic stratiform base metal deposits.

Many of the others are vein type deposits or have been structurally or metamorphically deformed to the extent that their original genesis is uncertain. However, the genetic association of all these deposits with volcanics is clear.

The tectonic, structural and lithological settings of these deposits are similar and the deposits in any one structural zone generally occur on or near the one stratigraphic horizon. In many cases, this stratigraphic horizon is the contact between a volcanic unit and an overlying sedimentary unit.

The deposits are characteristically polymetallic and fine grained though there is considerable variation within and between them. This is shown both in the relative proportions of the copper, lead, zinc and iron sulphides which constitute the bulk of the mineralization, and in the proportion of total sulphides to host rock.

Variations in base metal content and proportions have led to the identification of lead-zinc rich, copper-lead-zinc rich, copper rich, and barite rich types though there appears to be almost continuous gradation from one type to another. Distinctly different types can occur within the one deposit so that bulk assay grades for particular deposits may disguise considerable internal variation. Pyrite is ubiquitous and is generally the dominant sulphide; some deposits consist almost wholly of pyrite with only minor base metal sulphides.

Most deposits are wholly or partly massive sulphides, that is, sulphides constitute 50 per cent or more of the rock, but many are associated with disseminated mineralization. The form of the massive sulphide bodies is generally lenticular; the disseminated mineralization commonly has a greater lateral extent, and is tabular in form.

The most characteristic feature of the massive sulphides deposits is their depositional layering which is generally conformable with bedding in the host rock sequence. The disseminated mineralization is also generally conformable.

Vein type mineralization is also common throughout the eastern Lachlan Fold Belt. In many places, such as at Woodlawn, it is associated with the stratiform massive sulphides and is clearly part of the same volcanogenic mineralizing event. Such volcanogenic events can produce a wide range of mineralization types: massive sulphides; disseminated mineralization which may be in part due to clastic dilution of potentially massive sulphides; veins, either singly or as anastomosing networks; and widespread hydrothermal alteration, including intense chloritization, alkali depletion and silicification, with or without base metal mineralization. All of these are displayed at Woodlawn which, as well as being the largest, is the most completely developed and preserved of the known volcanogenic base metal deposits in the eastern Lachlan Fold Belt.

Lithologic association with acid volcanics is almost invariable; rare exceptions include the Currawang deposit where mineralogically (possibly depositionally) layered copper-lead-zinc-iron massive sulphides occur in a sequence of pillow lava, basalt and high alumina andesite.

The gross distribution of the deposits appears to be tectonically controlled, probably by virtue of tectonism controlling the loci of volcanism. The contribution of local structure to the localization of the deposits is less clear.

Many of the deposits occur near major faults and shear zones which are sub parallel to or bound the structural units. These faults possibly reflect zones of deep crustal fracturing which controlled the development of the structural units and provided conduits for volcanic extrusive activity.

Fig. 2 presents the same deposits as Fig. 1 but shows their distribution in relation to major structural units after Scheibner(1973 and 1974b), who proposed a plate tectonic model for the formation of these units. According to Scheibner, strong extension of the Australian plate took place within middle Silurian and caused the formation of marginal seas or volcanic rifts, the Cowra Trough, the Hill End Trough and the Captains Flat trough. These troughs were formed by subsidence of fault bounded blocks of the partly cratonized lower Palaeozoic sediments. Volcanics, volcanoclastics, and sediments were deposited in probably shallow seas occupying these troughs. Subsequent deformation, particularly east to west compression, established the synclinorial zones of Fig. 2. The mid to late Silurian sequence plus some early Devonian units are preserved in these synclinorial zones, hence the strong correlation between Figs. 1 and 2.

Figs. 1 and 2 also show a clear tendency for deposits to cluster in some areas while other areas of apparently similar structural and lithological setting contain no or few deposits. The cluster of deposits including the Peelwood, Cordillera and Mt. Costigan deposits are an example of this tendency. These deposits may be related to deep crustal fracturing in the hinge zone at the southern end of the Hill End Trough which opened up in the middle Silurian by rupture of the Molong Volcanic rise (Scheibner, 1973).

The volcanogenic deposits of the eastern Lachlan Fold Belt are representative of a group of deposits which are found in many parts of the world and in many ages of rocks, ranging from those in the Canadian Precambrian to possibly the best known, the Tertiary Kuroko deposits of Japan. A general theory of the genesis of these deposits has been developed (Matsukuma and Horikoshi, 1970; Stanton, 1972; Sangster, 1972; Hutchinson, 1973; Lambert and Sato, 1974; and others) and appears applicable with some modifications, to those in the Lachlan Fold Belt.

These deposits are regarded as being of volcanic, exhalative origin and associated with dominantly felsic volcanism in a marine environment. Some of the volcanism may be subaerial, thus providing the abundant volcanoclastic detritus of the mineralized sequence. The mineralization is compound, including deposition of stratiform ore at the depositional interface and operation of an extensive hydrothermal system in the volcanic pile beneath, adjacent to or remote from the site of stratiform deposition. Copper-rich vein and epigenetic mineralization predominate within the hydrothermally altered zone which is characterized by chlorite alteration, silicification and alkali depletion. The stratiform deposits in any one district generally occur on or near the one stratigraphic horizon, suggesting that some geological factors control the operation of discrete mineralizing events within a limited time span. The distribution of the deposits is clearly related to major structural units and clustering within these units may also relate to local structural control.

Bouguer Gravity-Eastern Lachlan Fold Belt

Bouguer gravity data for the Lachlan Fold Belt were obtained during the reconnaissance gravity survey of Australia (Fraser et al, 1976). This regional survey was completed using an average station spacing of 11 kilometers and contoured at an interval of 5 milligals.

According to Fraser (1976) the Lachlan Fold Belt falls within the much larger Lachlan Gravity Complex.

Fig. 3, covering the same area as Figs. 1 and 2, shows the Bouguer anomaly map of the eastern Lachlan Fold Belt. The mineral deposits discussed earlier are also shown on Fig. 3.

Present gravity trends in this region are generally northerly to northwesterly in common with the major structural units shown in Fig. 2. Apart from these definite trends the gravity pattern is generally complex, reflecting rapid and large lateral density variations within the Lachlan Fold Belt.

The structural elements represented in the region are described by Scheibner (1973, 1974a, b). As shown in Fig. 2, the southeastern Lachlan Fold Belt is composed of major synclinorial and anticlinorial zones. Many of these meridional structural elements are narrow relative to the station spacing of the reconnaissance gravity survey and, as a result, detailed correlations with the gravity are difficult.

In the north, stratiform mineralization occurs within the Hill End Synclinorial Zone and is most intense on the southern and western margins of this Zone. The regional gravity pattern in this Zone is dominated by a large gravity low (A, Fig. 3) associated with the Bathurst Granite Batholith (Fig. 1). The -50 milligals closure, possibly related to the root of the batholith, occurs just to the west of the town of Bathurst. This anomaly trends in a northerly direction from Bathurst and appears to be associated with smaller circular gravity lows (B, C, Fig. 3) near the northern margin of the Lachlan Fold Belt.

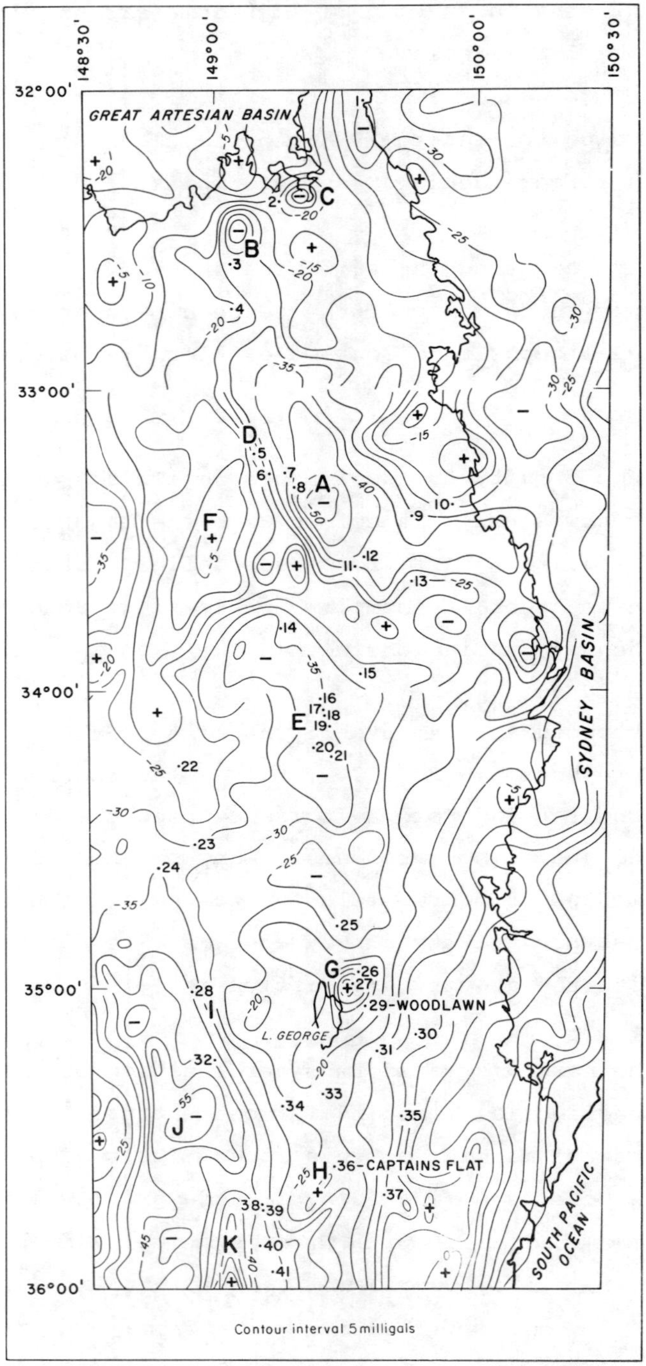

Figure 3. Bouguer gravity, eastern Lachlan Fold Belt.

South of Bathurst the anomaly swings abruptly to the east cutting across other major structural units and possibly following the Lachlan River Lineament described by Scheibner (1974b).

A steep gravity gradient (D, Fig. 3) on the western side of the Bathurst granite is probably associated with the Copperhannia Thrust which marks the boundary between the Hill End Synclinorial Zone and the Molong-South Coast Anticlinorial Zone in this region. The Mt. Bulga massive sulphide deposit occurs on this gravity gradient. This volcanogenic, stratiform deposit contains over 850,000 tonnes but the tonnage/grade combination is sub-economic (Stevens, 1974).

In the southern part of the Hill End Synclinorial Zone the Peelwood group (16 to 19,Table1) of mineral deposits (Stevens, 1974), occurs in a broad, northerly trending gravity low (E, Fig. 3).

The Molong and Rockley Anticlinoria, local units of the complex Molong-South Coast Anticlinorial Zone, occur to west and east of the Hill End Synclinorial Zone. The Rockley Anticlinorium, to the east of the Synclinorial Zone is relatively narrow compared to the station spacing used for the reconnaissance gravity survey, but appears to be associated with Bouguer gravity highs.

The most notable Bouguer anomaly within the Molong-South Coast Anticlinorial Zone is a northerly trending gravity high (F, Fig. 3) with a closure of -5mgls which occurs in the Molong Anticlinorium immediately west of the Bathurst Granite. An extension of this anomaly appears to trend easterly to the south of the Bathurst Granite and across the Hill End Synclinorial Zone

The Captains Flat-Goulburn Synclinorial Zone, at the southeastern end of the Lachlan Fold Belt, contains the two largest known stratiform volcanogenic deposits, the Captains Flat and Woodlawn deposits. This Zone trends north from Captains Flat to Woodlawn and then extends in a narrow belt north to northeast from Woodlawn. The Captains Flat-Woodlawn Region is discussed in more detail in the next section.

The most obvious Bouguer gravity anomalies in the Captains Flat-Goulburn Synclinorial Zone are the circular gravity high (G, Fig. 3) to the northwest of Woodlawn and a less well defined narrow northeasterly trending high (H, Fig. 3) just to the south of Captains Flat. Other closures in this Synclinorial Zone are gravity lows occurring near the northern extremities of the Zone, presumably associated with the Bathurst Granite.

The Cowra-Yass Synclinorial Zone occurs to the west and southwest of the Captains Flat Synclinorial Zone. This Zone is characterised by strong north-northwesterly trending gravity gradients and a complex gravity low (I and J, Fig. 3) near the northern end of the Cotter Anticlinorial Block.

Stratiform, volcanogenic mineralization (38-41, Fig. 3) occurs in the south of the area on the eastern margin of an intense northerly trending gravity high (K, Fig. 3) centred on the western boundary of the Cowra-Yass Synclinorial Zone.

The gravity pattern in the central Cowra-Yass Synclinorial Zone is characterized by low gravity gradients. The contours generally trend east-northeasterly across the Synclinorial Zone and the adjacent Molong-South Coast Anticlinorial Zone.

In the north of the Cowra-Yass Zone strong north trending gravity gradients occur to the east of the prominent northerly trending gravity high (F, Fig. 3) in the Molong-South Coast Anticlinorial Zone. A broad circular gravity high, centred just to the north of the town of Yeoval, is located at the northern end of the Cowra-Yass Synclinorial Zone.

The Cotter Anticlinorial Block, in the southwestern portion of the eastern Lachlan Fold Belt, is characterized by strong meridional gravity gradients. These are associated with major faulting and presumably basic igneous intrusions which produce complex gravity highs on the northwestern side of the Block.

The Woodlawn-Captains Flat Area

Regional Geology

Aspects of the regional geology of the Woodlawn-Captains Flat area have been discussed as part of the Southern Highlands Fold Belt by Packham (1969, Ch. 3), and more recently in New South Wales Geological Survey publications (Felton and Huleatt, 1976). A simplified geological map is shown on Fig. 4 modified after Gilligan (1974) and Felton (1974).

The oldest rocks in the area are Ordovician sediments, part of a thick distal flysch accumulation of quartz rich greywacke, shale, chert and siltstone deposited in a deep marine basin. These rocks, deformed and metamorphosed, crop out in anticlinorial zones, intruded by Silurian and Devonian granitoids Mid to late Silurian sediments and volcanics occupy synclinorial zones.

The earliest Silurian sediments, mudstone, siltstone and minor limestone, were deposited on a shallow shelf. Crustal rifting and subsidence initiated acid volcanism. Rhyolitic and dacitic flows, tuffs, agglomerate, welded tuffs, ashstone, volcanic derived sediments and shale accumulated in elongate troughs. Deeper rifting in the northern part of the area near the end of the Silurian resulted in basic volcanism succeeding the acid volcanism. Pillow lavas, basaltic tuff, chert, and keratophyre were deposited in the Currawang area. A thick sequence of quartz rich greywacke, slate and siltstone transgressed the volcanic sequence in the late Silurian to early Devonian. This was locally succeeded by Early Devonian terrestrial conglomerate and shallow marine sedimentation. Strong deformation in the early Devonian, particularly east to west compression, and intrusion of granitoid stocks into high levels of the crust in the anticlinorial zones virtually completed the cratonization of the region. The volcanic rift troughs were converted into fault bounded synclinoria, the narrower ones being steeply overturned so that all limbs dip west. The Silurian sequence, regionally metamorphosed to greenschist facies and generally exhibiting strong westerly dipping, axial plane cleavage, is preserved in the synclinorial zones. Mature topography of much of the region is responsible for extensive alluvial and colluvial deposits which limit the outcrop of the Palaeozoic rocks.

Mineral Deposits

Volcanogenic, stratiform base metal deposits occur in or are associated with the Silurian felsic volcanics in the region as discussed earlier. These include the Woodlawn and Captains Flat deposits. The geology and setting of the Woodlawn deposit is discussed by Malone (this volume p 49).

The massive sulphide deposits at Captains Flat are described by Glasson and Paine (1965) and further discussed by Davis (1975). These were discovered late last century and produced 4 million tonnes of lead-zinc-copper ore with some gold and silver before mining ceased in 1962. All production was derived from two narrow, lens like massive sulphides orebodies occurring in an overturned limb of folded volcanics in which there was the greatest concentration of shearing and wall rock alteration.

Volcanogenic, and apparently stratiform, base metal mineralization occurred in association with basic volcanics at Currawang. Two small lenses of ore were mined in the 1890s and sporadically during the early part of this century. Exploration drilling in the area has intersected other lenses of ore consisting of banded massive sulphides, including pyrite, pyrrhotite, sphalerite, chalcopyrite and galena. The ore appears to be associated with a chloritic basic tuff zone in a sequence of pillow lavas, basic flows and tuffs and some dolerite intrusives.

Bouguer Gravity

The Bouguer gravity pattern in this part of the Captains Flat Synclinorial Zone (Fig. 5) is quite complex. Of particular interest is the apparent association of the volcanogenic deposits at Captains Flat and Woodlawn with regional Bouguer gravity highs. As shown in Fig. 5 the Captains Flat deposit lies within a narrow gravity high (-20 mgl closure) which trends in a northeasterly to southwesterly direction.

The deposit at Woodlawn is on the southeastern margin of a prominent circular gravity high. This anomaly which has an amplitude of about 29 mgls is centred about 10 km north-northwest of the Woodlawn orebody and also has associated low amplitude gravity ridges which appear to trend to the northwest and southwest respectively.

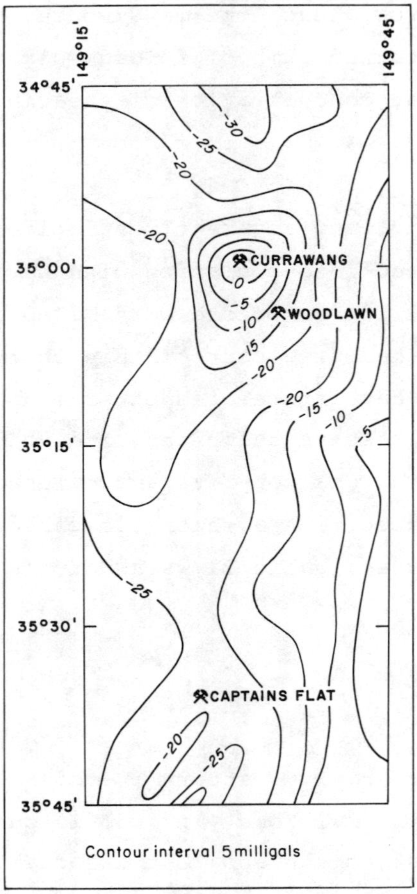

Figure 5. Bouguer gravity, Woodlawn - Captains Flat region.

Skeels (1963) method was used for depth interpretation along two orthogonal profiles through the centre of this circular anomaly. The source of the anomaly was assumed to be a vertical sided cylinder of denser material. Density contrasts of 300 and 400 Kg/m^3 were also assumed to give a range of interpreted depths.

Skeels method yielded a maximum depth to the top of between 1.8 and 2.4 km and a diameter of about 8 km. Since outcrops of basic intrusive material are common in the vicinity of the gravity high, the source of the anomaly is believed to be a shallow basic igneous plug of considerable size.

This anomaly is located in the Currawang area where the presence of basic volcanics suggests the deepest crustal rifting took place. The gravity anomaly coincides with the intersection of an east to west crossfold axis with north to south trending anticlinorial axes, which causes the two parallel limbs of the Captains Flat-Goulburn Synclinorial Zone to coalesce in this area. The Zone also changes strike in this area: it trends north to south to the south of the gravity anomaly and north-northeast to the north of the gravity anomaly.

It may be that the circular gravity anomaly occurs in a hinge zone where basic intrusion has occurred at the intersection of major crustal fractures, which also controlled the basic volcanism.

The effect of this gravity high on the detailed gravity surveys at Woodlawn (Templeton, this volume, p 485) was to create a complex gravity gradient with values increasing to the northwest.

Aeromagnetics

Regional aeromagnetic surveys of the Goulburn and Canberra map sheets were flown by the Bureau of Mineral Resources. These surveys were carried out at an average elevation of 150m above ground level along generally east to west lines with separations from 1.5 to 1.6 km.

Fig. 6 shows a simplified aeromagnetic map of the Woodlawn-Captains Flat area based on these data contoured at 25nT intervals. General meridional trends are again apparent

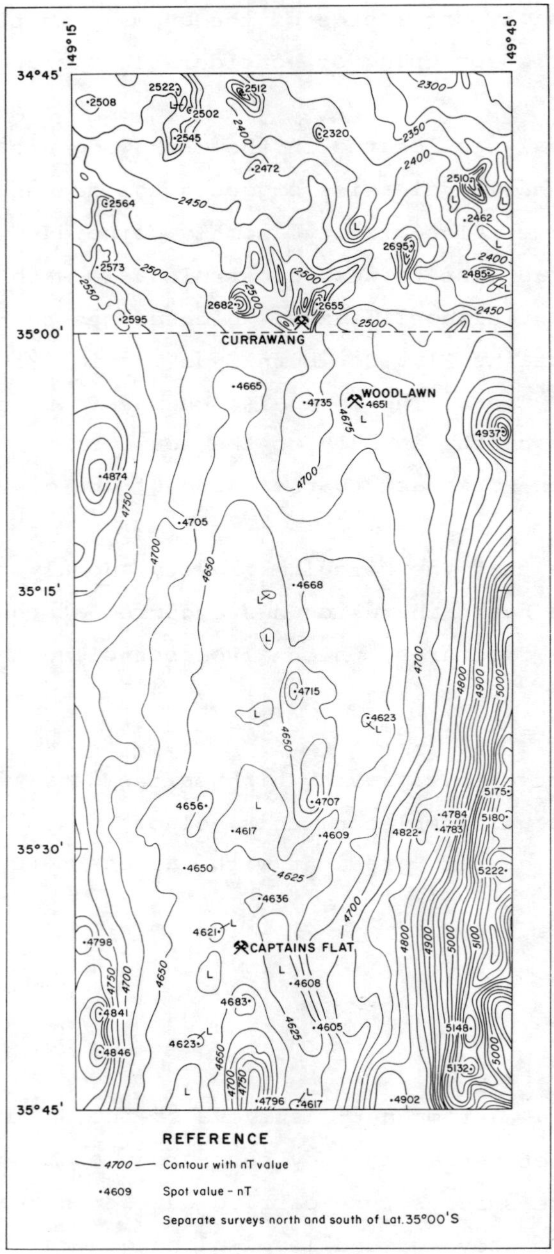

Figure 6. Aeromagnetic contours, Woodlawn - Captains Flat region.

in aeromagnetic data reflecting the strike of the major structural elements in this region as discussed earlier. Magnetic field values are generally lower in the central part of the area with prominent magnetic anomalies associated with granites to the east, southeast and south of the area. Magnetic anomalies occur over the Sutton, Tinderry and Urialla Granites and the Jerangle Granite Complex in the south. Very large magnetic anomalies occur above the heavily mineralized Braidwood Granite on the southeastern side of the area.

The Boro and Ellenden Granites are not obviously magnetic overall, although a narrow, linear, low amplitude magnetic anomaly trends through the centre of the area in a north to south direction along the western side of these granites. Limited ground magnetic measurements east of the town of Bungendore in the centre of the area showed this anomaly to be composed of a number of high amplitude anomalies which correlated with gabbroic dykes. This discontinuous series of anomalies is presumably due to a basic igneous complex (the Lockhart Igneous Complex) which consists largely of gabbro (Read, 1961) and has been extensively invaded by granite.

This linear anomaly belt extends from south of Captains Flat to just north of Woodlawn where the anomaly pattern becomes more complex and anomaly trends appear to swing northeasterly.

Local high amplitude magnetic anomalies also occur near the centre of the major circular gravity high near Woodlawn and are probably due to the basic igneous material shown in outcrop in Fig. 4.

The Wyangala Batholith in the northwest of the area is not obviously magnetic although a number of narrow intense anomalies occur near its eastern margin.

Magnetic profiles interpolated from the contour map were drawn in an east to west direction through both the Captains Flat and Woodlawn deposits. These are shown in Fig. 7 together with a simplified geological cross section through Captains Flat.

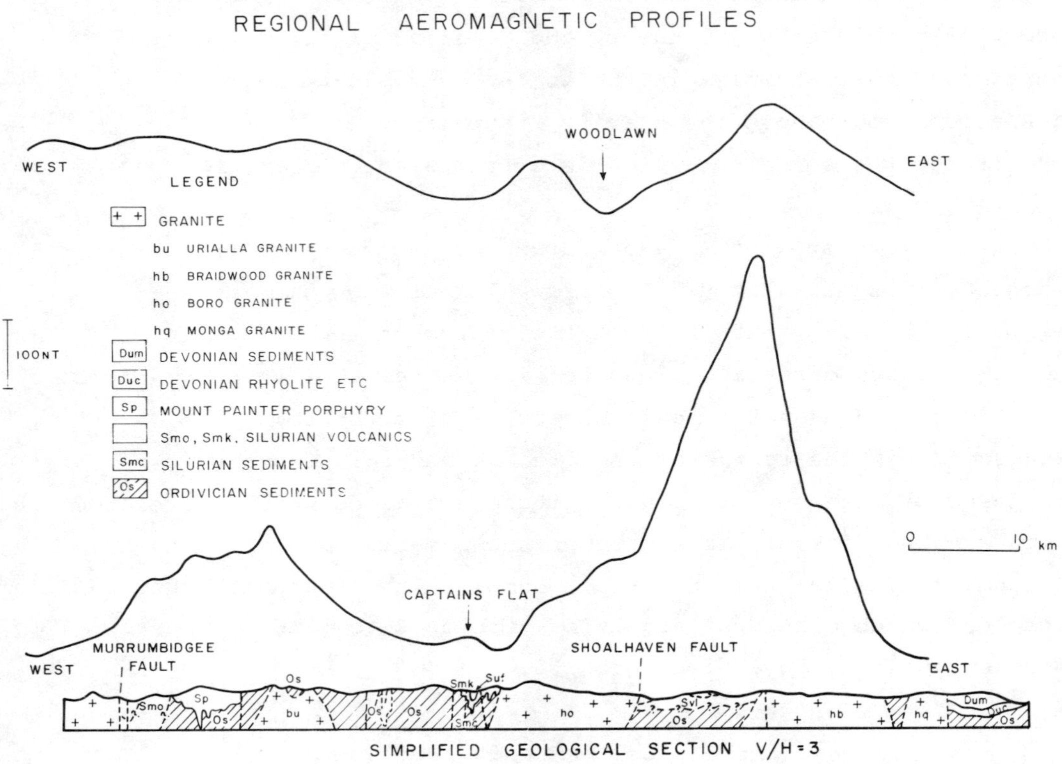

Figure 7. Aeromagnetic profiles and geological section.

The most striking feature of these profiles is their similarity in shape, although they are some 60 km apart. Anomaly amplitudes are generally much smaller in the profile through Woodlawn. Both sulphide deposits occur in broad magnetic anomaly lows near to a narrow low amplitude magnetic high and bounded on either side by large magnetic highs.

The magnetic data on the eastern side of the profile through Captains Flat show a large complex anomaly with an amplitude of about 600nT. Although from the shape of the anomaly more than one magnetic source is likely, it appears to be related to the Ordovician sediments and Silurian sediments and volcanics between the Braidwood and Boro Granites. The most intense anomaly does not seem to extend westward beyond the Shoalhaven Fault. This anomaly appears closely related to the contact between the western margin of the Braidwood Granite and the adjoining sediments and volcanics.

On the western side of the magnetic profile, a complex magnetic high with an amplitude of about 200nT is present. This anomaly occurs near the western margin of the Urialla Granite above Ordovician sediments and the Mount Painter Porphyry. It does not appear to extend westward beyond the Murrumbidgee Fault.

The intense anomalies on this profile are possibly due to the presence of magnetite near the contact of iron bearing sediments with the granites. Substantial magnetite deposits are known to occur in the area. Some contributions to these anomalies may also be due to the presence of basic igneous material.

The smaller magnetic anomalies occurring near the Captains Flat and Woodlawn deposits are believed due to narrow basic igneous intrusions.

Conclusions

Mineralization in the middle to late Silurian rocks of the eastern Lachlan Fold Belt is dominantly volcanogenic. Distribution of the mineralization reflects strong stratigraphic control and the influence of major structural units. The reasons for local clustering of deposits are not known.

Regional Bouguer gravity from the southeastern Lachlan Fold Belt shows considerable complexity which is difficult to correlate with major structural units, in other than a fairly broad sense. It is obvious that density variations in the Fold Belt are complex and mostly unresolved at the station spacing of the reconnaissance survey.

Regional gravity anomalies do not appear to be closely related to stratiform, volcanogenic mineralization throughout the Belt, although in the Captains Flat-Woodlawn region and further to the south, broad Bouguer gravity highs appear to be associated with significant sulphide deposits.

The most prominent gravity anomaly in the southern Captains Flat Synclinorial Zone is a large circular gravity high which is centred some 10km north-northwest of the Woodlawn deposit. This anomaly also occurs near the point where the Synclinorial Zone changes in strike direction and may represent an igneous centre in a hinge region. The effect of this circular gravity high is to produce complex gravity gradients at Woodlawn.

Aeromagnetic data from the Woodlawn-Captains Flat region also contains many complicated features. Major intense anomalies occur near the margins of many of the large granite bodies at their contact with older sediments and volcanics. These are believed to be mainly due to the presence of substantial magnetite concentrations. (Recently Connelly (1979) has attributed many of the aeromagnetic anomalies in the eastern Lachlan Fold Belt to the granites themselves. - Ed)

East to west aeromagnetic profiles through the major volcanogenic stratiform deposits at Woodlawn and Captains Flat are essentially similar with generally reduced amplitudes on the profile through the Woodlawn deposit. Both of these deposits occur in broad magnetic lows in the vicinity of narrow low amplitude highs and are bounded on either side by larger amplitude magnetic highs. The smaller magnetic anomalies in the vicinity of these deposits are believed to be due to narrow basic igneous intrusions.

References

Connelly, J.P. 1979. Interpretation of the Shape of Granites in the Eastern Lachlan Fold Belt using Aeromagnetic Data. Bull. Aust. Scc. Expl. Geophys. V.10, No. 1 March 1979. p 92-95.

Davis, L.W. 1975. Captains Flat lead-zinc ore body. In Knight, C.L. (ed). Economic Geology of Australia and Papua New Guinea, I. Metals (Australas. Inst. Min. Metall., Melbourne) pp. 694-700.

Felton, E.A. 1974. Goulburn 1:250,000 Metallogenic Map. Geol. Survey N.S.W., Sydney.

Felton, E.A. and Huleatt, M.B. 1976. Geology of the Braidwood 1:100,000 sheet. Geol. Survey N.S.W., Sydney.

Fraser, A.R., 1976. Gravity provinces and their nomenclature. BMR Journal of Geology and Geophysics 1, 1976, pp 350-352.

Fraser, A.R., Moss, F.J. and Turpie, A., 1976. Reconnaissance Gravity Survey of Australia. Geophysics Vol. 1, 6, p 1137-1145.

Gilligan, L.B., 1974. Canberra 1:250,000 Metallogenic Map. Geol. Survey N.S.W., Sydney.

Glasson, K.R. and Paine, V.R., 1965. Lead-zinc-copper ore deposits of Lake George Mines, Captain's Flat in Geology of Australian Ore Deposits 8th Comm. Min. Metall. Congr. Aust. N.Z., 1, pp 423-431.

Hutchinson, R.W., 1973. Volcanogenic sulfide deposits and their metallogenic significance. Econ. Geol. V. 68, pp 1223-1246.

Lambert, Ian B. and Sato, Takeo, 1974. The Kuroko and Associated Ore Deposits of Japan: A Review of Their Features and Metallogenesis. Econ. Geol. Vol. 69, pp 1215-1236.

McClatchie, L., 1970. Copper mineralization in the Lachlan Geosyncline, New South Wales. Proc. Australas. Inst. Min. Metall., 236, 1-16, 5 figs.

McPowell, C. and Johnson, B.D. (editors), 1976. Specialist Group Investigating the Genesis of Ore Deposits - Conference Macquarie University Dec 1975. Bull. Aust. Soc. of Expl. Geoph. Vol 7 No. 1 49 pp 682.

Matsukuma, T. and Horikoshi, E., 1970. Kuroko deposits in Japan, a review. in T. Tatsumi (Ed.) Volcanism and Ore Genesis. Univ. Tokyo Press.

Markham, N.L. and Basden, H. (editors) 1974. The Mineral Deposits of New South Wales. Geol. Survey of N.S.W. (publisher).

Packham, G.H. (editor), 1969. The Geology of New South Wales. Journal of the Geol. Soc. Aust. Vol 16 No. 1 654 pp.

Sangster, D.F., 1972. Precambrian volcanogenic massive sulphide deposits in Canada: a review. Canada Geol. Survey Paper 72-22, 44 pp.

Scheibner, E. 1973. A plate tectonic model of the Palaeozoic tectonic history of New South Wales. J. Geol. Soc. Aust., 20(4), 405-426.

Scheibner, E. 1974. a. An outline of the Tectonic Development of N.S.W. with special reference to mineralization in The Mineral Deposits of N.S.W. pp 1-47.

Scheibner, E. 1974 b. Definition and Review of Structural Elements in The Mineral Deposits of N.S.W. pp 109-113.

Skeels, D.C., 1963. An approximate solution of the problem of maximum depth in gravity interpretation. Geophysics, 28, 1, Oct pp 724-735.

Stanton, R.L., 1972. Ore Petrology. (McGraw Hill Book Co., New York) 713 pp.

Stevens, B.P.J., 1974. Hill End Synclinorial Zone in The Mineral Deposits of N.S.W. pp 277-293.

3.2. Geology, Geochemistry and Setting of the Woodlawn Deposit

E. J. MALONE

Contents

Abstract

The Woodlawn base metal deposit, situated in southeastern New South Wales, was discovered in 1970 by Jododex Australia Proprietary Limited. The orebody was originally located by soil geochemical prospecting and outlined as a geochemical anomaly. The first geophysical work was a frequency domain dipole-dipole induced polarization (IP) survey which was used to define potential drilling targets. Subsequent drilling proved up approximately 6.3 million tonnes of massive sulphides and 3.7 million tonnes of copper ore. The Woodlawn Mine was officially opened in December, 1978.

Geological and geochemical data relevant to an appreciation of the geophysical response of the Woodlawn deposit are reviewed and the setting and climatic conditions are also discussed. The data refer to the first discovered orebody which is a complex body comprising one large and several smaller lenses of polymetallic massive sulphides and an adjacent zone of mineralized volcanics containing mainly pyrite chalcopyrite mineralization.

The bulk of the massive sulphides occurs in one large continuous body folded into an open, asymmetrical, west plunging syncline. The southern limb strikes about 030°M and dips northwest at 40° to 60°; the northern limb strikes about 305°M and is near vertically dipping. The folded massive sulphide body has a strike length of about 300m. The plan width of the massive sulphides is about 25m although they lens out to the northwest. They consist of about 75 per cent sulphides, mainly pyrite, sphalerite, galena and chalcopyrite in decreasing order of abundance, with talc, chlorite, quartz and volcanic detritus as gangue. The massive sulphides are within 20m of the surface at the southern end of the body and within 10m of the surface at the northern end.

The mineralized volcanics generally contain less than 25% sulphides, partly disseminated but much of it in stringers and pods. In addition, quartz chlorite veinlets with 20% to 30% pyrite and chalcopyrite cut the mineralized volcanics in places and are accompanied by pervasive silicification.

Black shale units in the vicinity of the Woodlawn deposit, locally produced geophysical anomalies similar to those produced by the massive sulphide deposit. One such geophysically anomalous black shale locality was chosen for comparative study. The geology and geochemistry of this locality are briefly described.

Introduction

The Woodlawn deposit is at lat. $35^{0}04'$S, long. $149^{0}34'$E in the Canberra Goulburn district of New South Wales, 12km west of Tarago. (Fig.1). The deposit was indicated by geochemical surveys in 1968 and early 1969, and further defined by geological mapping and induced polariation (IP) surveys in late 1969 (Smith, this volume, p 79). A drilling programme to test coincident geochemical and induced polarization anomalies was commenced in November, 1969. The discovery hole was the second in the programme. In January, 1970, this hole penetrated 24.4m of massive sulphide ore averaging 1.45% copper, 7.0% lead and 18.0% zinc.

By the end of 1971, exploration and development drilling, and particularly grid drilling for metallurgical samples, had established reserves of 6.3 million tonnes of copper, lead, and zinc ore (1.7% Cu; 5.5% Pb; 14.4% Zn) and 3.7 million tonnes of copper ore (1.9% Cu). This ore was contained in one large and several small lenses of polymetallic massive sulphides and adjacent zones of dominantly copper mineralization. Subsequent exploration drilling has indicated a number of other lenses of massive sulphides but at greater depth.

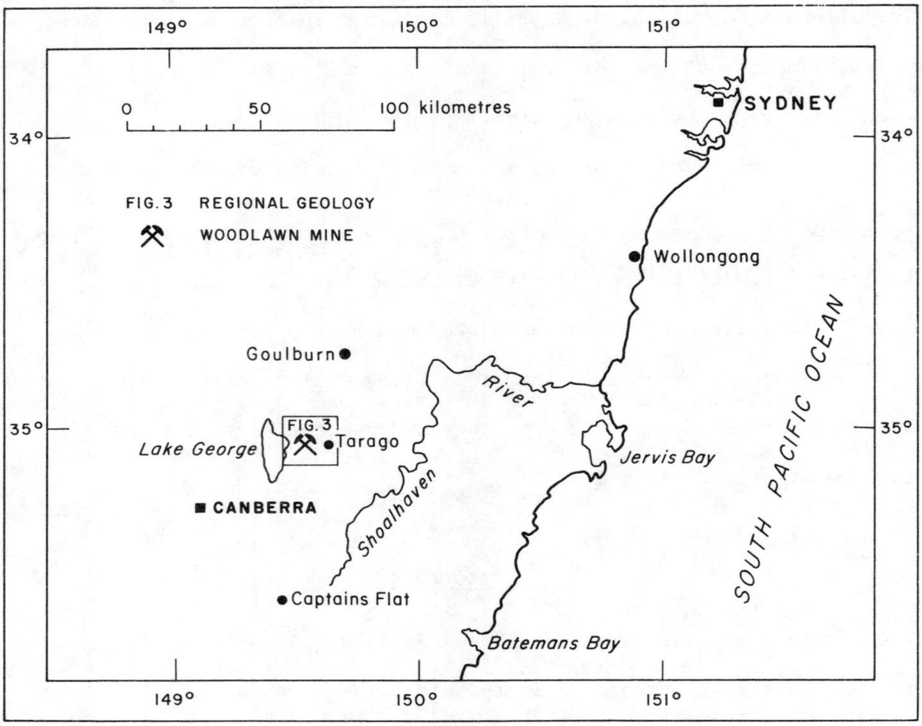

Figure 1 Locality map

Metallurgical testing in 1971 and 1972 indicated that a great deal of test work would be required to develop a process capable of satisfactorily recovering and separating the base metal sulphides. This implied that mining of the body would be delayed for some time. This provided time to carry out a wide range of geophysical surveys to specify the geophysical response of the orebody and thus assist exploration of the surrounding region. These surveys were commenced in 1971 and mostly completed late in 1976 just before the start of mining operations.

Only those attributes of the Woodlawn deposit which clearly contribute to the geophysical response are discussed in any detail. The geological setting of the orebody, the nature of the host sequence, and the structural and metamorphic regime are briefly described as some aspects of these matters may have bearing on the geophysical response.

Geological research into the Woodlawn deposit was commenced at about the same time as the geophysical research. Results of the geological research to date include seven papers published in Volume 26, parts 3 and 4, of the Journal of the Geological Society of Australia, issued in August, 1979. The authors and titles of these papers are:

L.B. Gilligan, E. Anne Felton & F. Olgers
The regional setting of the Woodlawn deposit

E.J. Malone
Nature, distribution and relationships of the mineralization at Woodlawn, New South Wales.

D.E. Ayres
The mineralogy and chemical composition of the Woodlawn massive sulphide orebody.

M.D. Petersen & I.B. Lambert
Mineralogical and chemical zonation around the Woodlawn Cu-Pb-Zn ore deposit, southeastern New South Wales.

W.R. Ryall & T. Nicholas
Surface geochemical and biogeochemical expression of base-metal mineralization at Woodlawn, New South Wales, Australia.

D.E. Ayres, M.S. Burns & J.W. Smith
Sulphur-isotope study of the massive sulphide orebody at Woodlawn, New South Wales.

B.L. Gulson
A lead-isotope study of the Pb-Zn-Cu deposit at Woodlawn, New South Wales.

These papers and Malone, et al. (1975) provide a much more detailed description of the Woodlawn orebody and consider many more aspects of its geology than appropriate within the present volume.

Black shale units are contained in the sequence in the vicinity of the Woodlawn deposit. It was recognised at an early stage in the exploration that some parts of these black shale units produce similar geophysical anomalies to those produced by the orebody. One such black shale locality was chosen for comparative study in the hope of developing ways of distinguishing the responses from these very different geological sources. The unit in the test area is referred to as the Black Shale.

Topography

Woodlawn is situated in a low part of the Great Dividing Range between the catchment areas of Lake George to the west, and the Shoalhaven and Wollondilly Rivers to the east. The orebody underlies a low, broad spur at the northern end of a prominent ridge. The massive sulphide body was completely concealed under grass and scattered tree cover growing on partly transported soils above a ferruginous clay gossan. Rubble of gossanous boulders occurred towards the south where massive sulphides grade into mineralized volcanics which produce higher and more rugged topography.

The Black Shale site is located in the Lake George catchment, on a dissected, moderately rugged hillside sloping towards the southwest. This shale is covered by thin rubbly soil and is generally poorly exposed except in creeks and gullies.

The topography of the area containing the Woodlawn deposit and the Black Shale site is illustrated in Fig. 2.

Climate

The climate in the area consists of a warm to hot summer and a long, cold winter; strong westerly winds are common throughout the year. The yearly rainfall is about 700mm (28in), some of it in sudden torrential downpours. The seasonal and annual distribution of rainfall varies widely. Annual totals in the past 40 years range from a low of 300mm (12in) to a high of 1330mm (52in). Rainless months are fairly common

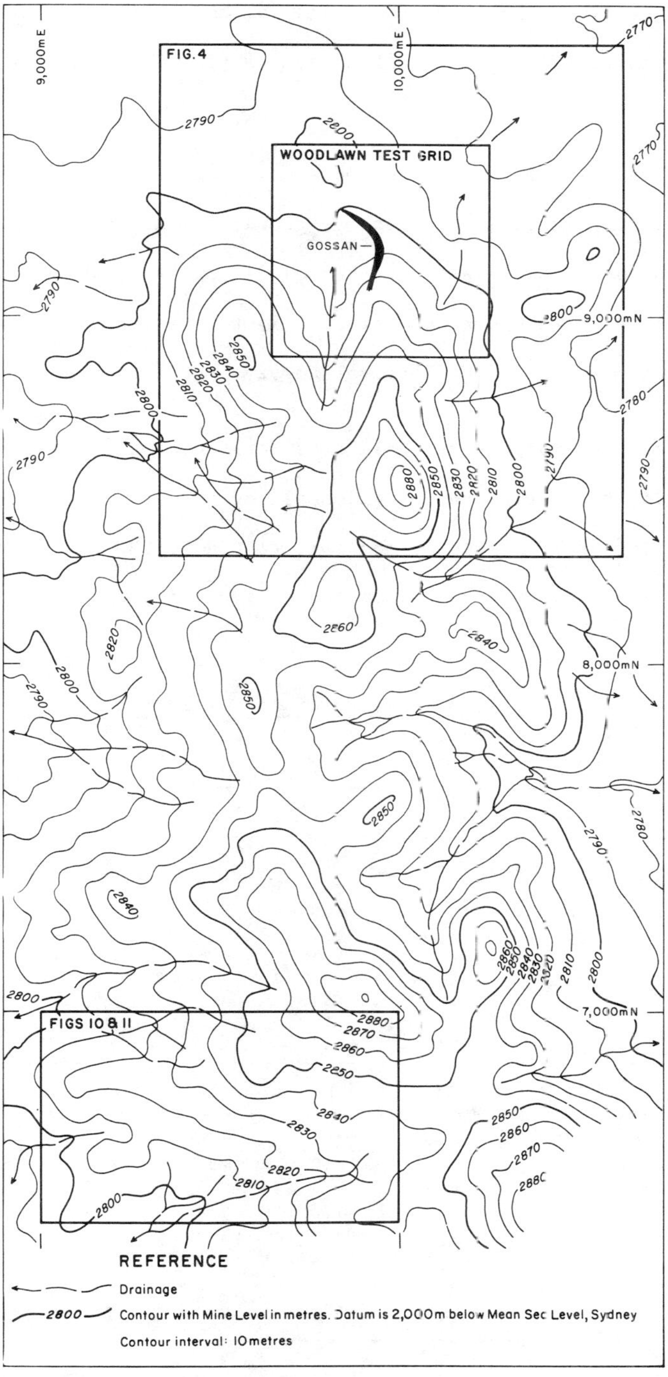

Figure 2 Woodlawn topographic map

and in the 1965 drought year only 100mm (4in) of rain fell in 8 months. By contrast, 770mm (30in) of rain fell in the wettest 8 months (April to November) of 1974.

Electrical and electromagnetic geophysical field data recorded at Woodlawn are, to some extent, dependent on ground conditions, particularly soil moisture content. Surveys run in the middle of 1974, when the ground was saturated to the surface, are likely to have produced somewhat different results from those of identical surveys run in dry weather.

Table 1 lists the monthly rainfall data for the Woodlawn area from 1969 to 1976, the period during which all of the geophysical data were gathered. This table makes it possible to deduce the likely ground conditions during the running of any particular geophysical survey.

TABLE 1 Monthly Rainfall Totals (cm)

	Jan.	Feb.	Mar.	Apr.	May	June
1969	.0	13.1	9.1	6.5	5.4	7.1
1970	7.0	7.1	6.2	7.2	5.9	2.7
1971	6.9	12.5	4.5	2.5	2.0	0.4
1972	24.1	3.0	1.9	6.2	2.6	0.6
1973	4.8	11.5	4.0	0.7	2.9	4.3
1974	8.0	2.7	4.9	16.0	7.7	5.0
1975	2.9	9.1	6.6	9.2	2.4	17.7
1976	13.3	7.2	10.1	2.1	0.7	5.2
	July	**Aug.**	**Sept.**	**Oct.**	**Nov.**	**Dec.**
1969	2.9	6.0	5.2	23.2	13.0	4.2
1970	0.5	4.0	13.0	4.0	8.3	10.4
1971	1.7	6.8	3.4	2.7	9.7	10.0
1972	1.0	6.1	1.4	7.3	4.7	0.5
1973	5.5	8.2	3.4	13.1	14.0	4.0
1974	4.7	19.7	6.8	11.8	5.4	0.0
1975	8.7	3.9	6.3	9.8	2.5	5.2
1976	4.0	2.1	5.1	18.9	7.4	2.0

Data for the years 1969 to 1971 are based on records kept at "Pylara", the adjoining property, data for later years are based on records kept at Woodlawn.

Geology, Geochemistry and Setting of the Deposit

Geological Setting

The regional geology is presented on Fig.3* and is briefly discussed here.

The oldest rocks in the Woodlawn area are a thick sequence of quartz rich greywacke, shale, chert and siltstone of Upper Ordovician age. These sediments represent a distal flysch accumulation, deposited in a deep marine basin. No Lower Silurian rocks have been identified in the area, though sedimentation may have continued into the Silurian.

During middle to late Silurian, a shallow shelf sequence of quartzose sandstone, mudstone, siltstone and minor limestone was deposited, transgressing onto the Ordovician rocks. This was succeeded by a period of acid volcanism, during which the Woodlawn Volcanics were deposited. The acid volcanics were succeeded by a sequence of basic volcanics, including lavas with well developed pillow structures, basaltic tuff, chert, spilite and keratophyre. Dolerite intruding the Woodlawn Volcanics may be related to this basic sequence. The acid and basic volcanic sequences are overlain or transgressed by a thick sequence of quartz rich sandstone, slate and siltstone with proximal flysch characteristics. Locally, quartz sandstone and basic volcanic beds interfinger at the contact. The flysch sediments are probably earliest Devonian.

The Ordovician and Silurian sequences were strongly deformed and regionally metamorphosed to lower greenschist facies during the Bowning Orogeny in the middle of the Early Devonian. At about this time, hornblende biotite granite stocks intruded into anticlinorial zones occupied mainly by Ordovician metasediments. The Silurian sequence largely occupied the

---

*The folded Figs. 3, 7, 8 and 9 can be found in the envelope accompanying the book.

intervening synclinorial zones, which were probably deformed depositional troughs or grabens. The Early Devonian deformation produced open, inclined folds accompanied by a ubiquitous, regional, westerly dipping, axial plane slatey cleavage. Westerly dipping, meridional thrust faults bound some of the synclinorial and anticlinorial zones. These possibly developed during the middle to late Silurian but further movement probably took place on them during the Early Devonian.

The terrestrial Tarago Conglomerate was deposited after the Bowning Orogeny and unconformably overlies all the older units. It consists of boulder and cobble conglomerate, lithic arenite and red siltstone and contains considerable amounts of volcanic detritus, probably derived from the Woodlawn Volcanics to the west. The conglomerate is overlain by calcareous sandstone and siltstone and thin bedded, fossiliferous limestone of late Lower Devonian age, deposited in shallow marine conditions.

The Tabberabberan Orogeny during the Middle Devonian terminated sedimentation in the area. The rocks were again affected by strong east-west compression and underwent low grade regional metamorphism.

Tertiary deep weathering and epeirogenic movements during the Quaternary accentuated the present horst-graben topography of the region, which largely reflects the structural configuration established in the Early Devonian.

Local Geology

The sequence containing the Woodlawn mineralization comprises coarsely porphyritic quartz feldspar porphyry, agglomerate, tuff, commonly foliated, rare welded tuff, tuffaceous shale, black shale and rare chert with some chlorite and talc chlorite rocks (Woodlawn Volcanics).

It overlies a dominantly sedimentary succession comprising quartzose sandstone, siltstone, shale, black shale, fossiliferous limestone and minor tuffaceous sediments. The contact between the two units at Woodlawn is faulted in part. The orebody is contained in the lower part of the Woodlawn Volcanics within fine grained, largely sedimentary rocks which are laterally equivalent to and partly derived from the acid extrusive rocks.

The local geology is presented in Fig. 4. This map also shows the location of the grid which was used for gradient array IP surveys (Quick and Cifali, this volume, p 333). This grid became the Woodlawn Test Grid and was used for locating most of the detailed geophysical surveys over the orebody.

The tuffs of the host sequence generally possess a welded or metamorphically foliated matrix containing clasts ranging in size from less than 1mm to more than 100mm across. The clasts include mainly acid volcanic material and some sedimentary and chert fragments. The coarse grained acid fragmental rocks commonly consist of large, flattened fragments of fine grained acid volcanics, mainly pyroclastics, in a fine grained tuff matrix. The fragments commonly show a different degree of alteration from that shown by the matrix. The rhyolitic flows are less abundant than the tuffs and generally occur higher in the sequence. Chlorite talc schist and chloritic tuff are common, particularly in the footwall zones beneath massive sulphide bodies. The chert horizons contain recrystallized radiolaria as well as volcanic and sedimentary detritus. Tuffaceous shale is the dominant sediment and grades into fine grained tuff. Black shale occurs at several horizons and locally contains poorly preserved graptolite remains. *Monograptus* sp., identified from black shale in the host sequence at Woodlawn, and *Monograptus bohemicus*, identified from an approximately

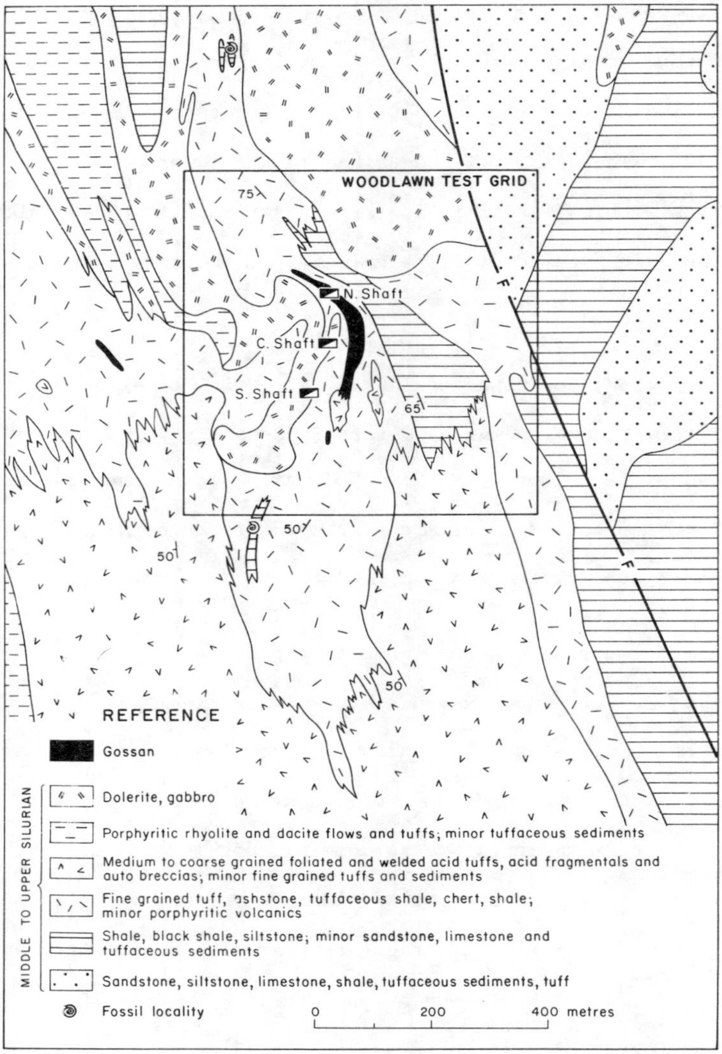

Figure 4 Woodlawn solid geology

equivalent horizon in the sequence southwest of Tarago, indicate a Middle to Upper Silurian age (Strusz and Nicoll, 1973).

The relationships between volcanics and sediments is marked by rapid facies changes with interfingering of rock type and gradational contacts. Minor strike slip and oblique slip faulting may account for some apparently discordant contacts between rock units. Overall, there is a gradation in the host sequence from dominantly acid fragmental volcanics in the south to dominantly sediments and fine grained volcanics in the north. In three dimensions, the fragmental volcanics appear to form a central pile fringed with finer grained volcanics and sediments. Minor base metal mineralization is common in the central volcanic pile. The lenses of massive sulphide ore are developed in the fine grained volcanic/sedimentary fringe and grade laterally through a stratigraphically equivalent zone of mineralized volcanics into the central volcanic pile.

The sequence has undergone low grade greenschist facies metamorphism reflected in the recrystallization of fine grained material and the development of sericite, muscovite, albite, tremolite, epidote, chlorite and talc. Structural deformation has produced ubiquitous slaty cleavage in the finer grained tuffs, generally parallel or subparallel to bedding. The tuffs, particularly the foliated tuffs, have undergone intense local silicification and alkali leaching. Soda depletion is particularly characteristic of the host sequence in the central volcanic pile and adjacent to mineralization in the fine grained volcanic/sedimentary fringe.

Factors Contributing to Geophysical Response

Three of the most important factors contributing to the geophysical response of the deposit are: the geometry of the deposit; the nature, degree of concentration and distribution of the mineralization; and the contrast between the orebody and the host sequence.

Geometry

The massive sulphides mainly occur in one large, continuous body, folded into an open asymmetrical westerly plunging syncline, but with marked changes of plunge in the trough of the syncline. The southern limb is fairly regular, striking about 030°M and dipping to the northwest at about 40° to 60°. The northern limb strikes about 305°M and dips steeply to the southwest though locally, near the surface, it is vertical or slightly overturned. The changes of plunge in the trough of the syncline have produced zones in which the lens of sulphides is near vertical on either side of a flat lying zone. The geometry is well displayed in the cross sections and plan of Fig. 5, (reproduced from Malone et al., 1975) which are based on the extensive drilling at Woodlawn. The strike length of the folded massive sulphide body is about 300m. The small strike length and complex shape pose considerable difficulties in planning and interpreting airborne geophysical surveys (Chapter 6) in particular.

Nature, concentration and distribution of Mineralization

Two styles of mineralization contribute to the geophysical responses. The most important contribution is from the massive sulphides; the other, minor contribution is from mineralized volcanics, located along strike to the south of the massive sulphide body. Other styles of mineralization present in the deposit include chlorite schist copper ore and small zones of chlorite schist complex ore. Both these types of ore occur beneath or downdip from the massive sulphide lens and either are too deep to produce a geophysical response at the surface or their response is masked by the response of the massive sulphide body. Only the massive sulphides and the mineralized volcanics are therefore described.

Massive Sulphide Mineralization

The main massive sulphide lens, apart from a few included lenses or tongues of volcanics, is a single continuous body of sulphides with

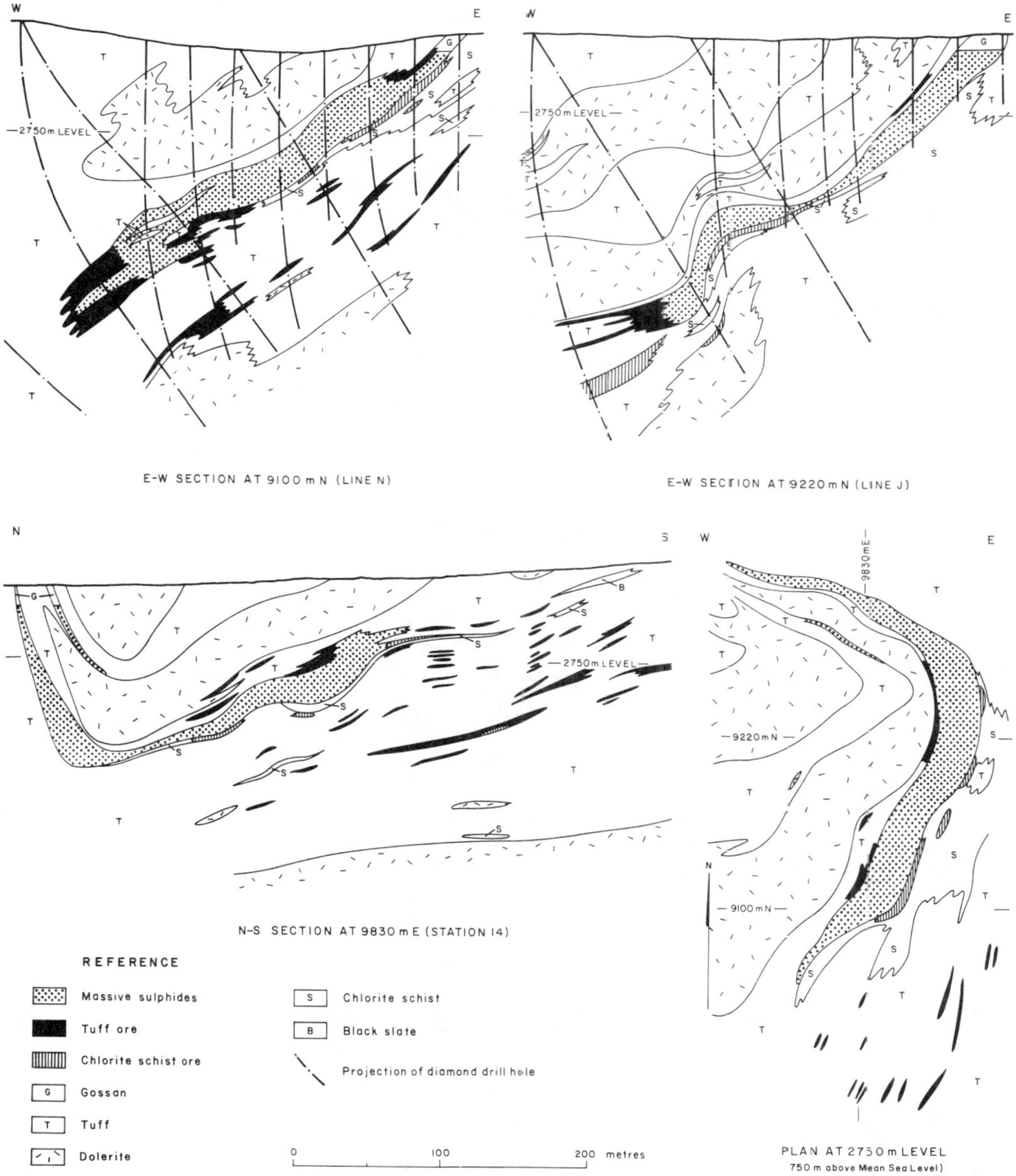

Figure 5 Geological sections and level plan

an unfolded maximum strike length of about 420m (1380 ft) and a downdip length of about 300m (980ft) (Fig.5). The thickness of the lens, apart from thinning at the edges, ranges from about 45m (150ft) to about 5m

(16ft). The more abrupt changes in thickness appear to be associated with folding and cross folding of the lens.

The upper surface of the main lens is in sharp, apparently concordant, contact with the overlying sequence. In a few places, the contact is cut by quartz veining which has locally carried remobilized sulphides into the adjacent tuff or dolerite. The bottom surface is generally in sharp contact with unmineralized footwall; it is marked by an abrupt drop in total sulphide content where chlorite schist ore underlies the lens.

The massive sulphide ore contains generally about 75% or more of total sulphides, consisting of pyrite, sphalerite, galena, and chalcopyrite in order of decreasing abundance, plus probably less than 1% of tetrahedrite-tennantite, arsenopyrite, pyrrhotite and stannite. The ore is fine grained, and generally finely to coarsely banded. The most conspicuous feature is the banding, particularly in the upper and lower parts of the body. The bands are regular, laterally extensive, and range in thickness from less than a millimetre to 20mm. They are generally concordant with the layering in the overlying and underlying tuffaceous sequence, apart from local kinking and folding. Plate 1 presents a polished section of typical banded ore, displaying minor folding of the sulphides as well as fine banding.

Individual bands are composed of the same components as make up the body as a whole, though the proportions of the individual sulphides vary from band to band; some bands consist solely of pyrite, sphalerite or, less commonly, galena.

The part of the orebody between the well banded top and bottom layers is either not banded or displays poorly developed wide banding.

Plate 1

Generally, this part of the orebody is more pyrite rich and contains less base metal sulphides than the banded ore.

The gangue minerals in the massive sulphide ore are talc, chlorite, barite, quartz and volcanic detritus; local thin lenses of tuffaceous shale are interbedded with the massive sulphides.

The grades of individual assay lengths of massive sulphide ore vary within wide limits, from 0.3 to 4% for copper, from 1 to 20% for lead and from 2 to 30% for zinc. Some of the local higher grades in the orebody may be due to remobilization of the minerals. In particular, remobilization may have produced the coarser grained, monomineralic layers.

The separate bodies of massive sulphides above and below the main lens are essentially similar to the main lens. Whether or not these bodies produce recognisable geophysical responses depends on their position relative to the main lens and the accuracy of the particular survey. For the most part, their responses will be masked by that from the main lens.

Mineralized Volcanics Ore

This style of mineralization occurs to the south of and downdip from the massive sulphide lens. The most important as regards geophysical response are those zones of ore occurring near the surface along strike to the south from the massive sulphide body. The locations of the zones reaching the surface are shown on Fig. 6 which shows the plan distribution of the massive sulphide lenses and mineralized volcanic ore zones at the base of the weathering profile. Other similar zones occur farther south and downdip but do not reach the surface.

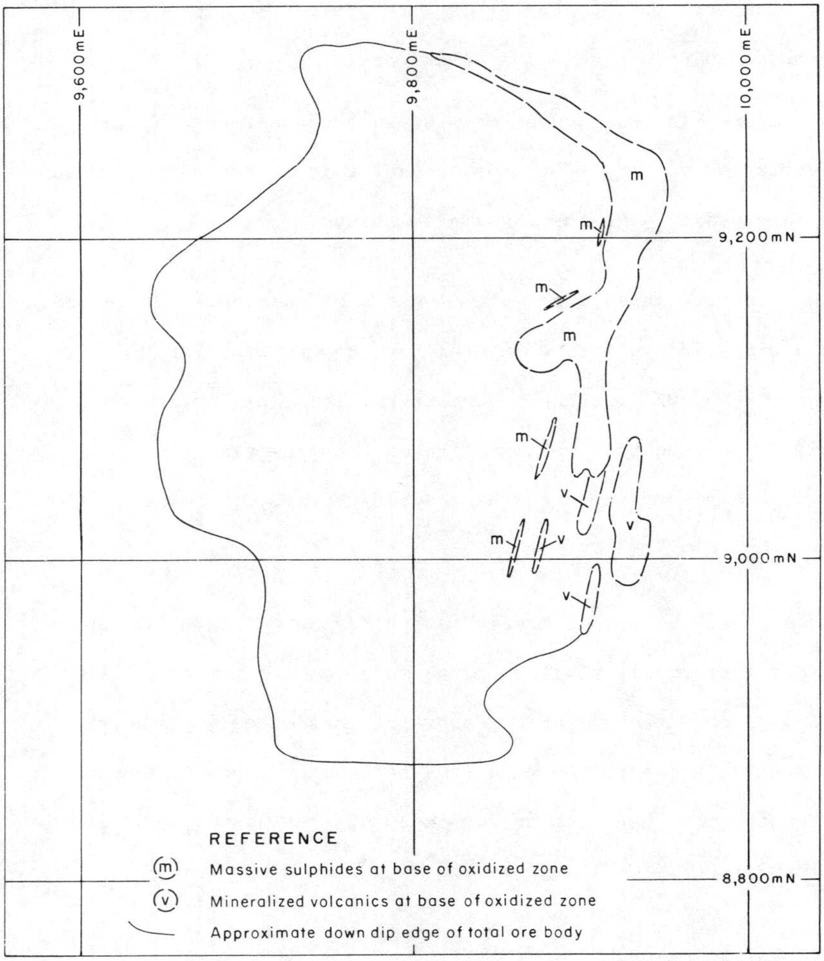

Figure 6 Distribution of mineralization at base of oxidized zone.

The mineralization in the volcanics achieves ore grade because of a greater concentration of stringers of massive sulphides and pyrite chalcopyrite veinlets than is present throughout the enclosing volcanics in which pyrite is the dominant or only sulphide. The mineralized and pyritic volcanics are commonly silicified and locally chloritized, and the silicification in particular may have a significant effect on their electrical conductivity.

The sulphide content of the mineralized volcanics ore is generally less than 25% on average; the enclosing volcanics probably contain 5% to 10% sulphides.

Fig. 6 shows the distribution of sulphides at the base of the weathering profile. The depth to fresh sulphides varies from place to place. Fresh sulphides occur in the mineralized volcanics generally within 5m (16ft) of the surface and is never deeper than 10m (33ft). The depth to solid massive sulphides is about 20m (66ft) at the southern end of the lens and reduces gradually to the north to about 10m (33ft). The massive sulphides are capped by a thin blanket of friable supergene sulphides, generally about 1m thick but locally much thicker. The supergene sulphides are overlain in turn by ferruginous clay gossan extending to near the surface. The supergene zone is porous and permeable and is saturated with acid, metal rich water. This relatively freely circulating fluid may have some influence on detailed electrical surveys.

Host Sequence

Most of the host sequence, particularly around the northern part of the deposit, contains very little mineralization and contrasts well with the orebody. Mineralization, mainly consisting of about 5% pyrite, is much more common in the volcanics at the southern end of the deposit, and in those volcanics below the ore zone or its lateral equivalents to the south. In addition, generally pyritic black shale occurs in the sequence below the central part of the ore zone and extends beyond it along strike to the north and south. Black, pyritic shale also occurs at about the same stratigraphic level as the main massive sulphide lens but some 50m (165ft) to the south. These black shale units and the mineralized volcanics produce locally quite strong electrical responses.

Superficial Cover

Soil and clay gossan overburden and the depth and degree of weathering of the massive sulphide orebody undoubtedly influenced the measured geophysical responses, particularly to detailed ground surveys.

A layer of partly elluvial soil, two to three metres thick, covered the subcrop of the massive sulphide body. Flat lying slabs about 0.5m thick, of ferruginous-cemented polymictic conglomerate occurred locally on the surface above the orebody and to the east. These were not directly related to the orebody but were dissected remnants of a once widespread Tertiary or Recent depositional profile.

The soils graded downwards into a ferruginous clay gossan formed by *in situ* weathering and leaching of the massive sulphides. This clay gossan was directly above a zone of supergene sulphides. The contact between the clay gossan and supergene zone, which apparently coincided with the water table, was extremely sharp generally but in places was marked by water filled cavernous zones.

The supergene zone, consisting of friable secondary sulphides, was permeable, and filled with acid, metal rich and probably very conductive ground water. This zone was generally about 0.5m to 1.0m thick overlying solid massive sulphides. However, towards the northern end of the orebody, the supergene zone was 10m to 15m thick in narrow channels incised into the massive sulphides. These channels of *in situ* alteration were presumably formed by percolating groundwater. Movement of conductive water through the supergene zone, and particularly through the relatively deep, narrow channels, may have had some effect on detailed electrical ground surveys and may have contributed to differences between electrical measurements at different times.

The contact between the supergene sulphides zone and the solid massive sulphides is fairly regular except for the local channels. The top of the massive sulphide orebody is about 6.5m below the surface at the northern end of the body. The depth to massive sulphides increases to a maximum of about 20m below the surface at the southern end, nearly in sympathy with the topography which rises 15m from the northern end to the southern end of the massive orebody. Geophysical responses undoubtedly have been affected by the different depths of burial of various parts of the Woodlawn orebody.

Geochemical Expression

The surface geochemical expression of the Woodlawn deposit is not directly relevant to a consideration of its geophysical response. However, in this instance, as is commonly the case, the geochemical anomaly was located first and helped to position the only exploratory geophysical survey, an induced polarization (IP) survey completed in 1969 (Smith, this volume,p 79). All subsequent geophysical work on Woodlawn was done after drilling had located the orebody.

The distributions of the copper, lead and zinc values contained in surface soils around the deposit are shown in Figs. 7, 8 and 9. Each of these figures also shows the location of the gossan cap of the orebody and of the Woodlawn Test Grid. The figures are based on resampling of the prospect carried out in 1970. The later samples were analysed by atomic absorption spectrometry yielding more accurate results than the emission spectrography analyses of the original samples. The geochemical anomalies located by the second survey agreed very closely with those of the original survey. Both surveys involved collection of B horizon samples wherever the B horizon was developed; C horizon samples were collected in areas of skeletal soils.

The distribution of the copper, lead and zinc values are all different. They neither singly nor in unison provide an unequivocal guide to the location of the mineralization. Peak values for lead and copper nearly coincide over part of the orebody and such a concurrence certainly indicates an area requiring further investigation. However other lead peaks of comparable magnitude occur to the northwest and northeast of the orebody. Consideration of the raw geochemical data and the surface geology indicate that these peaks are related to sources underlying the areas of peak geochemical values. On the basis of knowledge at the time, all of these anomalous areas required further exploration.

The zinc peaks are scattered over a wider area than the lead peaks, and none of the zinc peaks coincide with the orebody. Zinc is a mobile element and its wide dispersion was expected. However, the scattering of the zinc peaks coupled with the topography in the area indicated the possibility of multiple sources of the geochemical anomaly. This also tended to enlarge the area which required further exploration.

The main area of copper anomaly is located to the south of the massive sulphide body, partly over the area where the copper orebody was subsequently found but much of it over areas of insignificant mineralization. The distribution of the copper anomaly is particularly significant, bearing in mind the effect of topography on the distribution of geochemical anomalies. The expectation is that the anomalies will extend in the direction of ground water movement, which is generally roughly parallel to surface drainage.

At Woodlawn, the extension of the three areas of anomalous metal values to the northwest and to the northeast constitute hydromorphic tails in the directions of drainage. The bulk of the copper anomaly, however, is located topographically higher than the orebody (Fig.2). Hence, prior to the results of drilling being available, the copper anomaly distribution resulted in extending the area requiring further investigation much farther to the south.

The sources of most of the anomalous metal values are quite obvious once the data obtained by subsequent exploration and drilling are considered. Many sources contributed to the anomalies, although clearly, the orebody contributed most of the metal values. With only the geochemical and surface geological data as guides, it was not possible to delineate only the significant part of the geochemically anomalous area for further investigation.

A dipole-dipole IP survey was selected as likely to be the most effective technique for further testing the geochemical anomaly. It was expected that IP would be effective in identifying the best drilling targets. It was also felt that a strong IP anomaly would encourage management to allocate funds for a satisfactory drilling test. It must be remembered that, at the time, Woodlawn was merely a geochemical anomaly in a geological province in which no significant orebody had yet been found. The already mined Captains Flat orebody some 60km (37 miles) to the south (Malone and Whiteley, this volume, p 23) was regarded as being only marginally economic under current costs and conditions.

The IP survey was run by McPhar Gecphysics Australia Proprietary Limited in July and August of 1969 and covered the entire extent of geochemically anomalous ground. Results cf that survey are discussed by Smith (this volume,p 79).

In retrospect, it now appears that a self potential (SP) survey would have been the most cost effective geophysical technique to use in initial follow up of the soil geochemistry. The SP survey (Cifali and Whiteley, this volume, p 321) which was run after drilling had outlined the orebody, located a large anomaly coincident with the strongest centre of the geochemical anomaly and almost exactly outlined the top of the massive sulphide body. The SP survey also located five other anomalies,four of them comparable in magnitude to that over the orebody but lacking coincident geochemical anomalies which suggested sources other than significant mineralization. Black shale was suspected in most cases and this was subsequently confirmed. One such black shale locality which had been drilled to confirm the source of the anomalous responses was chosen as the site for comparative geophysical study. It was also realized that although the Woodlawn orebody would be mined, this Black Shale would remain for future study.

Black Shale Locality

The Black Shale occurs about 2.5km (1.6 miles) south of the Woodlawn deposit. The interpreted solid geology of this area is presented in Fig.10 and is briefly described below. Fig. 10 also shows the location of SP and strong IP anomalies. The location of the black shale and the orebody are shown in Fig.2.

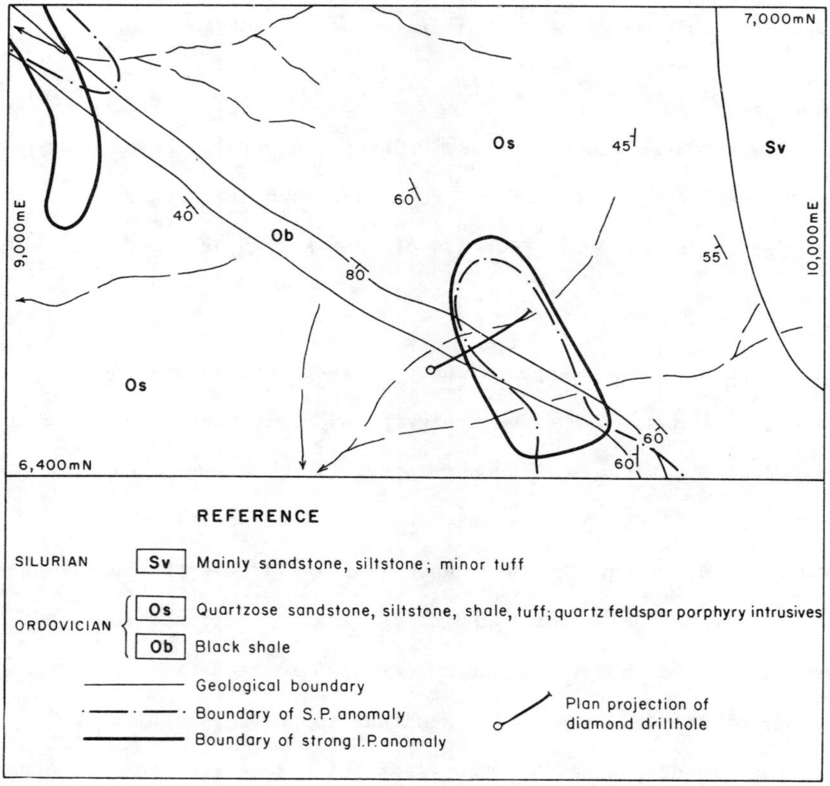

Figure 10 Interpreted solid geology, Black Shale locality

The Black Shale is about 40m (130ft) thick and crops out discontinuously for about 3km (1.9 miles) overlain and underlain by very fine to medium grained, locally feldspathic, quartzose sandstone and siltstone with probably intrusive quartz feldspar porphyry. The shale unit is black, displays ubiquitous slatey foliation, contains local silty interbeds and is interbedded with fine sandstone beds near the top and bottom contacts. Trace to common pyrite occurs on the foliation or infilling irregular fractures. The entire sequence dips to the southwest and is probably at least partly overturned. What appears to be the same black shale unit crops out farther to the northwest, dipping to the west but overlying Silurian volcanics. Graptolites from one exposure in this northern black shale indicate an Upper Ordovician age. This indicates an overturned relationship with the underlying Silurian rocks but whether this is due to overfolding, faulting or some combination is not known.

The foliation is the most prominent surface in outcrop but is not generally parallel to bedding; this is confirmed at the graptolite localities where the foliation cuts across the bedding planes on which the fossils are preserved.

Structural geology of the area is poorly known for several reasons: because the foliation masks bedding, because outcrop is poor, and because not a great deal of mapping has been done in the area. The extent and lack of repetition of the Black Shale suggests one black shale unit in a very thick sandstone siltstone sequence.

One diamond drillhole has penetrated the sequence. It drilled through quartz feldspar porphyry and shale to 73.2m (240ft) and through the Black Shale from there to total depth of 214.3m (703ft). The thickness in the drillhole is apparently due, at least in part, to folding and fault repetitions of the Black Shale although this is not reflected in outcrop. Minor amounts of pyrite occurred throughout the Black Shale. The only base metal sulphides consisted of minor amounts of sphalerite and galena accompanying pyrite in randomly oriented veinlets. The thick development of black shale in this vicinity may have contributed to the localized, strong geophysical response. Contact metamorphic effects caused by the porphyry intrusions may also have contributed to the geophysical response.

Geochemical expression

Copper, lead and zinc analyses of grid soil samples over the area of Fig.10 are presented on Fig.11. The analyses were done by

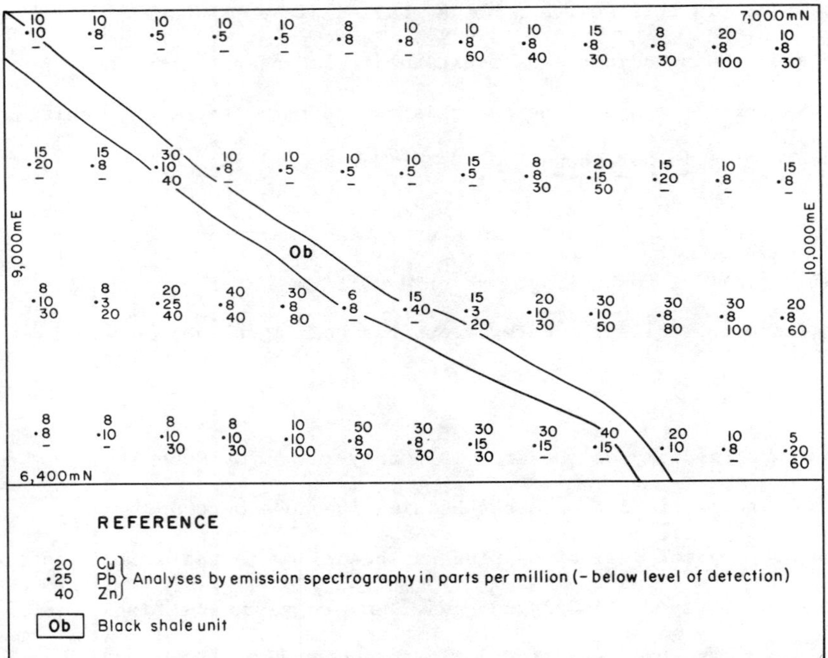

Figure 11 Surface geochemistry - Copper, Lead and Zinc, Black Shale locality.

emission spectrography which has a high lower limit of detection for zinc of about 20 parts per million and produces results which are accurate to about plus or minus 50%. Despite the limited analytical accuracy, the sampling was sufficiently close to show that no significant geochemical anomaly was associated with the Black Shale as a whole or with the part which produced an SP anomaly.

References

Malone, E.J., Olgers, F., Cucchi, F.G., Nicholas, T., and McKay, W.J., 1975. Woodlawn Copper, Lead and Zinc Orebody, in Economic Geology of Australia and Papua New Guinea, Volume 1, Metals. Australas. Inst. Min. Metall., Melbourne.

Strusz, D.L. and Nicoll, R.D., 1973. Palaeozoic fossils from the Tarago region, N.S.W. Rec. Bur. Miner. Resour. Geol. Geophys. Aust., 1973/15 (unpublished).

CHAPTER 4

PREDISCOVERY GEOPHYSICAL SURVEYS

4.1. Early Geophysical Exploration of the Woodlawn Orebody

R. J. SMITH

Contents

Abstract

Prior to the discovery of the Woodlawn orebody geophysical surveys were completed with the induced polarization and resistivity methods over an area selected by geological and geochemical reconnaissance. Supplementary vertical magnetic intensity measurements were also made.

Interpretation was hampered by interference from grounded fences in the centre of the surveyed area, however, resistivity and induced polarization anomalies were defined and significant subsurface mineralization was inferred. Subsequent drilling outlined the Woodlawn orebody.

Introduction

During July and August of 1969, geophysical test surveys were conducted at Woodlawn by McPhar Geophysics Proprietary Limited. This area had been selected as a base metal prospect following geological reconnaissance and geochemical sampling (O'Driscoll, this volume p 15) and the objective of the geophysical survey was to locate and define possible concentrations of either disseminated or massive metallic sulphide mineralization for testing by drilling.

The principal methods chosen were resistivity and induced polarization (IP) supplemented by measurements of vertical magnetic intensity. These were the only geophysical measurements undertaken prior to the discovery, by drilling, of the Woodlawn deposit.

Equipment and Field Procedures

Resistivity and IP measurements were made with a McPhar Model P660 variable frequency IP unit. The transmitter for this unit was powered by a 2.5kw motor generator capable of delivering a current up to 5 amps at frequencies of 0.125, 0.25, 1.25, 2.5 or 5.0Hz. At these frequencies, the transmitted waveform is square rather than sinusoidal.

The McPhar IP receiver filters all but the fundamental frequency in the transmitted waveform and measures voltages at any of the above frequencies and the percentage change in voltage between any pair of frequencies. These quantities are used to calculate apparent resistivity (ρa), percent frequency effect (PFE) and metal factor (MF).

For the survey at Woodlawn apparent resistivity was determined at 2.5Hz and PFE was measured over the frequency interval 0.3Hz to 2.5Hz according to the equation:

$$PFE = \frac{\rho 1 - \rho 2}{\rho 2} \times 100$$

where $\rho 1$ = apparent resistivity at 0.3Hz

$\rho 2$ = apparent resistivity at 2.5Hz

PFE is read directly from the IP receiver as a percentage. All measurements were made with a conventional dipole-dipole electrode array using dipole lengths (X) of 100ft (30.5m) or 200ft (61m) and dipole separations (n) of 1, 2, 3 and 4 times the dipole length.

At the time of the survey resistivity units of ohm ft/2π were widely used for convenience in field calculations although they are not common today. Since the original interpretation used these units they have been retained. (it is convenient to remember that 1 ohm ft/2π ≃ 0.5 ohm m:- Ed) Metal Factor (MF) was calculated by the following formula

$$MF = \frac{1000\ PFE}{\rho a}$$

In addition to the electrical survey, measurements of vertical magnetic intensity were made with a McPhar M700 fluxgate magnetometer.

<u>Traverse Locations</u>

A total of 12 resistivity and IP lines each approximately 3000 feet (900m) in length were completed using the dipole-dipole array. Only the 5 lines in the immediate vicinity of the Woodlawn deposit are discussed. The location of these lines is shown in Fig. 1. On all of the lines a dipole length of 200ft (61m) was used with an additional profile along 2000S using a dipole spacing of 100ft (30.5m).

Measurements of vertical magnetic intensity were also made at 100ft (30.5m) intervals along these lines.

It should be noted that all grid references in this paper are to the Imperial System grid in use during initial exploration. The base line of this grid was a rabbit proof fence, aligned approximately magnetic north, through the centre of the geochemically anomalous area. This fence was used as the base line for the initial resistivity and IP surveys.

As the exploration activity in the Woodlawn area was increased, a base line in the middle of the area became inconvenient and the origin was shifted 10,000 ft grid north and 10,000ft grid east. This placed the entire area of interest in the southwest quadrant. The baseline became 10,000W and the IP lines 1,500S to 3,500S became 11,500S to 13,500S. (Fig.1). This second grid was in use when the Woodlawn Test Grid was established. The lines and stations of the Woodlawn Test Grid are also shown on Fig.1 to facilitate correlation with the later geophysical work at Woodlawn reported in this volume.

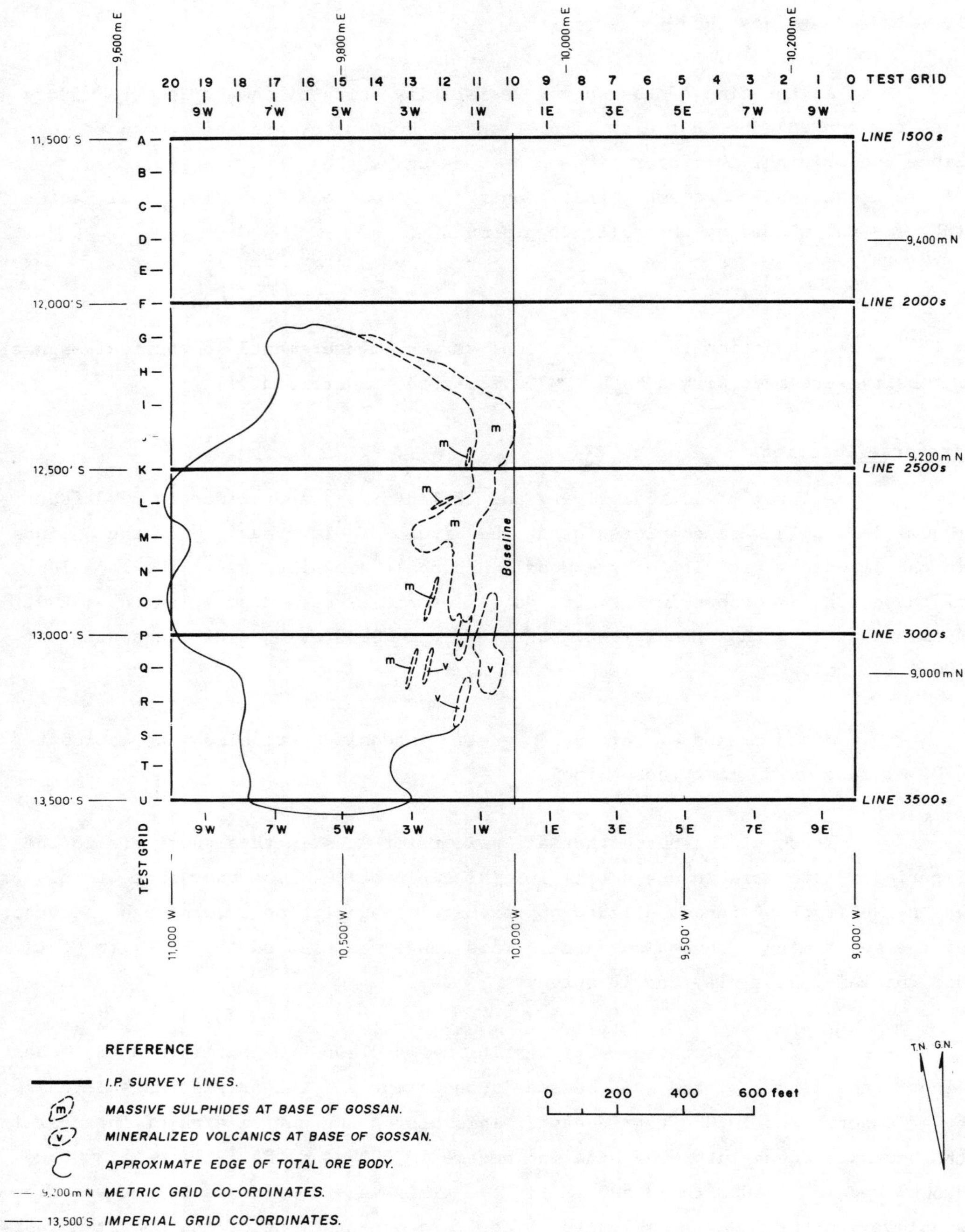

Figure 1. Imperial Grid Plan showing location of electrical traverses.

(Subsequently, Australia adopted the metric system and a metric grid was established at Woodlawn. The origin for the metric grid was placed at 13,000m grid south and 13,000m grid west from the origin of the second Imperial System grid, and consequently the area of interest lies in the northeast quadrant of the metric grid - Ed).

Discussion of Results

The magnetic intensity profiles are shown in Fig. 2. It should be noted that the McPhar M700 magnetometer treats the vertical field intensity as negative in the southern hemisphere. An increase in field intensity, therefore, results in a negative anomaly. It is usual to plot profiles of vertical field intensity with negative values upwards so that profiles retain a conventional appearance, as has been done in Fig. 2.

(The large magnetic anomaly trending in a south-southeasterly direction from 11W on line 500S to 11E on line 2500S is believed due to a dolerite sill (Agostini et al, this volume, p 475) and is not related to the Woodlawn orebody - Ed).

The resistivity and IP data are plotted as pseudosections of apparent resistivity, PFE and MF in Figs. 4 to 9. In the plotting convention used the pseudosection for apparent resistivity is inverted by plotting values for n = 1 at the bottom.

In the original interpretation anomalies were identified on the pseudosections as definite, probable or possible. Although the anomalies have been plotted above the MF pseudosection all three parameters were considered in the interpretation process.

The anomaly classifications are also shown on Figs. 4 to 9.

In general, the IP data were of good quality. Observed values of apparent resistivity were moderately high with no evidence of interference from electromagnetic coupling or masking due to conductive overburden. There was significant relief in both resistivity and PFE resulting in the identification of several anomalies.

A major difficulty was the presence of a rabbit proof fence (RPF), located near the baseline (OE, Fig.1) and extending through the entire grid

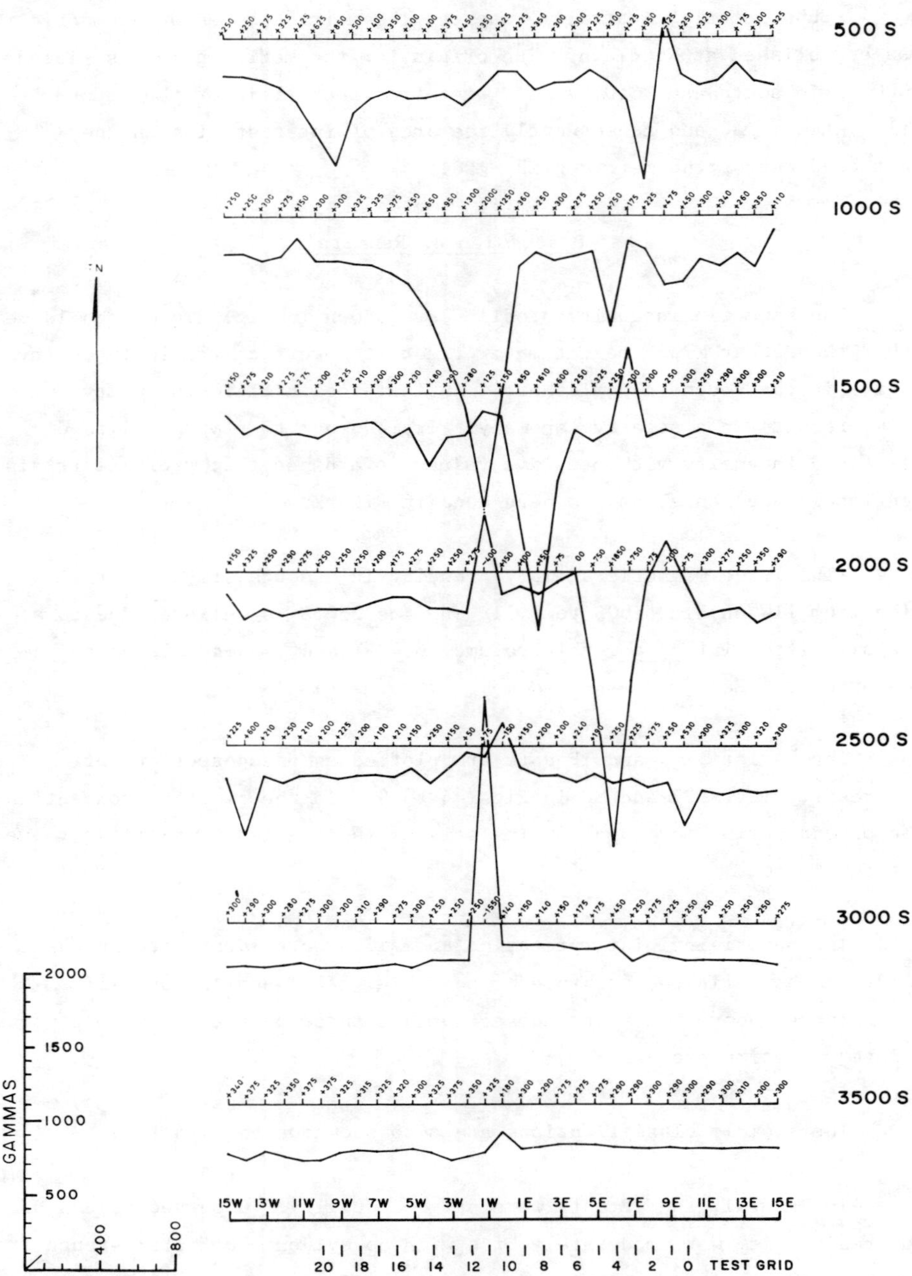

Figure 2. Vertical magnetic intensity profiles.

in a north to south direction. In addition, a branch of this fence extended west from the baseline close to line 2500S and subparallel with it. A rabbit proof fence is normally a mesh fence, buried at its base so that it represents a continuously earthed metallic conductor. As such it can cause pronounced resistivity and IP anomalies, which are dependent on orientation and location relative to the measuring electrodes (such fences are a common source of spurious electrical and electromagnetic anomalies in many agricultural areas of Australia, particularly in eastern Australia - Ed).

Examples of electrical anomalies to be expected from a continuously earthed conductor are shown in Figs. 3a, b, c. The responses shown are from scale model studies representing a shallowly buried water pipe or rabbit proof fence. It is apparent that a continuously earthed conductor can create pronounced apparent resistivity, PFE and MF anomalies. These anomalies are minimized if the conductor and survey lines are orthogonal and the conductor is midway between electrodes but, even so, a pronounced anomaly still occurs.

The north to south fence at Woodlawn was expected to cause pronounced electrical anomalies. Although care was taken to keep electrodes as far from it as possible, a response could not be avoided without complete removal of the fence. At Woodlawn the location of the fence with respect to the electrodes was noted carefully during the survey and the effect of the fence was taken into consideration during interpretation.

Line 3500S (Fig. 4)

Pronounced electrical anomalies are apparent near the baseline, coinciding closely with the rabbit proof fence (RPF). There is a decrease in apparent resistivity, a coincident increase in PFE and consequently an increase in MF. A generally high background PFE is present along the entire line, particularly west of the baseline suggesting widespread disseminated metallic mineralization. A second anomaly is suggested at 7W to 9W, primarily related to a local decrease in apparent resistivity.

The original interpretation identified a definite anomaly at 1E to 2W with a probable anomaly at 7W to 9W. While the precise limits of the anomaly may be debatable there is little doubt of a definite metallic source between 1E and 1W, possibly extending farther to the west.

Unfortunately this anomaly coincides with the rabbit proof fence so it is not possible to say with any certainty that an additional subsurface

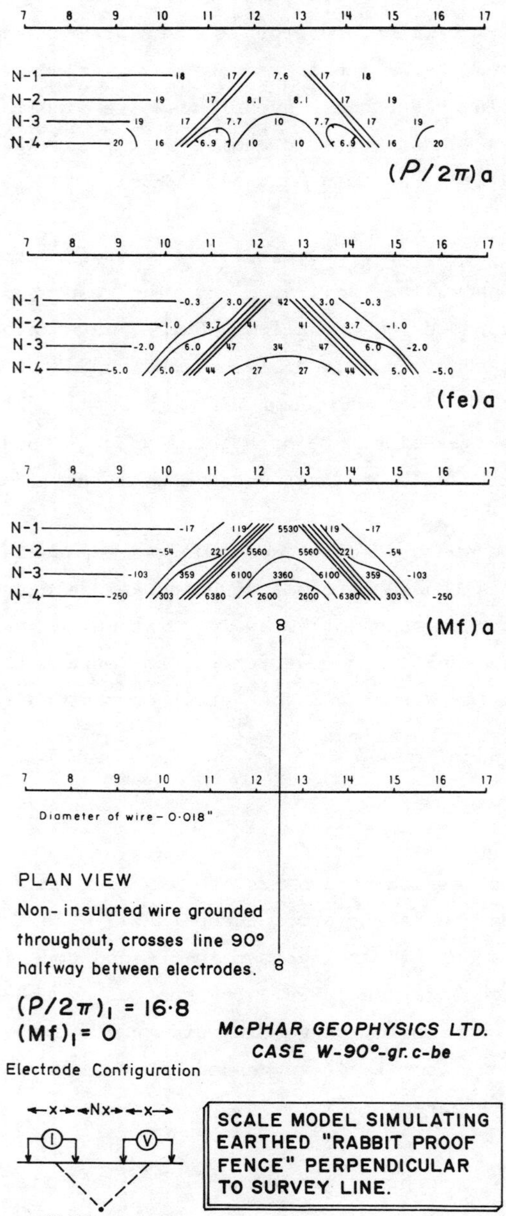

Figure 3(a). Scale model simulating a grounded rabbit proof fence orthogonal to the survey line.

metallic source is also present. The IP response on line 3500S is more pronounced than is associated with the rabbit proof fence on the lines to the south. Original interpretation did, therefore

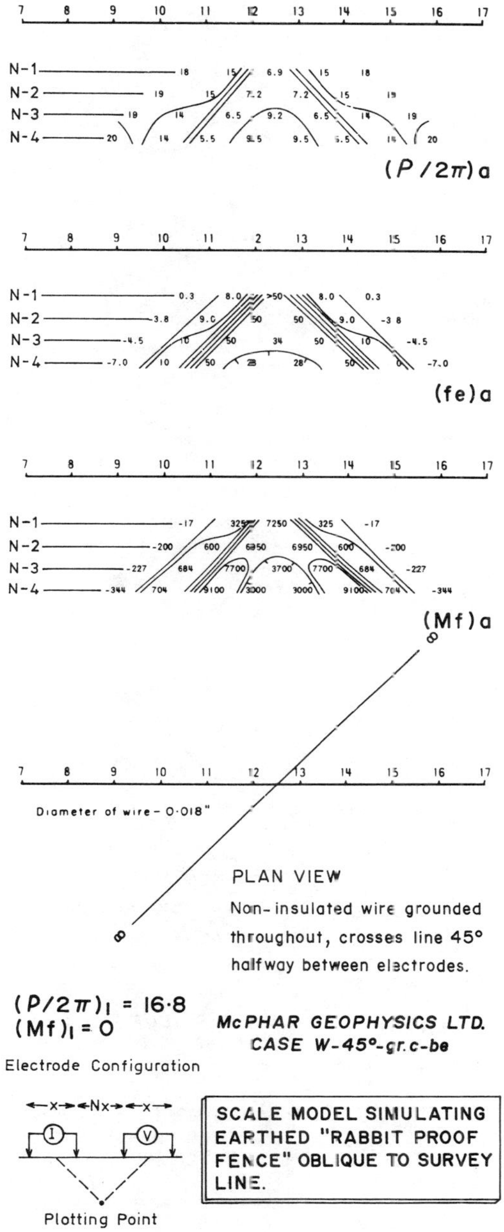

Figure 3(b). Scale model simulating a grounded rabbit proof fence oblique to the survey line.

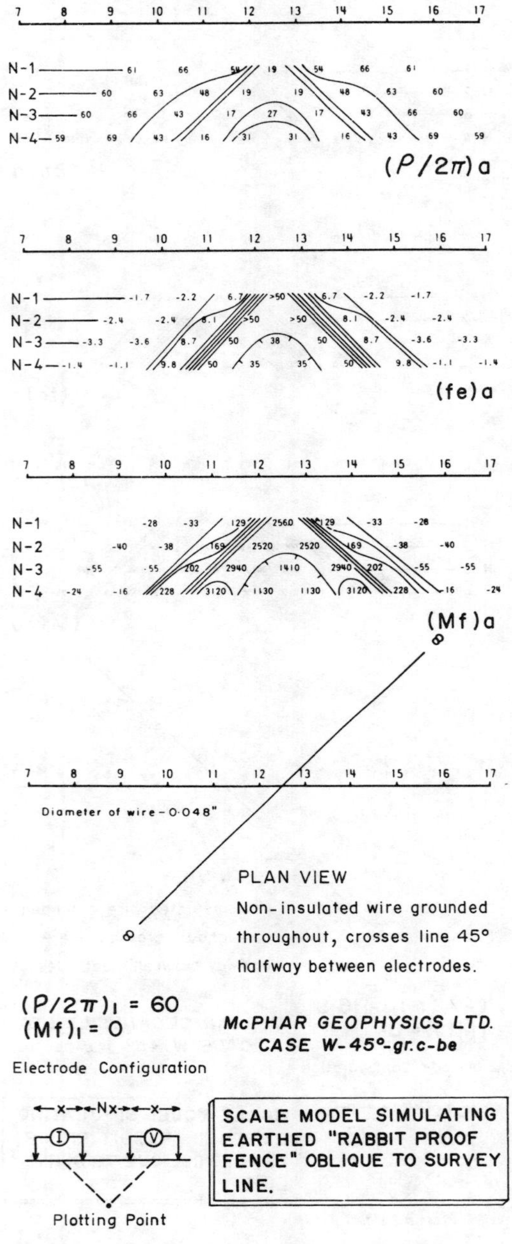

Figure 3(c). Scale model simulating a grounded rabbit proof oblique to the survey line.

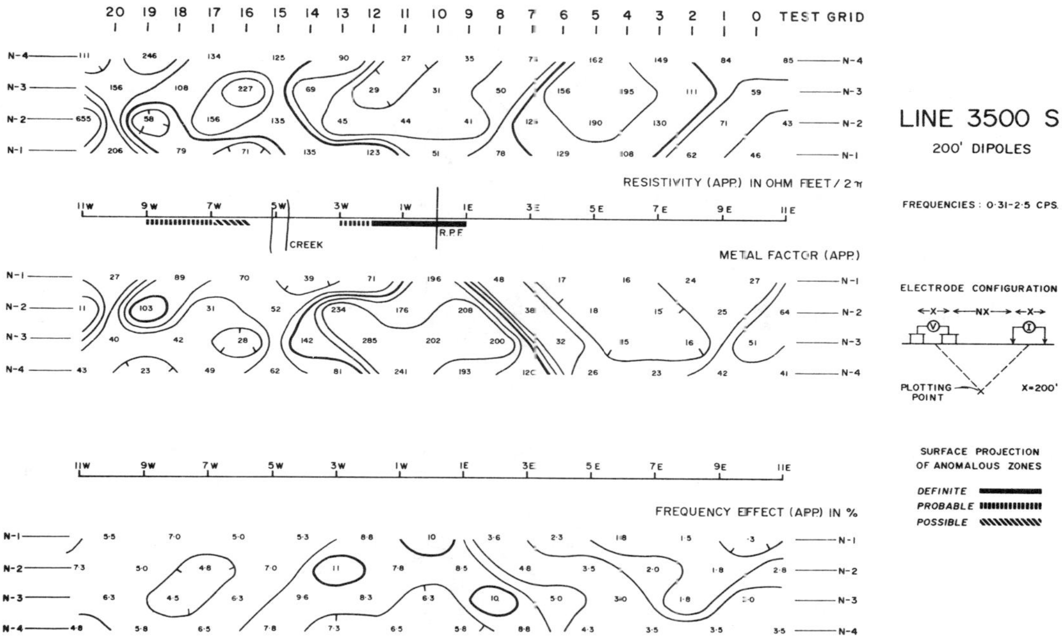

Figure 4. Electrical results, line 3500S (61m dipoles).

suggest a possible additional source at depth. Subsequent drilling suggests that this is unlikely, and that the high background PFE may be related to mineralized volcanics, particularly west of the baseline.

Magnetic measurements were widely spaced but a possible magnetic anomaly of about 100 nT appears to occur at OE. This was essentially a single station anomaly and would need to be confirmed by additional measurements at closer spacing. (This anomaly was associated with the fence since it did not appear in the later magnetic surveys with the fence removed, (Agostini et al, this volume, p 475 -Ed)).

Line 3000S (Fig. 5)

A pronounced electrical anomaly was evident again near the baseline extending to the west of the rabbit proof fence. This anomaly appears as a clearly defined zone of low apparent resistivity and increased PFE between 1E and 3W. Original interpretation identified a definite anomaly between 1E and 3W and, with hindsight, it would be difficult to disagree. Although there is a local increase in PFE the most marked feature of this

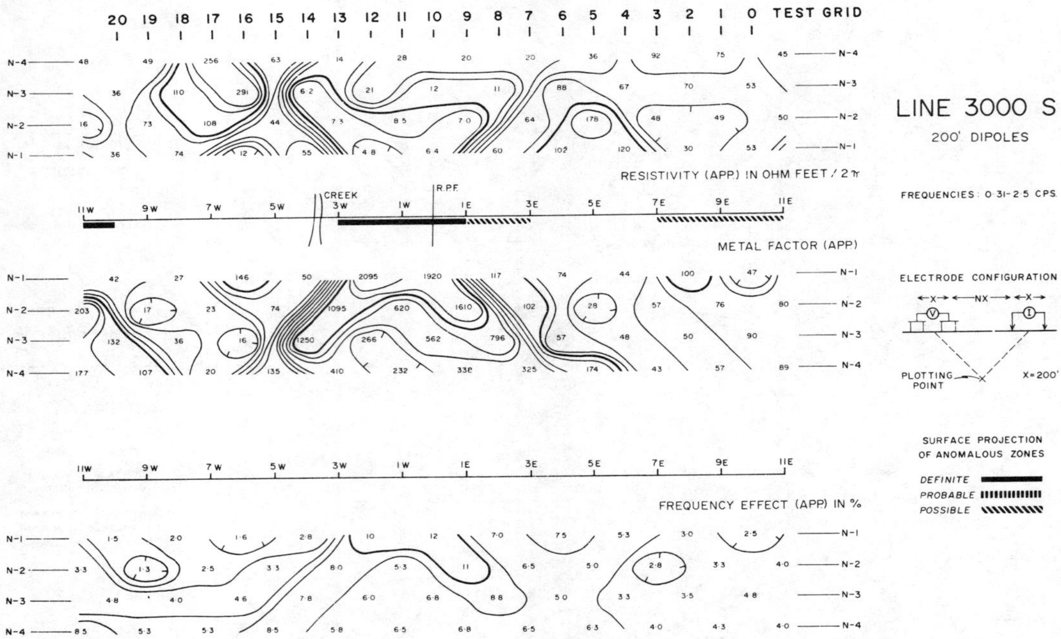

Figure 5. Electrical results, line 3000S (61m dipoles).

anomaly is the pronounced decrease in apparent resistivity between 1E and 3W. This is apparent for n = 1 suggesting a shallow source (or sources), presumably massive metallic mineralization or a cultural source. The rabbit proof fence at OE has probably contributed to this anomaly but it does appear to extend further west than would be expected if the fence was the only source. Measurements with 30m dipoles would probably have resolved this question and may have defined two separate anomalies, these however, were not made.

Magnetic measurements again detected a pronounced single station anomaly at 1W again due to the fence.

On this line subsequent drilling confirmed the massive sulphide mineralization of the Woodlawn orebody at a shallow depth beneath 1W to 2W, dipping to the west. It seems probable that the observed IP anomaly is due to a combination of the rabbit proof fence at OE and the orebody between 1W and 2W.

Line 2500S (Fig. 6)

Electrical data on line 2500S were considered unreliable due to the proximity of a parallel rabbit proof fence between OE and 21W.

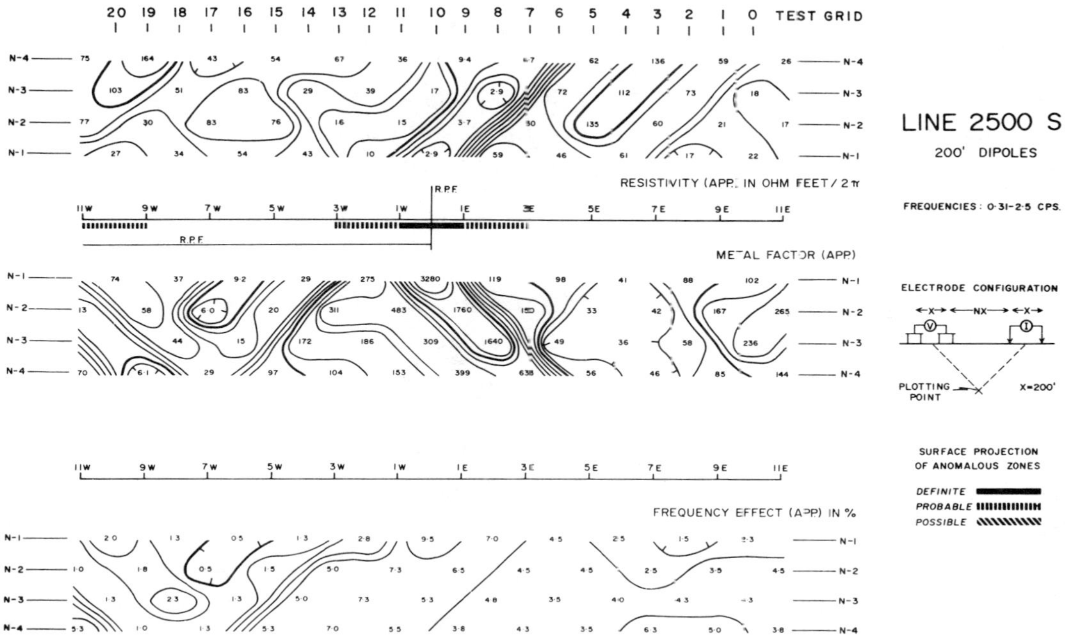

Figure 6. Electrical results, line 2500S (61m dipoles).

The IP response from this fence was difficult to predict since it depends on local soil conditions between the fence and the measuring electrodes. Discrete electrical anomalies on this line could however be interpreted and their possible significance discussed. The most pronounced anomaly on this line was again centred near the baseline. Original interpretation identified a definite anomaly at 1E to 1W with probable extensions to 3E and 3W. The eastern extension was mainly based on high background PFE values while the western extension was primarily based on a decrease in apparent resistivity.

The precise location and limits of the electrical anomalies are questionable. The zone of increased PFE extends from approximately 3W to 13E with a local peak between 1W and 3E. A sharply defined zone of low apparent resistivity between 3W and 1E is most prominent between 1W and 1E. This suggests massive mineralization or a cultural conductor between 3W and 1E with possible disseminated mineralization further to the east.

A possible magnetic anomaly was detected on the line adjacent to the fence between stations 1W and 2W with a maximum amplitude of 300nT at 0E. This anomaly is due to the proximity of the subparallel fence which crossed the survey line obliquely near 1W.

Subsequent drilling has shown massive sulphide mineralization at shallow depths between 0 and 1W, dipping to the west. The observed IP response is probably due to a combination of the rabbit proof fence and subsurface mineralization but they are so close to each other that it is unlikely they could be resolved by measurements with shorter dipoles.

Line 2000S (Fig. 7, 8)

IP measurements with 200ft (61m) dipoles detected a broad zone of increased PFE between 7W and 1E. There was also a pronounced zone of low apparent resistivity between 1W and 3W, with a less pronounced zone extending to 7W and 1E.

The original interpretation identified a definite anomaly at 1E to 1W, with a probable extension to 5W and a possible extension to 7W. This interpretation may be debatable, the main anomaly in all three parameters is between 3W and 1W but in any case the limits of the broad anomalous zone are clearly defined.

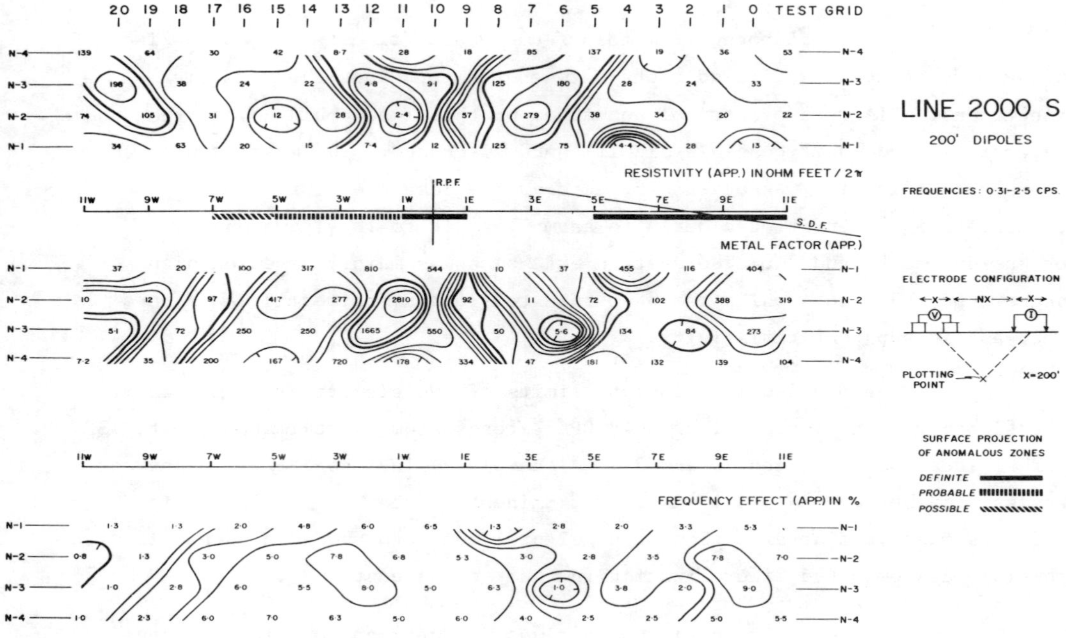

Figure 7. Electrical results, line 2000S (61m dipoles).

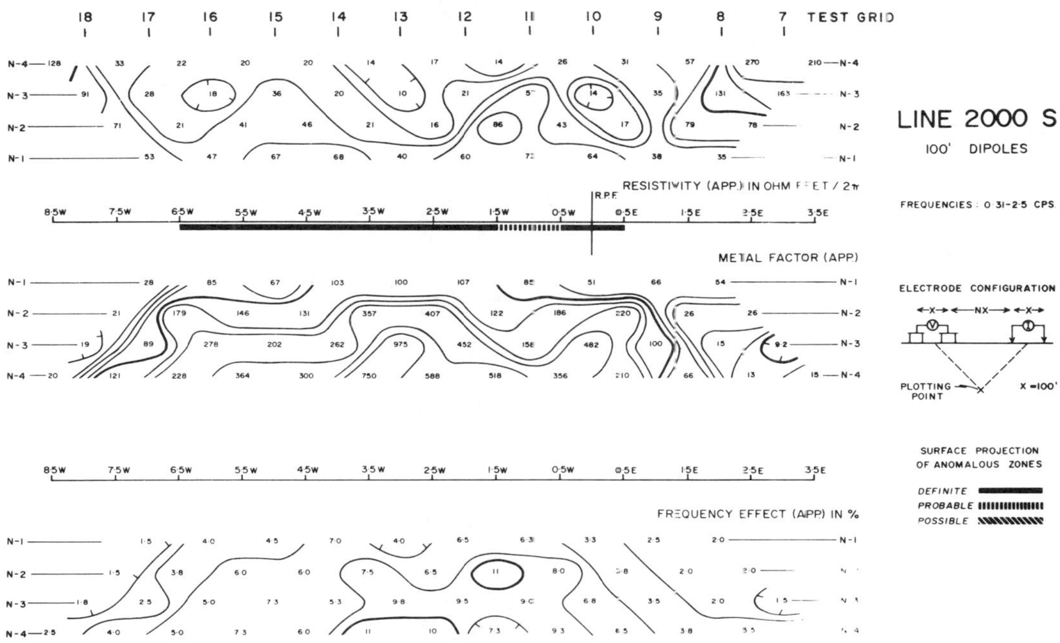

Figure 8. Electrical results, line 2000S (30.5m dipoles).

Further east a definite anomaly was interpreted at 5E to 15E, coincident with a regularly grounded, steel dropper fence (SDF) which crossed the line obliquely near 8E.

The rabbit proof fence at 0E was within the anomalous zone but the main anomaly was farther west and additional measurements were made with 100ft (30m) dipoles in an attempt to resolve it. Measurements with the shorter dipoles (Fig. 8) clearly defined a zone of anomalous electrical response between 0.5E and 6.5W. The original interpretation resolved separate anomalies at 0.5E to 0.5W, coincident with the fence, and 1.5W to 6.5W. In any case it is clear that the anomaly extends much further west than could be explained by the fence and subsurface metallic mineralization is suggested. The main response was apparent for n = 2, 3 and 4 suggesting either some depth to the source or an end effect from mineralization to the north or south of the surveyed line.

Magnetic measurements detected a single station anomaly of several hundred nT at 1W.

Subsequent drilling has confirmed massive sulphide mineralization to the south of line 2000S, subparallel to the survey line, between approximately 7W and 2W. This was in close agreement with the interpreted electrical anomalies.

The magnetic anomaly at 6E is part of the zone which trends northwesterly on several lines. It is not adequately defined by these early data but appears likely to be due to a dolerite which has been mapped in this area.

Line 1500S (Fig. 9)

A poorly defined IP anomaly was detected at 1E to 1W, coincident with the rabbit proof fence. This was mainly defined by a PFE anomaly with no significant decrease in apparent resistivity. A second zone of increased PFE was evident at about 5W to 9W but again without any clear decrease in apparent resistivity.

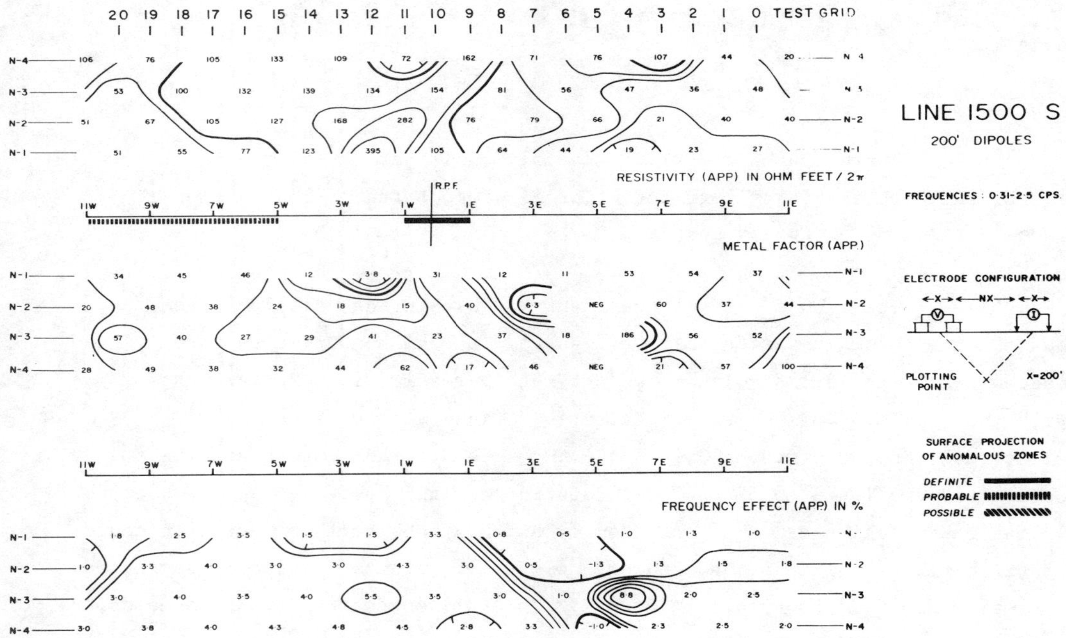

Figure 9. Electrical results, line 1500S (61m dipoles).

The original interpretation identified a definite anomaly from 1E to 1W and a probable anomaly at 5W to 11W. The anomaly at 1E to 1W coincided with the fence and no significant extensions were indicated. The probable anomaly at 5W to 11W is only poorly defined and it was checked by measurements with 100ft (30m) dipoles. A probable anomaly was identified centred near 7W but it is much less pronounced than on lines to the south.

Magnetic measurements detected a negative anomaly of approximately 2000 gamma centred near 2E, similar to anomalies on line 1000S and 2000S.

The electrical results show clearly that the main source of the anomaly on lines 2000S, 2500S and 3000S is absent on line 1500S. Drilling confirmed that the Woodlawn massive mineralization terminates south of line 2000S.

Summary and Conclusions

The original resistivity and IP surveys at Woodlawn were severely hampered by interference from cultural sources which were grounded rabbit proof fences. In spite of this interference there was good evidence to interpret the presence of massive mineralization near 0E on line 2000S, line 2500S and line 3000S. A broad anomaly was still present on line 3500S but the decreased amplitude suggested only disseminated mineralization. The IP response decreased sharply north of line 2000S.

A separate source was confirmed on line 2000S and suggested on line 3000S but interference on line 2500S was excessive.

Subsequent drilling at Woodlawn confirmed the presence of the massive sulphide deposit in good agreement with the original interpretation of these measurements. The original interpretation was inconclusive due to interference from the fences but later measurements, with fences removed, have substantiated the IP response of the sulphide mineralization. (eg Quick and Cifali, this volume, p 333).

Magnetic measurements at 100ft (30m) intervals failed to detect any significant anomaly which could be correlated with sulphide mineralization. A pronounced magnetic anomaly to the north of the mineralization appears to be due to basic intrusives.

The cultural interference experienced in this IP survey is a common problem in the more developed parts of Australia, particularly eastern N.S.W. In general, by careful choice of line and electrode locations it is possible to minimize cultural interference but often these cannot be avoided. At a reconnaissance stage of an exploration program it is rarely practical or possible to remove fences and this survey illustrates the importance of detailed measurements with short dipoles in order to resolve responses from neighbouring sources. It would have been tempting to attribute these anomalies to the fences and downgrade them if detailed measurements on line 2000S had not confirmed a separate source.

CHAPTER 5

PHYSICAL PROPERTY MEASUREMENTS

5.1. The Physical Properties of the Woodlawn Orebody and Surrounding Rocks

E. D. TYNE, M. IDNURM AND E. J. MALONE

Contents

Abstract

Physical properties of the Woodlawn orebody and surrounding rocks were obtained from measurements on rock samples and drill core and in drill holes. The Woodlawn style of mineralization is clearly an excellent electrical and electromagnetic geophysical target. The massive sulphide ore exhibits resistivities in the range 0.01 to 5.0 ohm m and a contrast with surrounding rocks of up to 1 to 10,000. The orebody zone is also very polarizable with typical frequency effects of up to 30%, compared with background effects of 2% to 10%. The frequency effects measured for the orebody may not, however, be representative of the massive orebody as a whole.

The Woodlawn massive sulphide ore has an average specific gravity of 3900 kg/m^3, exceeding the average value of the country rocks by about 1000 kg/m^3. This indicates the potential for gravity surveying in delineating near surface deposits of this type.

Measurements of magnetic susceptibility and natural gamma radiation on core samples and in drill holes show that the massive sulphides are essentially non-magnetic and non-radioactive.

Introduction

No single comprehensive study of the physical properties of the Woodlawn orebody and country rocks has been undertaken. However, Jododex Australia Proprietary Limited (Jododex), the Bureau of Mineral Resources (BMR) and the Geological Survey of New South Wales (NSWGS) have carried out independent physical property measurements on both drill core samples and in drill holes. The results obtained by these three organizations are summarized and the contrasts between the physical properties of the ore and host rocks are discussed.

The very high conductivity of the Woodlawn massive sulphides ore is the most significant petrophysical property of the rocks at Woodlawn. Therefore, although a number of physical properties were measured, work on drill core samples and in drill holes has been concentrated on measuring the electrical properties of the ore and host rocks.

Physical Property Measurements on Drill Core

Measurement Techniques

Jododex carried out limited resistivity and induced polarization (IP) laboratory testing on selected drill core samples during early exploration at Woodlawn, in an attempt to elucidate some of the complexities encountered in interpretation of the surface electrical surveys (Quick and Cifali, this volume, p 333). Very simple techniques were used for core measurements and these involved placing current electrode pads in contact with the ends of faced drill core which was wet but not evacuated, and positioning potential electrodes at known separations along the core axis. A Heinrichs Geoex Mark 2 Lab-Sender was used at frequencies of 0.3 and 3.0Hz to supply several microamps of current to the samples. Resistivity and IP measurements were taken with a Heinrichs Geoex Mark 4C receiver.

The NSWGS also carried out some laboratory testing on core in conjunction with drill hole detection logging (Tyne, this volume, p 375). Susceptibility measurements on core from W276, which intersects the main ore lens, were obtained using a Bison 3101 susceptibility meter. Resistivity measurements on the same core which was wet but not evacuated, were made using the recommended laboratory procedures for a four electrode galvanic system (Emerson, 1969) and a current source frequency of 1.0Hz.

A well controlled study of a range of physical properties of drill core samples from W54 (collar location 9500mN, 9400mE) was undertaken by the BMR. This hole does not intersect the main Woodlawn orebody but penetrated two massive sulphide lenses about 400m to the northwest. Magnetic susceptibility was determined with a Bison 3101 susceptibility meter and magnetic remanence with a Digico Complete Results magnetometer. The samples for the susceptibility and remanence measurements were cylinders 2.5cm in diameter and 10cm and 2.5cm long respectively.

A four electrode horizontal cell was used for electrical property measurements (Fig. 1). The samples were saturated under

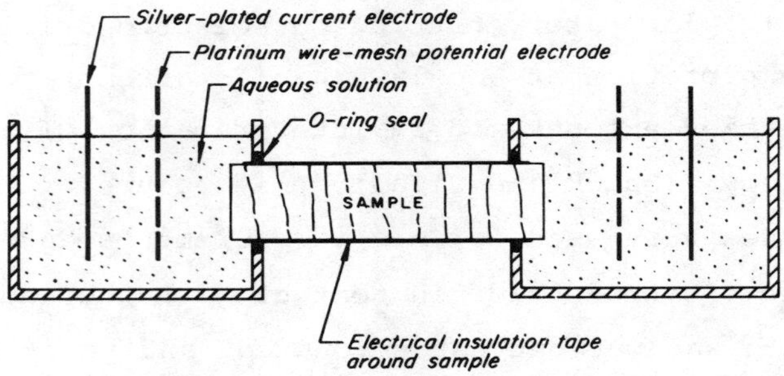

Fig. 1 Schematic diagram of conductivity cell

vacuum with a saline solution of 0.02 siemens/m conductivity. The same solution filled the cell. Resistivities of the samples were determined at 1kHz and the percent frequency effect (PFE) (Keller and Frischknecht, 1966) was obtained by measuring resistivities at two frequencies 1.0 and 0.1Hz thus,

$$PFE = \frac{\rho_{0.1} - \rho_{1.0}}{(\rho_{0.1} \; \rho_{1.0})^{\frac{1}{2}}} \times 100\% \qquad (1)$$

where $\rho_{0.1}$ and $\rho_{1.0}$ represent the resistivities of the sample at frequencies 0.1Hz and 1.0Hz respectively.

A time domain square waveform (2s on, 2s off) was used for chargeability measurements. The IP decay transients were recorded by a storage oscilloscope. A numerical integration of the decay curve yielded the chargeability value for the interval 100ms to 1000ms. The samples for the electrical measurements were in the form of cylinders 2.5cm in diameter and approximately 8cm long.

Core Measurement Results

The results of core sample testing are summarized in Table 1. Only the measurements of Jododex and the NSWGS refer to the main massive sulphide orebody. Ranges of resistivity and PFE values are given in Table 1(a) because the measurements were not sufficiently well controlled to list individual sample results. Although the results in Table 1(c) refer to mineralization separate from the near surface orebody, a comparison of various petrophysical properties such as specific gravity, susceptibility and resistivity between groups (a), (b) and (c) on Table 1 indicate that the measurements on core from W54 are representative of the Woodlawn orebody and country rocks.

Resistivity and IP Properties

The electrical property results obtained by Jododex vary over a wide range. The resistivity values obtained for ore

Table 1.

Results of Physical Property Measurements on Drill Core from Woodlawn

(a) Jododex - Electrical Property Measurements on Core Samples

Rock Type	Resistivity (ohm m)	Per Cent Frequency Effect (%)
Massive Sulphides	5 - 0.01	0 - 200

(b) NSWGS - Magnetic Susceptibility & Resistivity of Non Evacuated Core Samples - W276

Depth (m)	Rock Type	Magnetic Susceptibility (SI x 10^{-6})	Resistivity (ohm metres)
27.4	Tuff		279
29.3	Tuff		178
56.7	Tuffaceous Shale		677
93.5	Dolerite		7,438
114.0	Dolerite		7,430
134.9	Tuffaceous Shale	251	110,000
135.9	Massive Sulphide with intercalated Tuff	0	0.1
145.8	Massive Sulphide	0	0.01
179.0	Tuff	327	28,779
191.0	Tuff	138	43,188

Table 1. (cont.)

(c) BMR - Physical Properties of Core Samples - W54

Sample No	Depth (m)	Rock Type	Specific Gravity (kg/m^3)	Resistivity (ohm m)	FE (%)	Chargeability (ms)	Porosity (%)	Susceptibility (10^{-6} SI) units)	Remanence (mA/m)
1	28.30-28.50	Dolerite	2890	4900	3	21	0.4	580	1.3
2	64.10-64.25	Welded Tuff	2680	9100	0.3	4	0.3	60	0.05
3	96.05-96.25	Rhyolite	2650	9700	0.5	5	0.7	50	0.04
4	168.55-168.65	Dolerite	2820	1400	1	7	0.8	550	5.4
5	183.14-183.34	Massive sulphides	4860	0.9	25	400	0.2	60	0.1
6	242.53-242.68	Tuffaceous shale	2770	7500	2	8	0.7	80	0.2
7	265.46-265.60	Massive sulphides	3340	16	18	370	0.3	50	1.0
8	267.60-267.70	Massive sulphides	3050	520	15	140	0.5	50	0.2
9	280.35-280.45	Tuffaceous shale	2780	13000	0.3	4	0.6	150	0.2

samples vary from about 0.01 ohm m to approximately 5 ohm m and are generally inversely proportional to the sulphide content of the samples. Resistivity of the samples tested, averaged 0.3 ohm m. A resistivity contrast of up to 10,000 to 1 was noted between samples from hanging wall rocks and from the orebody. These resistivity contrasts are confirmed by the results of the NSWGS and BMR. The resistivity values reported in Table 1(b) represent the results of rapid exploratory tests on non-evacuated core samples and therefore the resistivities of very siliceous samples could be as much as an order of magnitude greater than the true in situ resistivities of the samples because normal electrolytic conduction throughout the rock is absent. However, the resistivities of the metallic conducting samples may be considered to accurately portray the true sample resistivity.

PFE measurements on the massive ore by Jododex gave widely varying values. An average PFE could not be determined because it appeared to depend critically on the ore resistivity and consequently the gangue content. PFE and chargeability values in Table 1(c) confirm that there is a very strong polarizability contrast between massive sulphide and country rock samples. However, the wide ranging IP properties obtained from these small core samples may not be representative when considered in the light of the entire polarization process in the ground. It is likely that the dominant polarization effect will occur at the upper and lower boundary regions of the orebody where the sulphide/electrolyte interfaces necessary to the IP process are present. There should be very little internal polarization within the massive sulphides because their fine grained nature and low porosity (see Table 1c) would largely preclude the development of internal polarization interfaces. It may therefore be misleading to attempt to estimate the overall intrinsic IP effect or PFE of the orebody from measurements on small core samples because the intrinsic PFE will probably depend more on the gross or macroscopic polarization effects of the massive sulphide/host rock contact zones rather than on any microscopic polarization effects distributed throughout the massive sulphide ore itself.

Magnetic Susceptibility

Table 1(b) and (c) indicate that the massive sulphides and tuffaceous country rocks have a very low to unmeasureable susceptibility. Fig. 2 shows the susceptibility of 25 drill core samples from W276. The hole intersected about 30m of massive

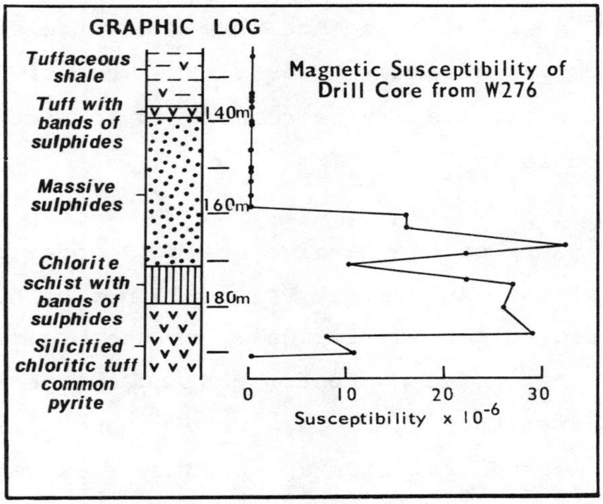

Fig. 2 Magnetic Susceptibility of Drill Core from W276

sulphides. The upper section of the massive sulphides, consisting predominantly of lead and zinc sulphides, is non-magnetic. The lower section of the massive sulphides, the underlying chlorite schists and the upper part of the silicified chloritic tuff are weakly magnetic with susceptibilities ranging from 250 to 375 x 10^{-6} SI units (20 to 30 x 10^{-6} cgs units). Dolerite is the only major magnetic rock type at Woodlawn and this varies in susceptibility from 500 to 600 x 10^{-6} (SI) for the hanging wall dolerite sills (Table 1(c)) to an interpreted susceptibility of 110 to 120 x 10^{-3} (SI) for the dolerite northeast of the orebody (Agostini <u>et al</u>., this volume, p 475).

In Situ Physical Property Measurements in Drill Holes

The details of *in situ* measurements in drill holes at Woodlawn are given elsewhere in this volume (Hone and Young, p 519 ; Tyne, p 541). Hone and Young show comprehensive geophysical logs of PFE, resistivity, SP, point resistance and gamma radiation in 4 holes, two of which (W256, W276) intersect the main massive sulphide lens. Gamma radiation logs show that the orebody is generally very low in radiation levels, and that the logs are only useful in identifying tuffs and tuffaceous shales which exhibit levels of 0.02 to 0.03 millirontgens/hour (mR/h) (1mR = 2.58×10^{-7} coulombs/kg (SI)). SP logs confirm that the orebody is actively oxidizing above the water table.

IP and resistivity logs in W256 and W276 substantiate the large resistivity and PFE contrasts obtained by measurements on drill core (Table 1). Hone and Young conclude that the massive sulphides have frequency effects of over 30% and resistivities as low as 0.1 ohm m. The country rocks exhibit background frequency effects of 2% to 10% and resistivities of the order of 1,000 to 10,000 ohm m, with dolerite being more resistive and polarizable than acid country rocks.

A resistivity log of W276 is compared with the results of core measurements from Table 1(b) (Tyne, this volume, p 545) and this comparison shows a satisfactory correlation between core and *in situ* measurements of resistivity. Measurements in W276, corrected for the effect of the bore hole fluid on the resistivity logs, yield resistivity values of about 0.1 ohm m for the massive sulphides, 100 ohm m for felsic tuff, 1000 ohm m for tuffaceous shale and 6800 ohm m for dolerite. There is a maximum resistivity contrast of 1:100,000 between the massive sulphides and host rocks intersected in W276.

Resistivity of Overburden

A discussion on the electrical properties of rock types at Woodlawn is not complete without comment on the overburden resistivity and thickness. The soil and weathered rock layer is about 5 to 10 metres thick at the northern end of the orebody and increases to 20 metres at the southern end (Malone, this volume, p 49). Overburden resistivity obtained from analysis of dipole dipole apparent resistivity pseudosections (Quick and Cifali, this volume, p333 ; Whiteley and Tyne, this volume, p 423) generally varies between a lower limit of about 40 to more than 100 ohm m, except over the gossanous zone where resistivities may be low.

Specific Gravity

A high specific gravity is an obvious physical property distinguishing the massive sulphides from the host rocks at Woodlawn. However, the specific gravity of the massive sulphides varies through a wide range of values depending on the total amount of sulphides present in particular samples and the proportions of base metal sulphides present.

The Woodlawn deposit also includes mineralized volcanics which locally are of ore grade because of their copper content. These rocks contain a much lower proportion of sulphides than the massive sulphide ore and have a much lower average specific gravity.

The host rocks include felsic volcanics which is locally silicified and/or pyritic as well as shale and dolerite. The specific gravities of these rock types vary through a considerable range but are much less than those of the massive sulphides.

The specific gravity of many samples of the gossan overlying the massive sulphides have been measured and an average value determined. The soils, alluvial deposits and decomposed rocks overlying the host sequence are generally comparable in specific gravity to the gossan and since these are locally extensive and thick, they may influence the results of gravity surveys.

Table 2 presents the specific gravities determined for a selection of samples to indicate the ranges of values for the various ore and rock types. The grades of mineralized samples are also shown in the table. A bulk sample of mineralized volcanics provides a better indication of the average value than do the individual samples.

The values of specific gravity used in ore reserve calculations are 3.9 for massive sulphide ore, 2.9 for mineralized volcanics (both copper ore and uneconomic material) and 2.0 for gossan. The figure of 3.9 for the massive sulphides appears low compared to the individual sample values on Table 2. Mining experience at Woodlawn has confirmed that it is a reliable average figure; its value reflects the practical necessity to incorporate small amounts of low grade or unmineralized rocks with the massive sulphides for mining purposes.

Conclusions

Fig. 3 graphically summarizes the electrical properties obtained from both core and drill hole measurements at Woodlawn. It is clear from these results that Woodlawn type mineralization is an excellent electrical and electromagnetic geophysical target.

The large density contrast of about 1000 Kg/m^3 between the massive sulphides and country rock indicate the potential for gravity surveying in delineating near surface deposits of this

Table 2.

Specific Gravity Determinations

Rock Type	Grade Cu %	Pb %	Zn %	Fe %	Specific Gravity kg/m^3
Massive Sulphides	3.80	0.89	3.55	36.20	4310
	0.28	3.30	7.15	26.50	4150
	0.64	6.60	13.70	28.70	4520
	0.88	7.05	14.60	24.00	4390
	0.72	5.30	10.60	24.20	4050
	0.74	5.00	9.80	17.40	3600
	1.23	5.40	13.60	28.70	4460
	2.30	1.35	6.00	33.00	4290
	0.50	4.50	10.20	20.20	3930
	4.30	0.47	16.50	32.00	4350
	0.78	17.70	24.00	6.70	4090
	0.86	8.55	16.90	19.80	4100
	1.33	11.00	22.00	12.00	3850
	1.70	3.45	23.00	20.40	4130
	0.96	9.80	20.50	17.00	4230
	1.86	1.61	19.30	18.40	4070
			Average Value		4160
Mineralized Volcanics	0.47	0.03	0.13	7.65	2830
	0.37	0.03	0.23	6.55	2810
	1.03	0.09	0.30	6.50	2770
	0.45	-	0.39	11.20	2950
	0.54	-	0.05	7.00	2820
	0.81	0.04	0.13	7.80	2850
	1.89	0.04	0.43	12.40	3040
	1.37	0.02	0.37	6.20	2820
	0.62	0.01	0.08	10.00	2910
	0.96	0.13	1.13	9.05	2940
	10.90	0.20	0.63	20.20	3330
	2.55	0.02	0.11	16.80	3080
	6.60	0.01	0.37	16.00	3150
			Average Value		2950
Mineralized Volcanics (Bulk Sample)	1.7	0.14	0.84	9.40	2900
Gossan	Average of 28 determinations				2130
	Range of values				1570 to 2800
Dolerite					3000
Dolerite					2860
Shale					2750
Tuff					2730
Tuff					2760

type. The orebody is neither magnetic nor radioactive and so these petrophysical properties cannot be exploited as a direct aid to exploring for similar orebodies.

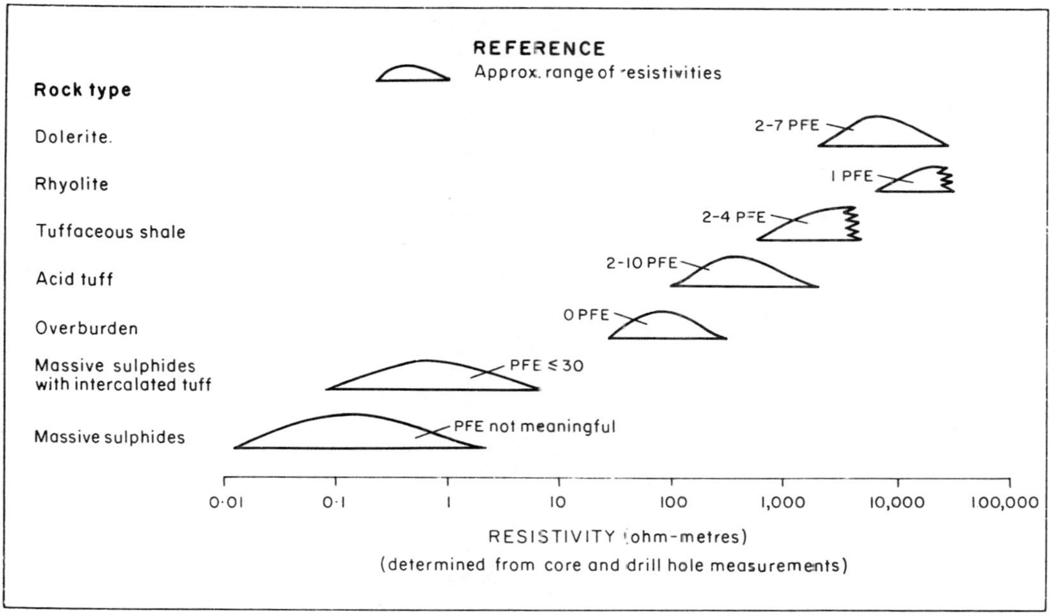

Fig. 3 Resistivity and PFE of Woodlawn massive sulphides country rocks and overburden.

Acknowledgements

This paper is published with the permission of the Under Secretary, New South Wales Department of Mineral Resources and Development, Sydney and the Director, Bureau of Mineral Resources, Canberra.

References

Emerson, D.W., 1969. Laboratory Electrical Resistivity Measurement of Rocks. Proc. Australas. Inst. Min. Metall., 230, 51-62.

Keller, G.V. and Frischknecht, F.C., 1966. Electrical Methods in Geophysical Prospecting, Pergamon Press, Oxford.

CHAPTER 6

AIRBORNE GEOPHYSICAL SURVEYS

6.1. Review of Helicopter Electromagnetic Data over the Woodlawn Orebody

R. J. WHITELEY

Contents

Abstract

A detailed helicopter electromagnetic (HEM) survey with the Scintrex HEM701 system revealed a large number of anomalies in the vicinity of the Woodlawn orebody. Many of these are of considerable size with generally low in-phase to out-of-phase response ratios indicating moderate to poor conductors. It is likely that these anomalies are due to rock types in the area, particularly conductive black shales.

A single HEM anomaly attributed to the Woodlawn orebody is characterized by small amplitude, complex shape and a very large in-phase to out-of-phase response ratio. All these features can be qualitatively explained as being due to a single, shallow, highly conductive, thick, tabular dipping conductor.

The results of this survey indicate that the HEM method would have difficulty in detecting and delineating targets similar to the Woodlawn orebody. However, the most diagnostic feature would be a very high in-phase to out-of-phase ratio.

Introduction

From January 22 to February 11, 1971, a helicopter electromagnetic (HEM) and magnetic survey was flown over a substantial area which included the Woodlawn orebody. The HEM survey was completed by Siegel Associates Australia Proprietary Limited (Pty. Ltd.) and utilized the Scintrex HEM701 system. This well known system consists of coaxial transmitter and receiver coils mounted on a rigid boom 9.1m (30 ft) apart. Coil axes are horizontal and aligned in the direction of flight. The boom is towed approximately 30m (100 ft) below the helicopter. The system operates at 1600 Hz measuring both in-phase and out-of-phase components of the oscillating magnetic field at the receiver coil. These components are expressed in parts per million (ppm) of the normal undistorted primary field at the receiver. Both in-phase and out-of-phase components are displayed on a chart recorder generally having a full scale width of 500 ppm. The scale sensitivity of the system is periodically checked during the flight by means of calibration signals.

The HEM701 compares favourably with similar EM systems, Ward (1967), having a low inherent noise level with comparable limits of exploration.

The survey at Woodlawn was flown for saturation coverage with a line spacing of generally 185m (600 ft), and boom terrain clearance of between 30 and 60 m. Positioning was by means of a strip film camera.

Interpretation

Interpretation of the HEM survey was carried out by Seigel Associates Australia Pty. Ltd. in the following manner. The location of the peak in-phase response of an anomaly, believed to be due to a geological conductor, was plotted as a circle at the appropriate position on the flight line derived by linear interpolation between successive fiducials. The distance between half maximum points on the anomaly i.e. the half width, was also plotted as a bar along the flight line. This parameter is simply related to depth in the case of a narrow steeply dipping conductor, but the relationship becomes complicated if thicker or deeper targets with a conductive overburden produce anomalies of large half width. The scaled in-phase and out-of-phase peak amplitudes also contain information on the nature of the target and are shown on the flight plan in ppm opposite the peak location. The ratio of these amplitudes is indicative of a conductivity-size factor for the target. In general, highly conductive bodies such as massive sulphides can be expected to have ratios of 3 or more. Bodies of moderate conductivity-size products have ratios between 1 and 3, and bodies of poor conductivity have ratios less than 1. In some areas these ratios are quite diagnostic whereas in others, the ranges overlap and they cannot be reliably used.

In the interpretation, anomalies were also graded into categories 1, 2 and 3 based on in-phase amplitudes, taking terrain clearance into account. Assuming a 15m thickness of overburden, a category 1 anomaly has a peak in-phase response equivalent to 350 ppm or greater at a 30m bird terrain clearance. In similar conditions, a category 2 anomaly has a peak in-phase response between 100 and 350 ppm, while a category 3 anomaly has less than 100 ppm. On the flight plan, the respective peak circles are fully shaded for a category 1 anomaly and partly shaded for category 2 and unshaded for category 3.

Magnetic anomalies coincident with EM anomalies are indicated on the flight plan as a double concentric circle with the amplitude in nanoteslas (nT) also printed.

Survey at Woodlawn

A segment of the interpreted HEM data in the vicinity of the Woodlawn orebody is shown in Fig. 1. Also

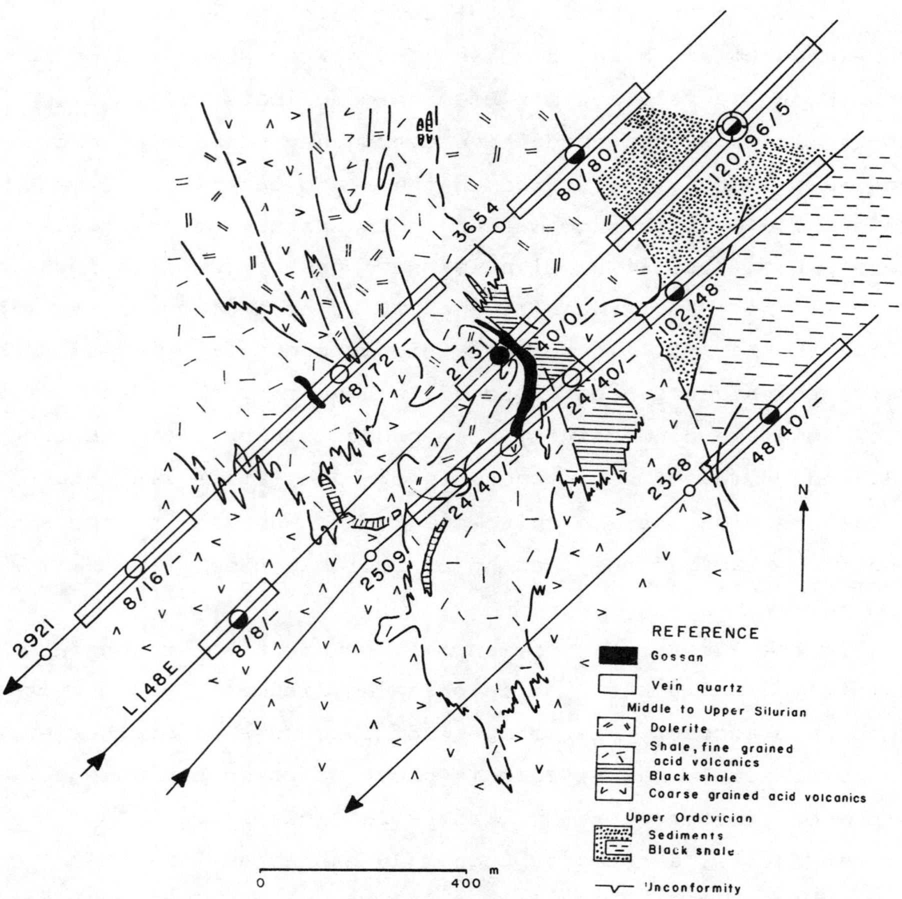

Figure 1 HEM Data over Woodlawn with simplified solid geology.

shown is a simplified solid geology map after Malone (this volume, p 49).

A large number of airborne EM anomalies were detected in the Woodlawn area. Some of these were due to cultural sources but the bulk were believed to be due to geological features. The majority of anomalies were graded into categories 2 and 3 with peak in-phase to out-of-phase ratios from about 3 to less than 1. Many of these anomalies had substantial half-widths and some coincided with magnetic anomalies.

A single line (line 148E) passed the northern part of the orebody. This line was flown in a southwesterly to northeasterly direction, i.e. up dip. Fig. 2 shows the section of line 148E crossing the orebody and the relationship of this line to other airborne EM surveys (Pederson, this volume, p 123).

A narrow category 3 anomaly was detected on this line near fiducial 2731 close to the hanging wall of the orebody. This anomaly has a peak in-phase response of 40 ppm, a very large in-phase to out-of-phase ratio and was later attributed to the Woodlawn orebody. Fig. 3 shows a section of recorded data from line 148E. The anomaly shows a complex double peak on the in-phase response and no out-of-phase anomaly.

The character of the in-phase anomaly is difficult to explain and would have been made more complex by the altitude of the boom varying across the orebody. The presence of a double peak anomaly may be mostly due to the large effective width of the orebody relative to the size of the coil system (which enhances edge effects) and the flight line orientation.

A larger amplitude category 2 anomaly is also present on this portion of line 148E. This anomaly has a

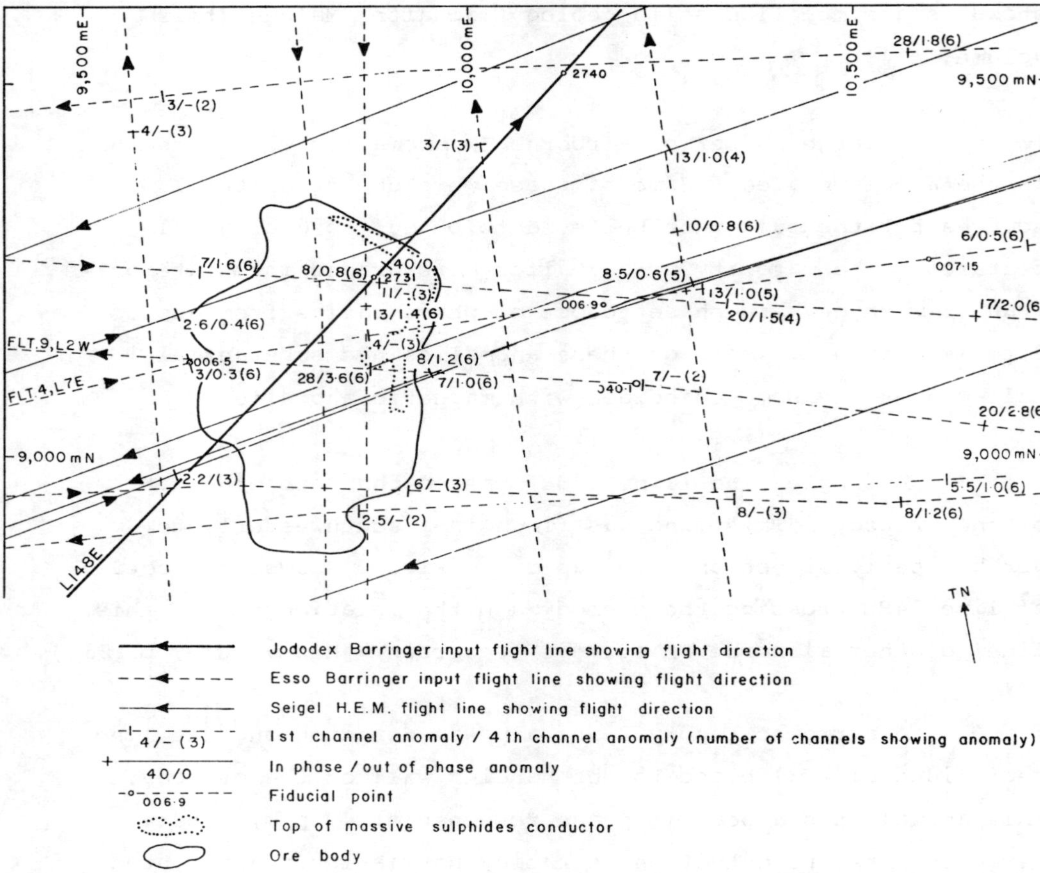

Figure 2 Relationship of HEM line 148E to other airborne EM lines.

peak response at fiducial 2747 and is produced by a much broader feature. The anomaly has a low in-phase to out-of-phase response ratio suggesting a moderate to poor conductor. This coincides with a weak magnetic anomaly. The solid geology map indicates the presence of sediments at the location of this EM anomaly. Since the general position of the EM anomaly is within a broad topographic depression enhancement of the EM response by wet ground cannot be ignored.

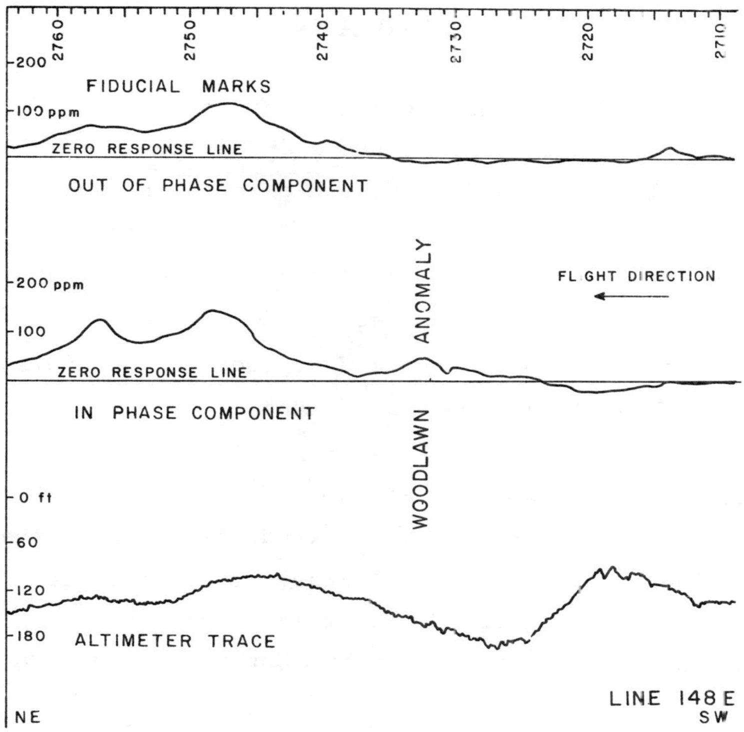

Figure 3 HEM anomalies on line 148E in the vicinity of the Woodlawn orebody.

An anomaly immediately to the north of this anomaly with a peak in-phase response at about fiducial 2757 is believed due to cultural sources.

On the flight line immediately to the south of line 148E, two weak category 1 anomalies with low in-phase to out-of-phase ratios suggest the presence of broad poor conductors. As these occur on either side of the ridge on which the southern part of the Woodlawn orebody occurs, it is likely that they are due to moist ground possibly enhanced by the host rock lithologies present.

Conclusions

A large number of HEM anomalies occur in the vicinity of the Woodlawn orebody. Many of these are of considerable size with generally low in-phase to out-of-phase response ratios, indicating moderate to poor conductors. It is likely that many of these anomalies are due to some rock types in the area particularly conductive black shales. Undoubtedly a portion of these have been caused or enhanced by local occurrences of wet ground, and flight altitude variations.

The single HEM anomaly attributed to the Woodlawn orebody is characterized by small amplitude and complex shape but a very large in-phase to out-of-phase response ratio. All of these features can be explained, at least qualitatively, by a shallow, highly conductive, thick tabular dipping conductor.

The most diagnostic feature of the HEM response over the Woodlawn orebody is the very high in-phase to out-of-phase ratio.

However, this detailed HEM survey indicates that detection and delineation of similar targets in this area would be difficult due to complexity of shape, small strike length, the likelihood of nearby cultural and host rock sources of EM anomalies and abrupt variations in topography.

Reference

Ward, S.H., 1967. Airborne electromagnetic methods. Canad. Mining and Groundwater Geophysics, 1967. pp. 81-108.

6.2. INPUT Results over Woodlawn

R. N. PEDERSEN

Contents

Abstract

Two sets of airborne electromagnetic data were gathered with the INPUT method over the Woodlawn deposit. The orebody displays the typical response from a dipping conductor with a much larger anomaly when flown in the direction of dip. It is concluded that a similar orebody could be discovered with the INPUT system, provided flight lines were sufficiently closely spaced.

Introduction

Two separate sets of INPUT airborne electromagnetic (EM) data were gathered over the Woodlawn orebody. In the first set of data, the area of interest was included in a regional survey. In the second, the main motive was an orientation test to establish a signature as a guide in planning further exploration in similar geological environments.
A Mark V INPUT system, together with the necessary ancillary equipment, was used in both surveys.

Equipment and Field Procedures

The INPUT (Induced Pulse Transient) system was first developed and constructed some twenty years ago. For many years INPUT has been the most widely used airborne EM system in the search for base metals. In 1975, it accounted for roughly half of the total flown in this category (Geophysical Activity, 1975). It is based upon the study of the decay of secondary EM fields created in the ground by short pulses generated from a large horizontal loop of wire surrounding a fixed wing aircraft.

The basic principle of the system is illustrated in Figs. 1 and 2. The transmitted primary field is discontinuous in nature (Fig. 1a) with each pulse lasting one millisecond. The pulse repetition rate is 288 per second. The axis of the receiving coil is horizontal and parallel to the flight direction. It is housed in a bird towed behind and below the aircraft. In between successive primary pulses (Fig. 1b) the INPUT transient voltages are measured. These are produced (Fig. 1c) by the decaying eddy currents in the ground.

The transient voltage decay curve is sampled at six points or gates (Fig. 1d). Signals received at each sampling gate are processed and recorded as continuous analog traces (Fig. 2). Each trace represents the integration of transient voltages within a sample window. When a significant highly conductive geological feature is present the induced eddy currents are stronger and more persistent and induce larger transient voltages in the receiving coil.

Other data recorded during flight include fiducial marks, terrain clearance, total magnetic field intensity and a 50Hz monitor for defining cultural EM sources such as power lines.

During survey the aircraft is maintained at a mean terrain clearance of 120 meters (400 feet) with the bird about 55 meters (180 feet) above the ground. Line spacing may be varied. Adjacent lines are usually flown in opposite directions.

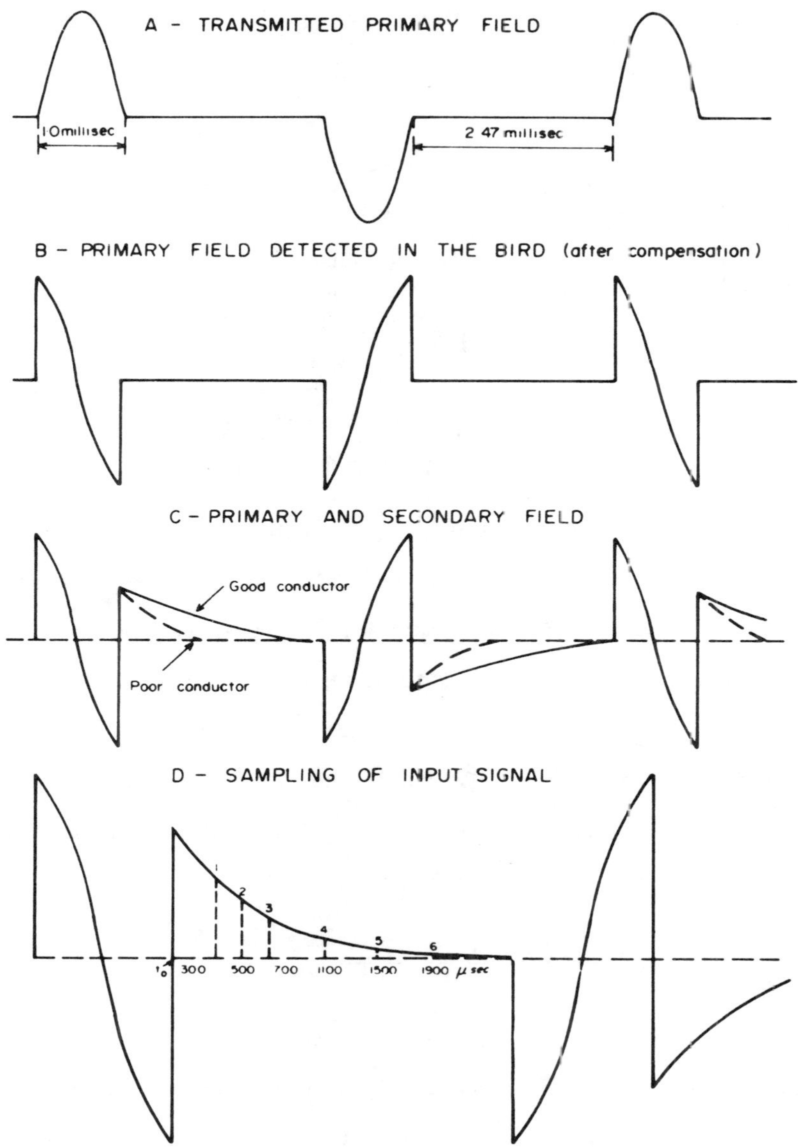

Fig. 1. INPUT Signal

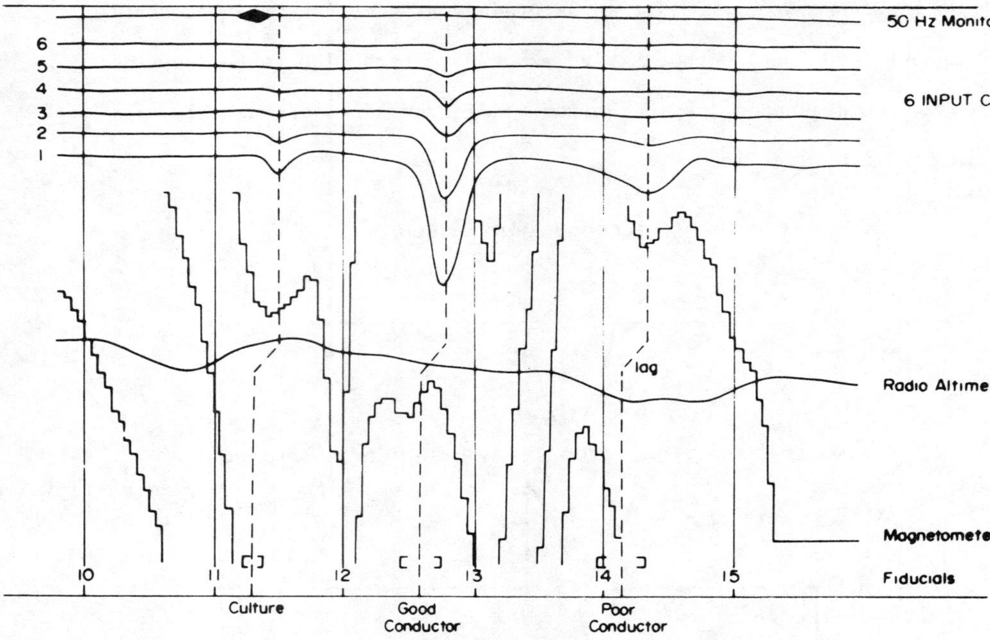

Fig. 2. Typical INPUT Recording

At the end of each flight, all records and film are developed and edited. The flight path is recovered and the INPUT anomalies are plotted on a plan of the survey grid. These anomalies are represented by means of symbols that display their most significant characteristics; the center and half peak width, the number of channels showing deflection, and the peak amplitudes of the first and fourth channels. Also shown are terrain clearance of the aircraft, the amplitude of any coincident magnetic feature, and any associated response on the 50Hz monitor.

Discussion of Results

Fig. 3 shows a portion of the EM map from the first survey which was flown at a 200m line spacing in April 1973 on behalf of Jododex Australia Proprietary Limited. The zone identified as P-1 is over the Woodlawn orebody.

Two simple criteria are used for selecting bedrock conductors which are considered to have a good chance of being due to massive sulphides.

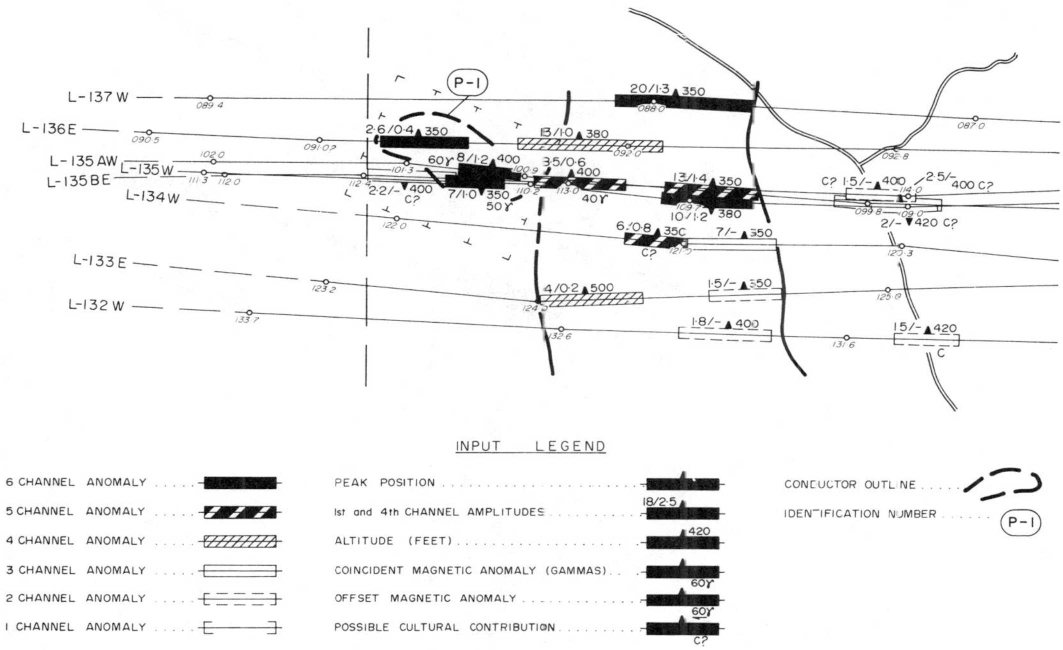

Fig. 3. EM Plan Map of Woodlawn Area (Flown for Jododex Australia Proprietary Limited)

They are (1) short strike length, and (2) isolation, or as in this case, a discordance from the main formational trend.

In Fig. 4 the area of interest is enlarged and also shows the Woodlawn Test Grid together with a simplified plan of the orebody. Results from lines flown over the orebody are illustrated in Figs. 5 to 8.

It is evident that the character of the INPUT response is affected by the geometry of the conductive target. The phenomenon of updip versus downdip anomaly variation has been established by model studies (Palacky, 1973) and was known at least as early as 1969. The Woodlawn INPUT anomalies are an excellent example of this effect of a dipping conductor, and only the westbound lines (Figs. 6 and 7) record well developed, sharp, slowly decaying anomalies which are consistent with a dipping highly conductive target.

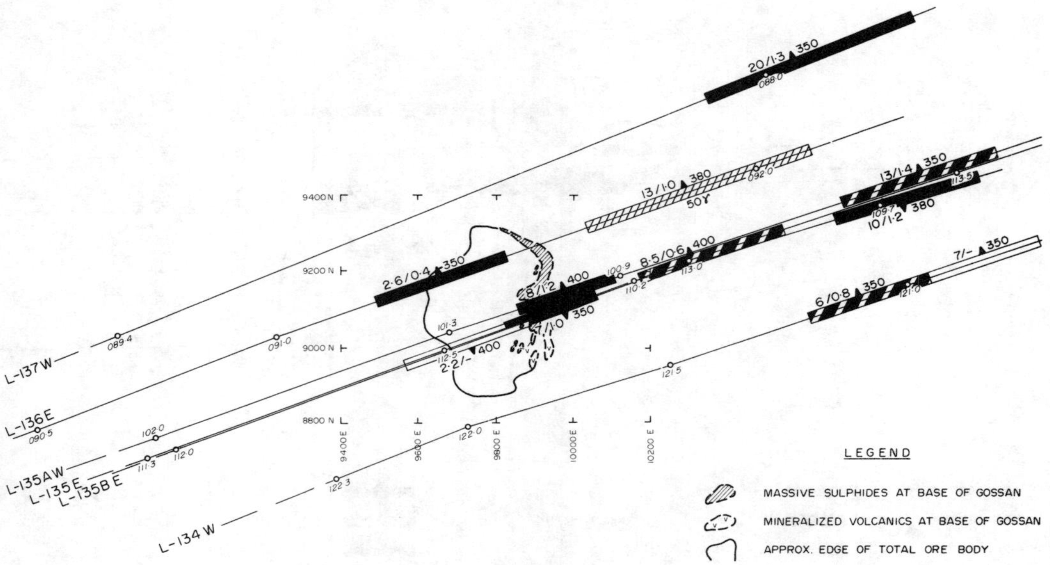

Fig. 4. Jododex EM Plan Map

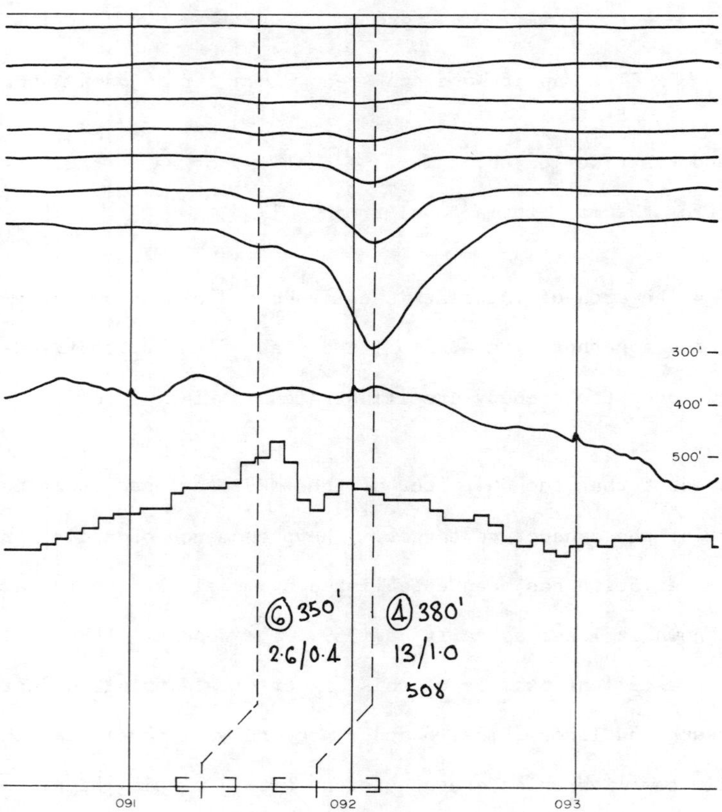

Fig. 5. Airborne geophysical record, Line 136-E

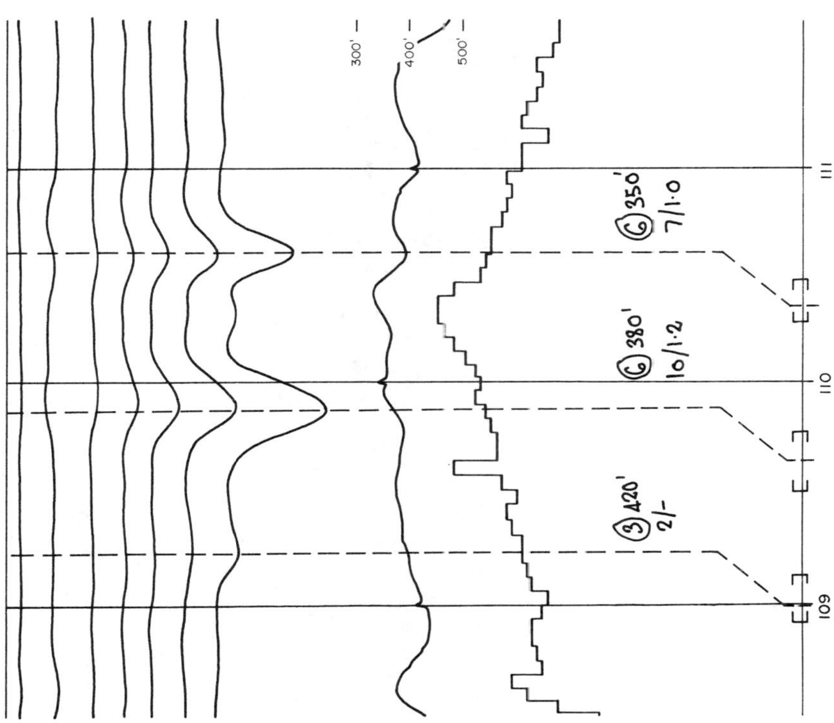

Fig. 6. Airborne geophysical record, Line 135B-W

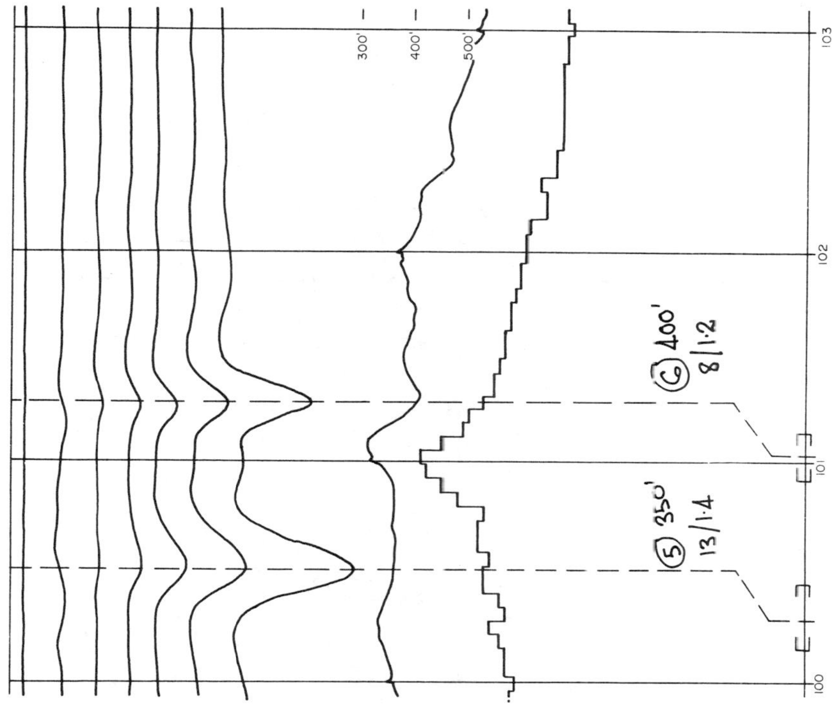

Fig. 7. Airborne geophysical record, Line 135A-W

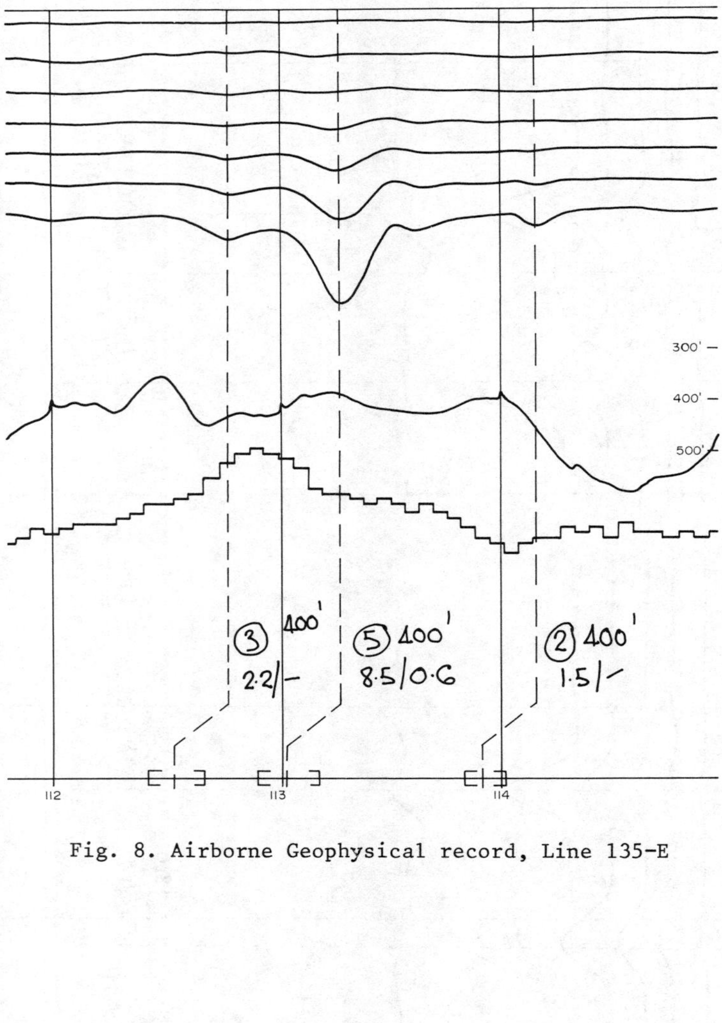

Fig. 8. Airborne Geophysical record, Line 135-E

Data from the second INPUT survey flown for Esso Australia Proprietary Limited are similar to those from the first survey. Differences, however, do occur as can be seen by comparing in detail both sets of data. Variations in both line orientation and position relative to the conductive zones are the main factors producing these differences.

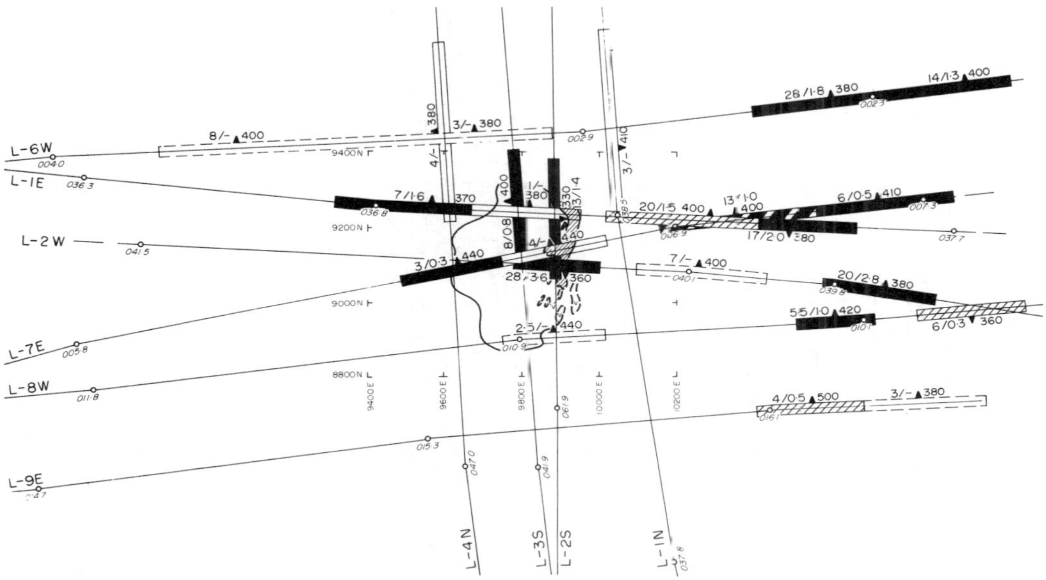

Fig. 9. EM Plan Map of Woodlawn Area (flown for Esso Australia Proprietary Limited).

Data from two lines flown in a southerly direction over the ore deposit are shown in Figs. 14 and 15. The responses on these lines in the vicinity of the orebody are typical of a highly conductive target.

Summary and Conclusions

The Woodlawn orebody is detectable with the INPUT method. INPUT data from the two separate surveys in the area are similar with

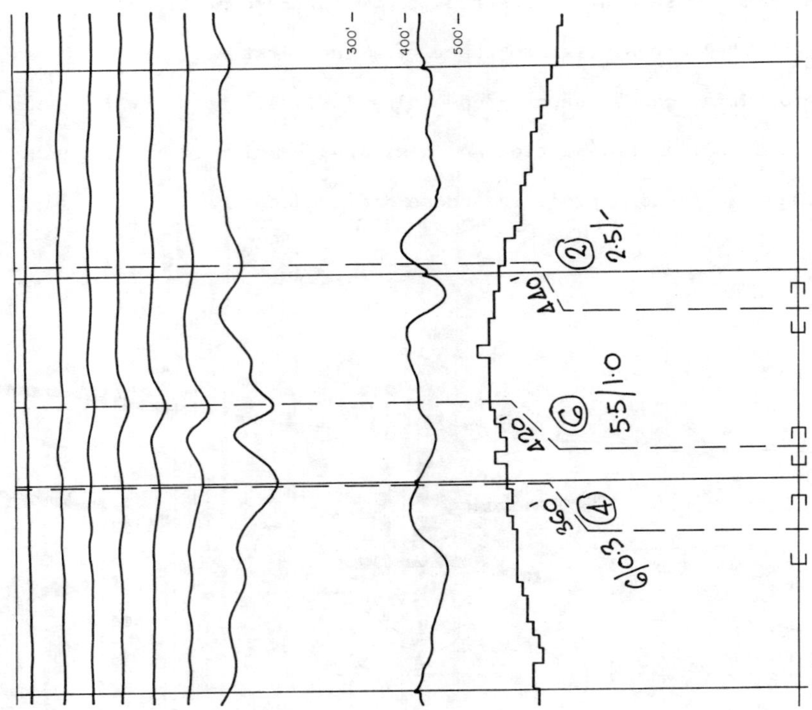

Fig. 11. Airborne geophysical record, Line 8-W

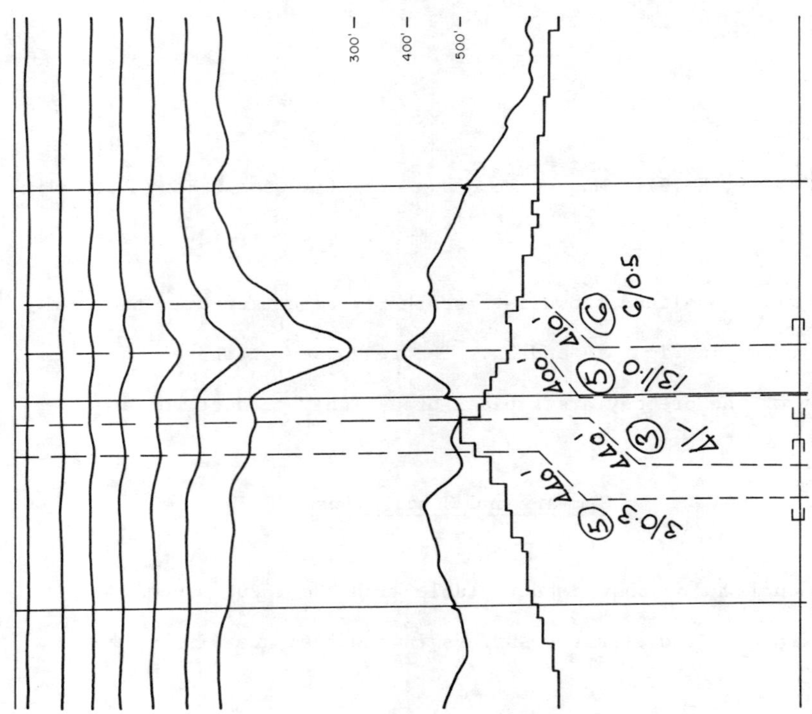

Fig. 10. Airborne geophysical record, Line 7-E

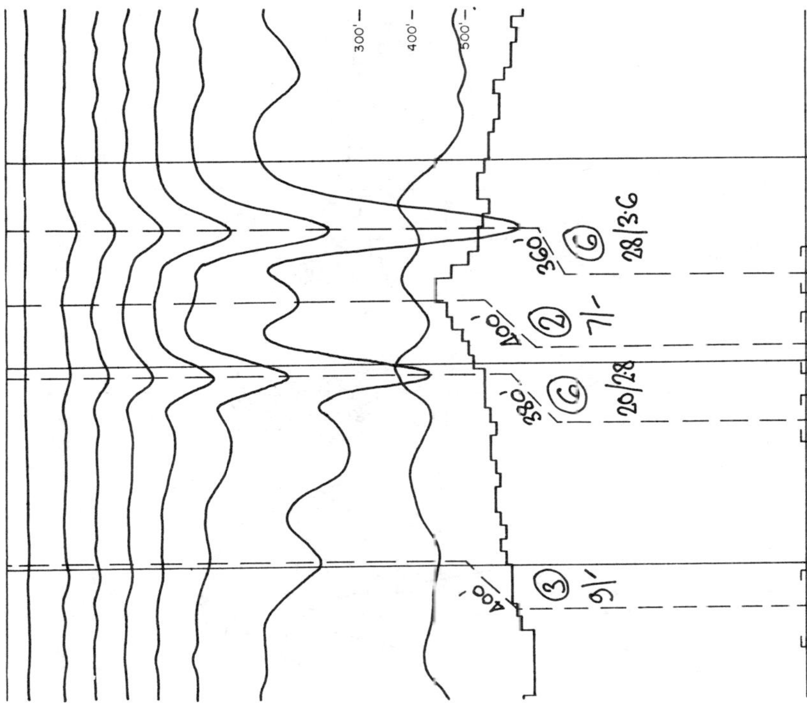

Fig. 13. Airborne geophysical record, Line 2-W

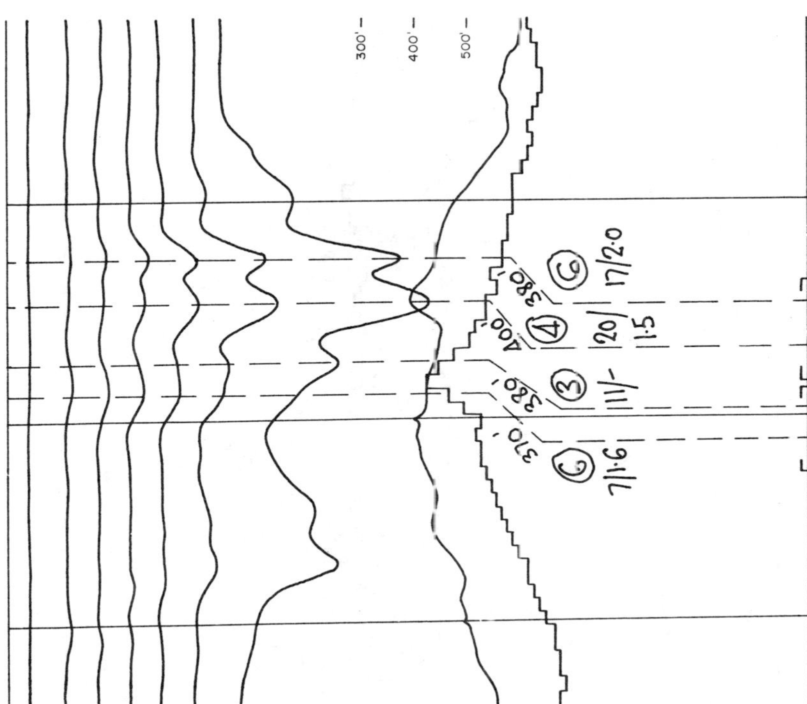

Fig. 12. Airborne geophysical record, Line 1-E

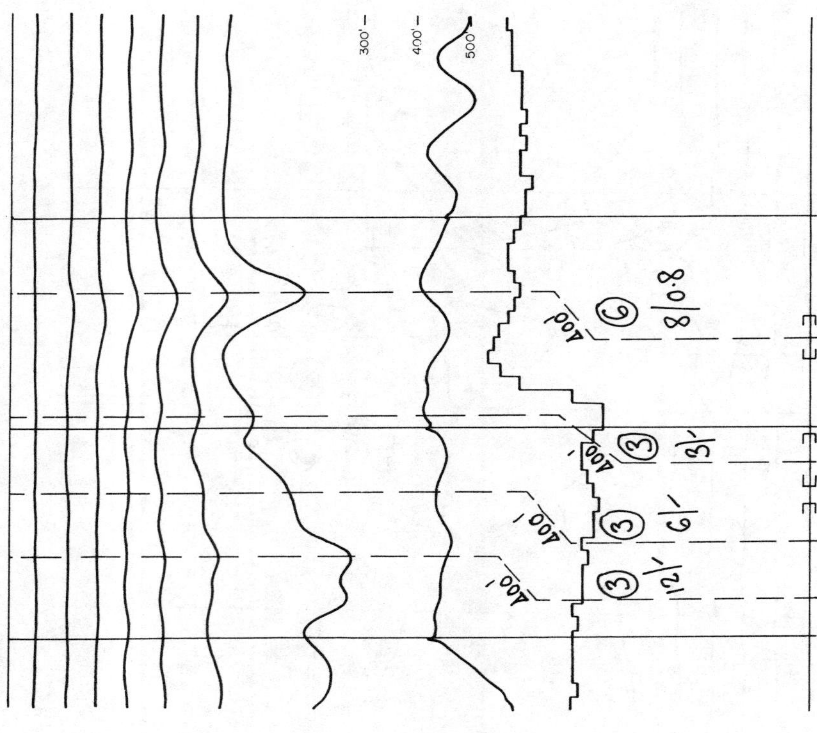

Fig. 14. Airborne geophysical record, Line 3-S

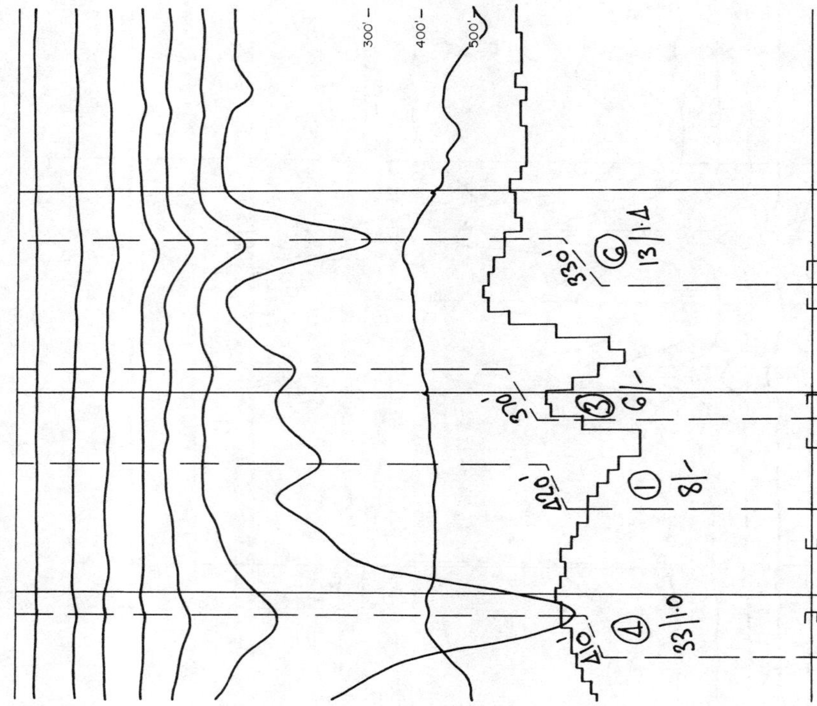

Fig. 15. Airborne geophysical record, Line 2-S

with differences in responses attributed to variations in flight line location and heading. The Woodlawn results are typical of those from a dipping conductor of limited strike length. In order to detect a similar orebody flight line spacings should not exceed 300 metres in each direction.

References

Geophysical activity in 1975, 1976, Special Report: Geophysics, V. 41 P. 780 to 794.

Palacky, G.J., 1975. Interpretation of INPUT AEM measurements in areas of conductive overburden: Geophysics, V. 40, p. 490 to 502.

Palacky, G.J. and West, G.F., 1973. Quantitative interpretation of INPUT AEM measurements: Geophysics, V. 38, p. 1145 to 1158.

Wagg, D.M., 1970, Airborne electromagnetic systems: Ottawa, Geoterrex Limited.

6.3. Response of the Woodlawn Orebody to the DIGHEM^{II} System

D. C. FRASER

Contents

Abstract

The helicopter towed DIGHEM^{II} electromagnetic (EM) system consists of two independent coil pairs contained within a 9m bird. The vertical coaxial coil pair is common to all other helicopter EM systems, and yields two EM channels (i.e., one each of in-phase and quadrature or out-of-phase components). The horizontal coplanar coil pair is a feature of the DIGHEM^{II} system, and provides a resistivity data channel in addition to two EM channels for this coil pair. Three other mathematically derived EM channels are also produced, and so the system yields a total of eight EM channels.

Survey flying over the Woodlawn area detected EM anomalies from the orebody as well as from broad, conductive zones and cultural sources. The Woodlawn deposit could not be separately identified from cultural and other responses on the vertical coaxial channels. However, the deposit was readily identified on the resistivity channel computed from the horizontal coplanar coil pair data. The resistivity channel is thus of major importance in the exploration for massive sulphides in the Woodlawn environment.

Introduction

The original DIGHEM system was introduced to the mining exploration industry in 1970, and the new DIGHEM^{II} system in 1976. Each is a multicoil system contained in a 9m bird which is towed 30m below a helicopter. A frequency of 900 Hz is commonly used because it provides a bias towards high conductivity targets by minimizing interference from the usual geologic noise sources such as conductive overburden.

The short separation between transmitter and receiver coils produces high resolution data. This resolution allows smaller sulphide bodies to be more readily detected than would be the case if the transmitter-receiver coil separation was considerably larger. The advantage of small scale was demonstrated by the original system's discoveries in Canada of the New Insco copper orebody (Fraser, 1974) and the Montcalm copper and nickel orebody (Fraser, 1978b). The strike lengths of these two deposits were, respectively, 100m and 200m.

The original DIGHEM system had a single transmitter coil and three orthogonal receiver coils, whereas the DIGHEM^{II} system has two orthogonal transmitter coils and two corresponding receiver coils. The original system had two shortcomings which have been largely overcome with the new system. These shortcomings were as follows:

(1) Conductive overburden diminished the ability of the original system to effectively explore the bedrock, even in cases where the bedrock conductor was energized. This was because a high degree of activity on the recorded EM channels from all sources including geological and cultural features, reduced the likelihood of recognizing that a particular EM anomaly might be caused by a bedrock conductor. This problem is inherent in all previously developed airborne EM systems.

(2) It was not possible to measure the width of a steeply dipping bedrock conductor with the original system. It could not differentiate between a broad anomaly produced by a wide body and a similar anomaly produced by several closely spaced thinner bodies. Again, this ambiguity exists in all previously developed airborne EM systems.

The above limitations have been largely overcome with the introduction of the multicoil DIGHEM^{II} system. Such systems have been in operation in Europe since the middle of 1976, in North America since early 1977 and in Australia since July 1979. A detailed description of the new system is given in Fraser (1979). This system was used at Woodlawn.

The DIGHEM^{II} Multicoil System

The multicoil system consists of two independent and orthogonal coil pairs contained within a 9m bird (Fig. 1). The vertical, coaxial coil pair is common to all other helicopter EM systems. The horizontal, coplanar coil pair is a feature of the DIGHEM^{II} system. Both coil pairs usually operate at frequencies close to 900 Hz although other frequencies have occasionally been

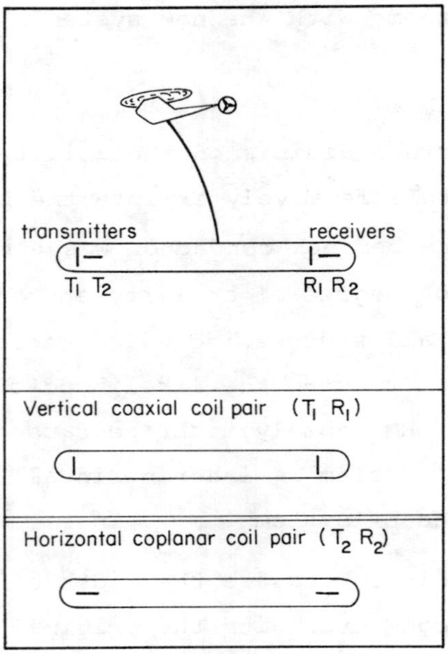

Fig. 1 DIGHEM^{II} system.

used. The two orthogonal transmitter coils energize conductors in two orthogonal directions and the resulting anomalies provide information on the geometry of the conductors. This information aids considerably in the recognition of bedrock conductors from conductive overburden and cultural conductors (e.g. fences).

The DIGHEM^{II} data are recorded digitally as well as displayed on a chart recorder. The computer software is an integral part of the system because the data handling procedures are too complex to be performed manually. Briefly, the procedures and display are as follows:-

1. Difference channels show the difference between the two in-phase responses and the two quadrature (out-of-phase) responses which are generated by the two independent coil pairs, i.e.,

   difference in-phase = coaxial in-phase - 0.5 (coplanar in-phase), and

   difference quadrature = coaxial quad. - 0.5 (coplanar quad.)

The difference channels are displayed as profiles together with the raw data, as in Fig. 2 with channels 33 and 34. The difference channels tend to be free of the geological noise caused by conductive overburden. This is significant, since this geological noise may mask the response of bedrock conductors as may occur with all previously developed airborne EM systems. The difference channels may contain as much as an order of magnitude increase in the signal to noise ratio for bedrock conductors in a geologically noisy environment (Fraser, 1979).

2. The conductance of anomaly sources is computed and displayed, as in Fig. 2, channel 37. The conductance channel is essentially an automatic anomaly selector. The channel will show zero except where discrete conductors occur. The existence of a conductor is indicated by a spike whose amplitude is calibrated in conductance units of siemens and the computed conductance represents the conductivity-thickness product of a vertical halfplane model. The EM anomalies are selected by computer. This allows the EM anomaly map to be plotted directly in one pass through the digital data.

3. Resistivity (Fig. 2, channel 40) is computed from the horizontal coplanar channels using a technique described by Fraser (1978a, 1979), based on a pseudo-layer halfspace model. The resistivity channel yields data which can be contoured, and so a DIGHEM^{II} survey provides a resistivity contour map as well as an EM anomaly map.

4. Contour maps of the total magnetic field and the enhanced magnetic field are also produced by computer. The enhanced magnetic map is generated by a digital filter operator which has characteristics similar to a downward continuation operator.

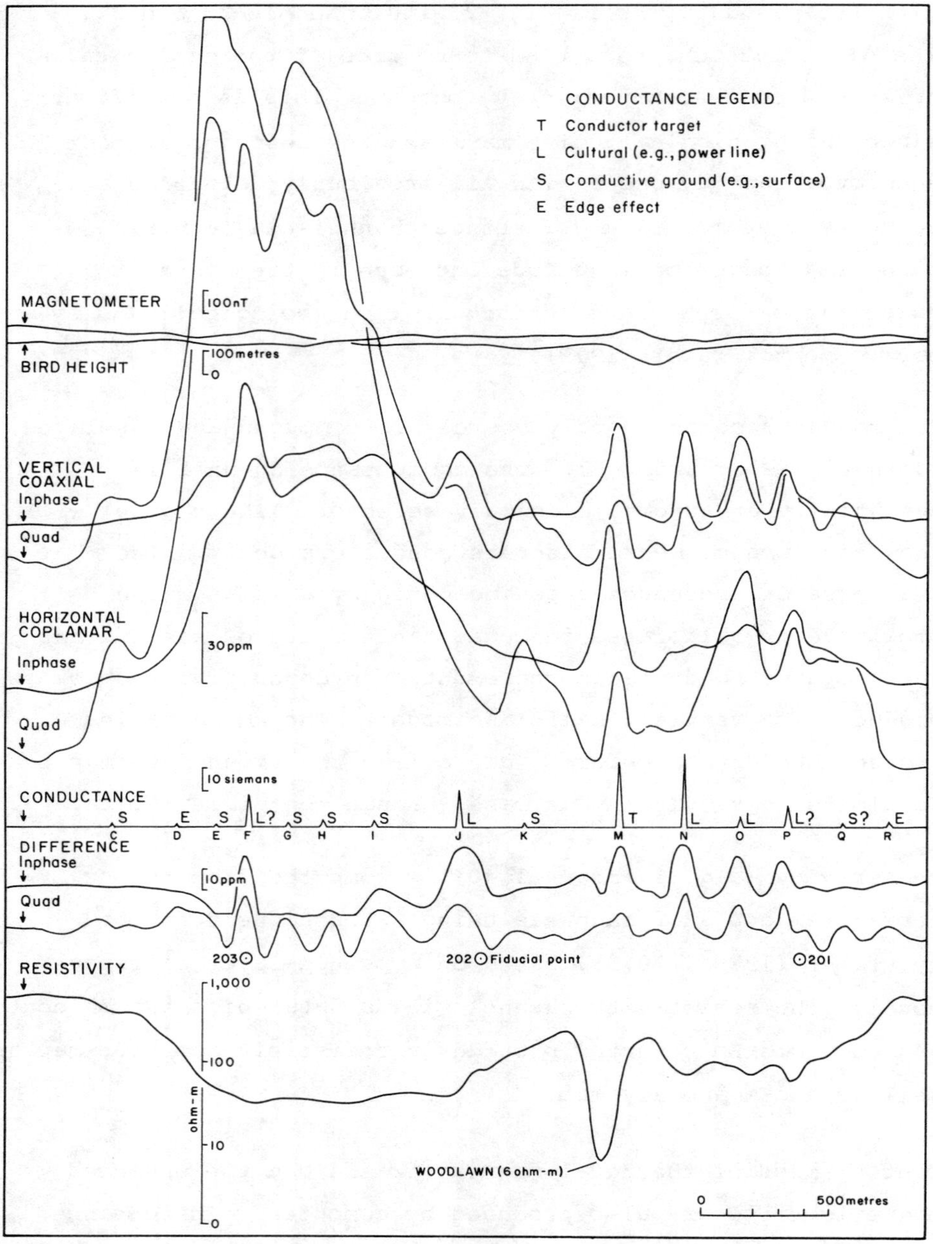

Fig. 2 DIGHEM^{II} Data over the Woodlawn orebody.

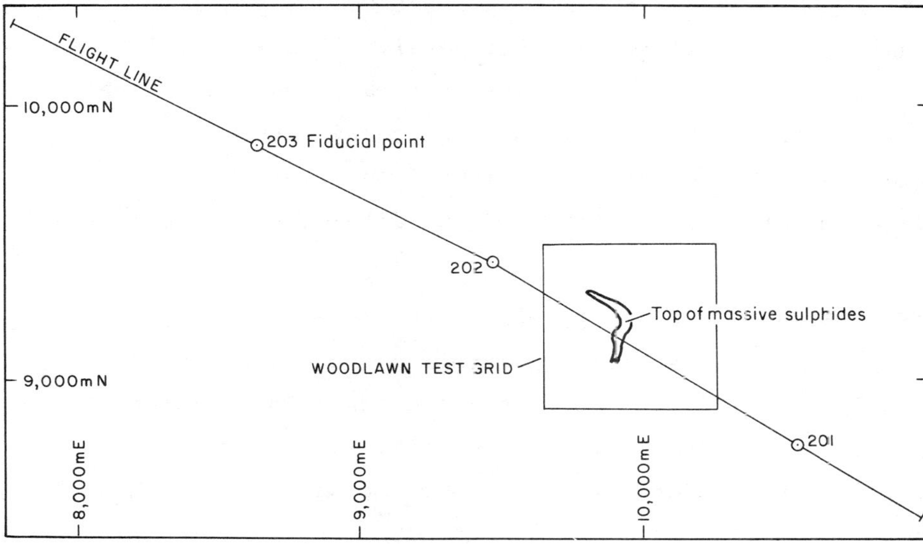

Fig. 3 Location of DIGHEM^{II} flight line.

Results

Fig. 3 shows one line which was flown in a northwesterly direction over the Woodlawn orebody. Since DIGHEM^{II} is a symmetrical system results are independent of whether the system is flown in an updip or downdip direction across the deposit. The anomaly shapes and amplitudes shown in Fig. 2 are fairly representative of the 3 lines which were flown both updip and downdip at 50m intervals over the orebody.

Mining of the deposit had already commenced at the time of the survey in July 1979, and this resulted in a high density of cultural influences including electrification. The system responded to these cultural influences as well as to the geology. However, the two can be readily separated as may be seen from a study of the computer generated profiles of Fig. 2, in which the following should be noted:-

1. The EM channels 22 and 23 represent the in-phase and quadrature anomalies from the vertical, coaxial coil pair. The Woodlawn anomaly on these vertical, coaxial channels is similar to the many anomalies produced by cultural features. Geological and cultural sources cannot be discriminated on the basis of the data from this coil pair.

2. EM channels 24 and 25 represent the in-phase and quadrature components of the horizontal coplanar coil pair. The Woodlawn anomaly on the in-phase channel 24 is larger than the cultural anomalies. The deposit is therefore more readily identifiable from the horizontal coplanar channel data.

3. The resistivity data (channel 40) clearly identifies the Woodlawn deposit which has the lowest resistivity observed on the traverse. Consequently, it can be concluded that the resistivity parameter is of major importance in surveys for highly conductive massive sulphide deposits such as Woodlawn. It is also important to note that cultural influences do not greatly distort the resistivity profile.

4. The largest EM response obtained occurred to the northwest of the Woodlawn orebody. This anomaly is very broad and is presumably caused by a broad zone of higher conductivity than normal and yielded an anomaly of up to 300 ppm on the horizontal, coplanar quadrature channel 25. In contrast, the Woodlawn deposit on this same channel gave an anomaly of only 30 ppm. Similarly, the broad conductive zone yielded a horizontal coplanar in-phase anomaly (channel 24) of 100 ppm which is larger than the 60 ppm for the Woodlawn orebody. In spite of the dominance on the EM profiles of the broad anomaly, the resistivity channel 40 shows the Woodlawn orebody to have a higher conductivity or lower resistivity (6 ohm m) than the broad zone (30 ohm m). Consequently, the Woodlawn

deposit is identified more readily on the resistivity channel than on the EM channels.

5. From the data over Woodlawn itself, the following may be inferred regarding the orebody:-
   (a) the deposit dips to the west because the horizontal, coplanar in-phase channel 24 peaks to the left of the vertical, coaxial in-phase channel 22.
   (b) the deposit appears to be more conductive on the hanging wall side because the horizontal, coplanar in-phase channel 24 peaks to the left of the horizontal, coplanar quadrature channel 25.

6. The conductance channel 37 which is the automatic anomaly selector calibrated in siemens, indicates a conductance of 28 siemens for the Woodlawn deposit. This channel also selected all cultural features and several inhomogeneities within the broad conducting zone discussed above. This channel has been designed to indicate all conductive features and one of the interpreter's tasks is to discriminate between anomalies produced by different sources, as shown by the L, S and E symbols of Fig. 2.

Conclusions

The DIGHEM^{II} system was shown to be capable of yielding substantial anomaly amplitudes over the Woodlawn massive sulphide orebody. It is also capable of identifying the massive sulphide deposit in an environment with a high density of cultural influences. The survey showed that the resistivity channel is of major importance in identifying significant conductors in areas containing conductive country rocks and/or cultural features. The Woodlawn deposit yielded the most anomalous resistivity response recorded during the survey, although larger EM amplitudes were obtained from other broad conducting zones. Cultural features such as power lines produced EM anomalies similar in amplitude to those from the Woodlawn deposit, but the resistivity channel proved to be largely unaffected by these. Consequently, the deposit was well defined by the resistivity channel, in spite of the cluttered anomalies which were present on the EM channels.

(At the time of this survey, mining operations had already removed the overburden and highly conducting shallow supergene zone and top of the orebody down to a depth of about 20m below the original ground surface above the orebody. This may have had the effect of reducing the response of the orebody to this method - Ed.).

References

Fraser, D.C., 1974, Survey experience with the DIGHEM AEM system: Cdn. Inst. Mng., Bulletin, April, p. 97-103.

Fraser, D.C., 1978a, Resistivity mapping with an airborne multicoil electromagnetic system: Geophysics, v. 43, p. 144-172.

Fraser, D.C., 1978b, Geophysics of the Montcalm Township copper-nickel discovery: Cdn. Inst. Mng., Bull., January, p. 99-104.

Fraser, D.C., 1979, The multicoil^{II} airborne electromagnetic system: Geophysics, v. 44, p. 1367-1394.

CHAPTER 7

GROUND GEOPHYSICAL SURVEYS: ELECTROMAGNETIC METHODS

PART 1

CONTINUOUS WAVE, REMOTE SOURCE METHODS

7.1. Very Low Frequency Electromagnetic Profiles at Woodlawn

R. J. WHITELEY

Contents

Abstract

The Woodlawn orebody produces large Very Low Frequency electromagnetic (VLF EM) anomalies. These anomalies demonstrate the need to explore with at least two VLF stations whose primary magnetic field directions are approximately orthogonal.

Large VLF anomalies of comparable magnitude and character to those over the orebody were detected both near the orebody and over the weakly mineralized Black Shale area about 2.5km to the south. These anomalies suggest that it is unlikely that the VLF response of the Woodlawn orebody would be sufficiently diagnostic to distinguish it from other conductive geological formations in the region.

Introduction

The Very Low Frequency electromagnetic (VLF EM) method is one of several radio frequency (RF) methods which utilize as an energy source the EM field produced by a distant radio transmitter.

Although the exploration possibilities resulting from the use of radio waves had been known for many years (Heiland, 1946) widespread development of techniques was hampered because of the relatively high frequency of most conventional radio transmitter (above 500kHz) and their low power. As a result, early methods suffered from poor penetration and a lack of ability to discriminate geological conductors.

Some thirteen years ago a major impetus to the use of remote radio sources in geophysical exploration followed the construction of a world wide network of powerful VLF transmitters for navigation and communication purposes. Since their introduction there have been many successful VLF EM surveys for base metal sulphide deposits e.g. Paal (1965), Paterson and Ronka (1969), Phillips and Richards (1975) and Caven (1976). These VLF stations operate in the frequency range from about 10 to 26kHz and can be used as remote electromagnetic sources for distance up to about 5000km.

A VLF transmitting antenna is effectively a grounded vertical wire approximately 100m high. There are several possible modes of electromagnetic radiation from this antenna although at distances exceeding about 100km, VLF waves appear to propagate in a waveguide between spherical reflecting shells formed by the earth's surface and the lower ionosphere with little energy loss at these boundaries.

The VLF signal is essentially a transverse electromagnetic (TM) mode with the main field polarized roughly cylindrically about the antenna. The magnetic field component carries the bulk of the signal energy within the earth and the vector assumes an attitude approximately parallel to the average ground surface. At distances from the station exceeding about 500 km, rectangular field components can be assumed and in most cases the primary field is practically uniform over several square kilometers of ground surface.

As the earth has a finite conductivity, the primary VLF field is attenuated with depth and undergoes a phase shift proportional to the average ground conductivity. When the primary magnetic field lines encounter a contrast in ground conductivity secondary magnetic fields are produced which generally differ in amplitude, phase and direction relative to the normal primary field. Ideally, the strike of the geological conductor should be at right angles to the direction of the primary field lines. The resultant of primary and secondary fields is elliptically polarized in an essentially vertical plane, and the nature of the conductivity contrast can be determined by measurement of parameters of the polarization ellipse (Ward, 1966). To some extent the VLF polarization ellipse in the vicinity of a geological conductor is characteristic of that conductor.

Survey at Woodlawn

Two VLF transmitters, NDT Tokyo and NWC Western Australia are sufficiently powerful to be used at Woodlawn. The characteristics of these stations are listed on Table 1. This table shows that NDT is almost due north of Woodlawn while NWC is approximately west-northwest. These stations differ in azimuth by approximately 60 degrees which, although not the ideal 90 degrees, is sufficient to obtain a response from most conductors with the variety of strikes in the Woodlawn area.

Table 1 Characteristics of VLF Stations used at Woodlawn.

STATION	FREQUENCY (KH_z)	RADIATED POWER (KW)	APPROX. BEARING FROM WOODLAWN
TOKYO JAPAN NST,	17.40	500	352°
NWC NORTH WEST CAPE W.A.	22.30	1000	292°

Fig. 1 shows the angular relationships between the strike directions of the Woodlawn orebody, the primary VLF field directions and the traverse directions used.

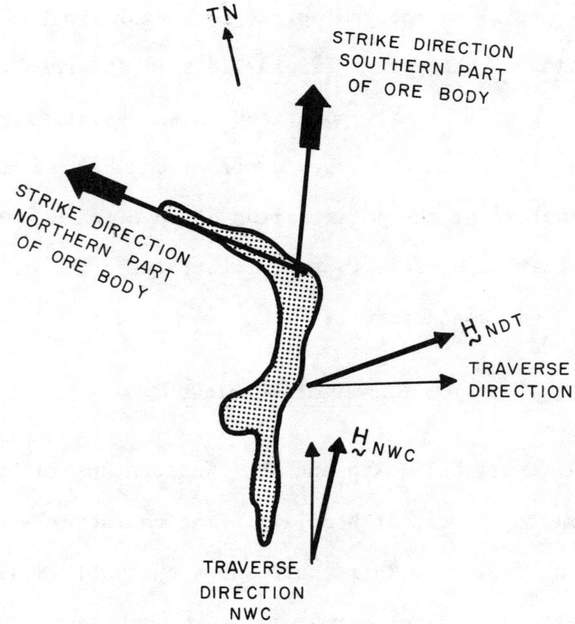

Figure 1 Angular relationships of Woodlawn orebody to Primary VLF magnetic Fields (H_{NWC} and H_{NDT})

The Woodlawn orebody varies considerably in strike direction with the northern part of the body striking at about 305°M and the southern part at about 030°M. The primary magnetic field for station NDT ($\underline{H}_{NDT}$) bears at 030°M and the traverse direction for this station along the east to west Woodlawn Test Grid was within about 18° of this direction. The primary field for station NWC ($\underline{H}_{NWC}$) bears at about 022°M and the traverse direction along grid north to south is within about 22° of this direction.

Equipment and Field Procedures

An EM16 VLF Electromagnetic Unit (Geonics Ltd., Toronto) was used at Woodlawn. This instrument has individual tuned receiver coils for station NDT and NWC respectively. The EM16 unit, operating along or close to the direction of the respective primary fields approximately determines inclinations from the vertical and the eccentricities of the polarization ellipse, by measuring in-phase and out-of-phase vertical field strengths. (Paterson and Ronka, 1969). These are expressed as a percentage of the horizontal primary field and obtained by physically tilting the instrument in a vertical plane to a minimum signal position or null, and reading the inclination for the in-phase measurements. The out-of-phase measurements are then made by attempting to improve the null by varying phase with a calibrated dial. Measurements with the EM16 are performed very rapidly. Highly conductive sources produce reverse crossover anomalies in which the in-phase and out-of-phase profiles move in an opposite sense near the conductor and intersect near its axis. The measured in-phase component generally increases in the direction of the primary field as the conductor is approached.

Readings were taken at from 5 to 15m intervals in a west to east direction with station NDT across the Woodlawn Test Grid on Lines H, K and M (9280mN, 9190mN and 9130mN respectively) and from south to north along the grid lines 11, 12, 13 with station NWC (Fig.1).

During the survey strong winds at Woodlawn caused some problems in obtaining accurate audio nulls using the monotonic speaker mounted on the instrument and necessitated the use of the earplug supplied. To ensure reliability of measurement most stations were repeated for verification and results averaged.

An additional VLF traverse was also carried out using NDT on the Black Shale some 2.5km south of the orebody which had a strong self potential anomaly and was drilled on the basis of that anomaly. (Cifali and Whiteley, this volume, p321; Malone, this volume, p 49).

The VLF survey was carried out during May, 1975.

Results: Station NDT

In-phase and out-of-phase profiles plotted above the orebody plan for station NDT are shown in Fig. 2.

Line H (9280mN)

The northernmost profile surveyed (Line H) shows a prominent reverse crossover anomaly in the vicinity of the orebody, which is typical of a shallow, highly conductive body. In this region, the orebody strikes at an azimuth of 305°M (Fig. 1) and is essentially vertical with a width of approximately 30m. It is capped by about 8m of gossan.

The reverse crossover is located near the centre of the orebody. The in-phase anomaly has a minimum amplitude of -25% on the eastern side of the body and a maximum of +22% on the western side. The large angle (about 45°) between the strike of the northern part of the orebody and the primary magnetic field direction causes the in-phase anomaly to be elongated in a westerly direction and characterized by a more prominent bowl shaped minimum to the east. These effects have been noted by Ragowsky and Bowes (1970) on

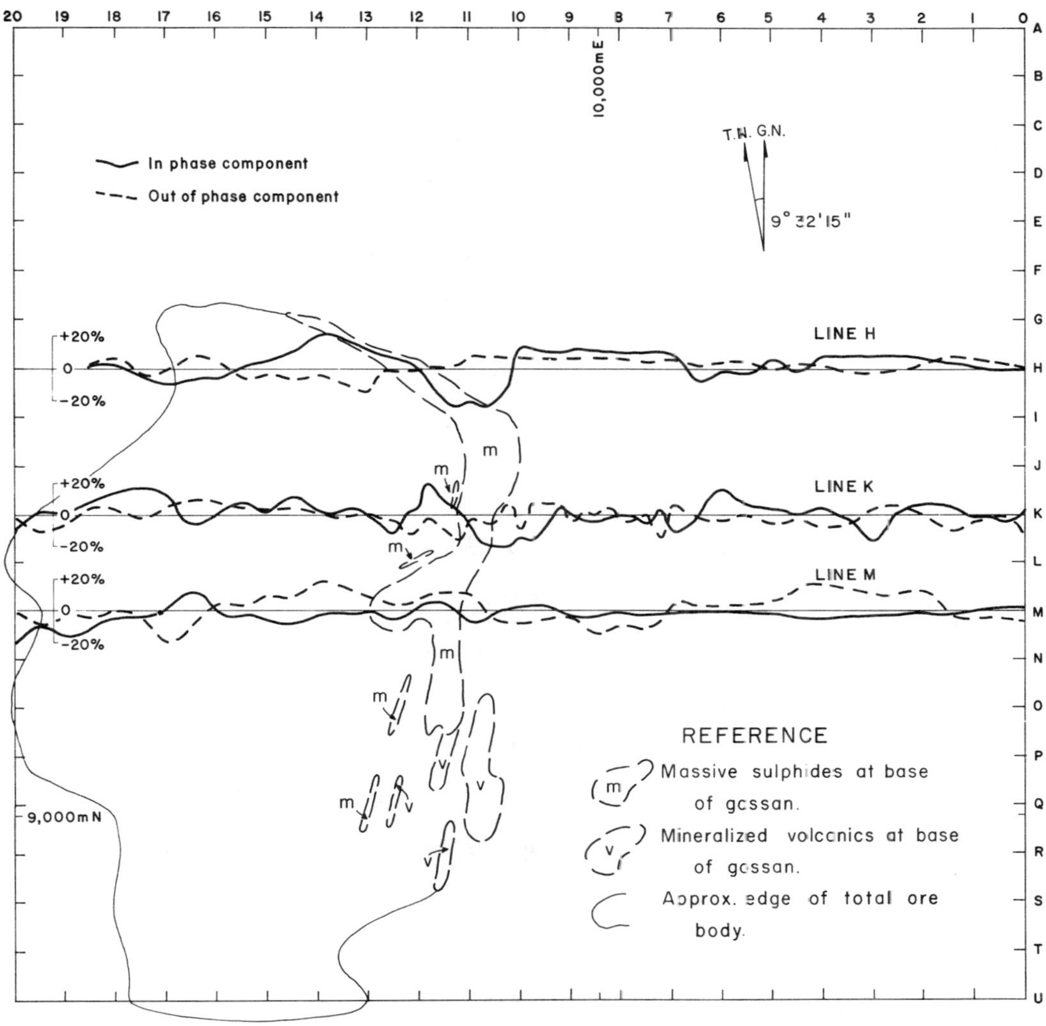

Figure 2 VLF Profiles, Station NDT Lines H, K, M, Woodlawn Test Grid.

scale model experiments for oblique VLF profiles over thin, vertical, tabular conductors. Similar asymmetric in-phase anomalies can be caused by conductors dipping at a low angle (say 30°). In the case of the Woodlawn deposit, unless the strike of the northern part of the orebody was known an erroneous dip to the east may have been inferred from this VLF profile.

The out-of-phase response is about one third the amplitude of the in-phase response. This feature and its reversed nature substantiate the highly conductive nature of the orebody.

Line K (9190mN)

The VLF profile on Line K near the centre of the orebody again shows a prominent reverse crossover anomaly in the vicinity of the body, indicating the presence of a good electrical conductor.

In this region, the orebody strikes approximately north to south with a dip of about 45° to the west and is approximately 25m wide and is capped by about 13m of gossan.

Fig. 3 shows a portion of this VLF profile plotted above a geological cross section based on drilling. The in-phase response on Line K above the orebody is similar in amplitude to that on Line H but contains many more localized disturbances. Prior to the survey a deep pit was dug between lines J and K for metallurgical testing purposes and subsequently filled in. This may account for many of the localized, small amplitude disturbances on this line. The minimum on the in-phase anomaly is immediately above the gossan and the crossover point appears to be migrated westward towards the hanging wall. Because of the local disturbances it is difficult to attribute any asymmetry in the anomaly to dip although the westward migration of the crossover point is probably a dip effect.

An additional reverse, crossover anomaly near station K17 on Line K, with a somewhat smaller amplitude than the orebody anomaly, could be correlated with the contact between tuff and dolerite on the western side of the orebody.

Line M (9130mN)

The VLF profile on Line M towards the southern part of the body shows no crossover anomaly in the vicinity of the orebody.

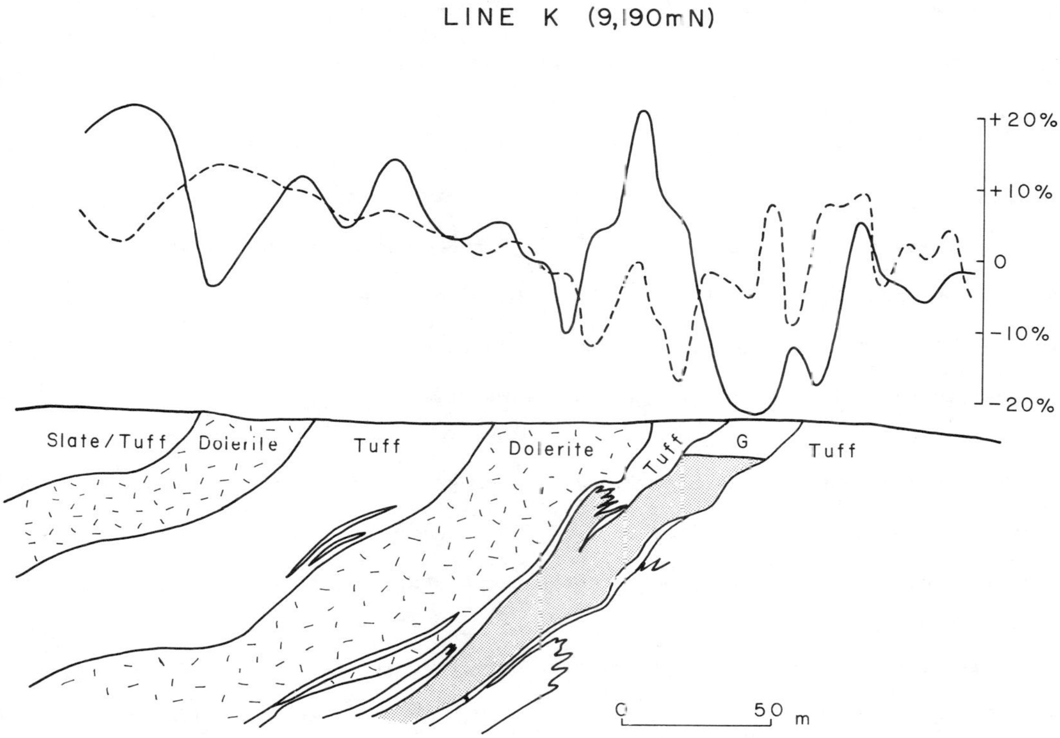

Figure 3 VLF Anomaly and the Woodlawn orebody using Station NDT.

In this region the orebody is more complex in shape, strikes at 030°M with a dip to the west of about 45° and an average width of approximately 45m. It is capped by about 18m of gossan. There is somewhat greater topographic relief along Line M which is located on the side of a hill marking the southern end of the massive orebody.

The only obvious anomaly on Line M is a reverse crossover near station M16 (Fig.2) to the west of the orebody. This could be correlated with a similar anomaly on Line K and is believed due to a contact between tuff and dolerite.

The lack of any anomaly in the vicinity of the orebody on Line M is difficult to explain since an additional thickness of about 5m of gossan on this line relative to that on Line K would seem an unlikely cause for such rapid attenuation of the large anomaly measured on Line K.

Prior to the VLF survey, a gradient array resistivity and IP survey was completed over the Woodlawn Test Grid. (Quick and Cifali, this volume p 333). These data indicate practically no difference in the conductivity of the body from the centre Line K to Line M and give bulk conductivity values for the near surface region of the orebody and gossan of about 10^{-2} S/m although the actual conductivity of the ore material is much higher.

The gradient array data also yield an average conductivity over $0.4 km^2$, which includes the orebody, of about 1.6×10^{-3} S/m. This shows the substantial conductivity contrast between the orebody and the host rocks.

This average conductivity can also be used to calculate a skin depth by applying the appropriate plane wave equation. At Woodlawn this gives a skin depth of about 90m at the frequency of station NDT.

The low conductivity of the host rocks with the resulting relatively large skin depth, make it unlikely that attenuation alone accounts for the lack of anomaly on Line M.

Results: Station NWC

Line 11, 12, 13

VLF profiles for the south to north lines 11, 12 and 13 obtained using station NWC are shown in Fig. 4. These profiles are basically similar and show reversed crossover anomalies near the northern part of the orebody where it strikes at a considerable angle to the primary magnetic field direction. No large responses were obtained over the central and southern part of the body where it essentially parallels the primary magnetic field direction and is therefore poorly coupled to the primary field.

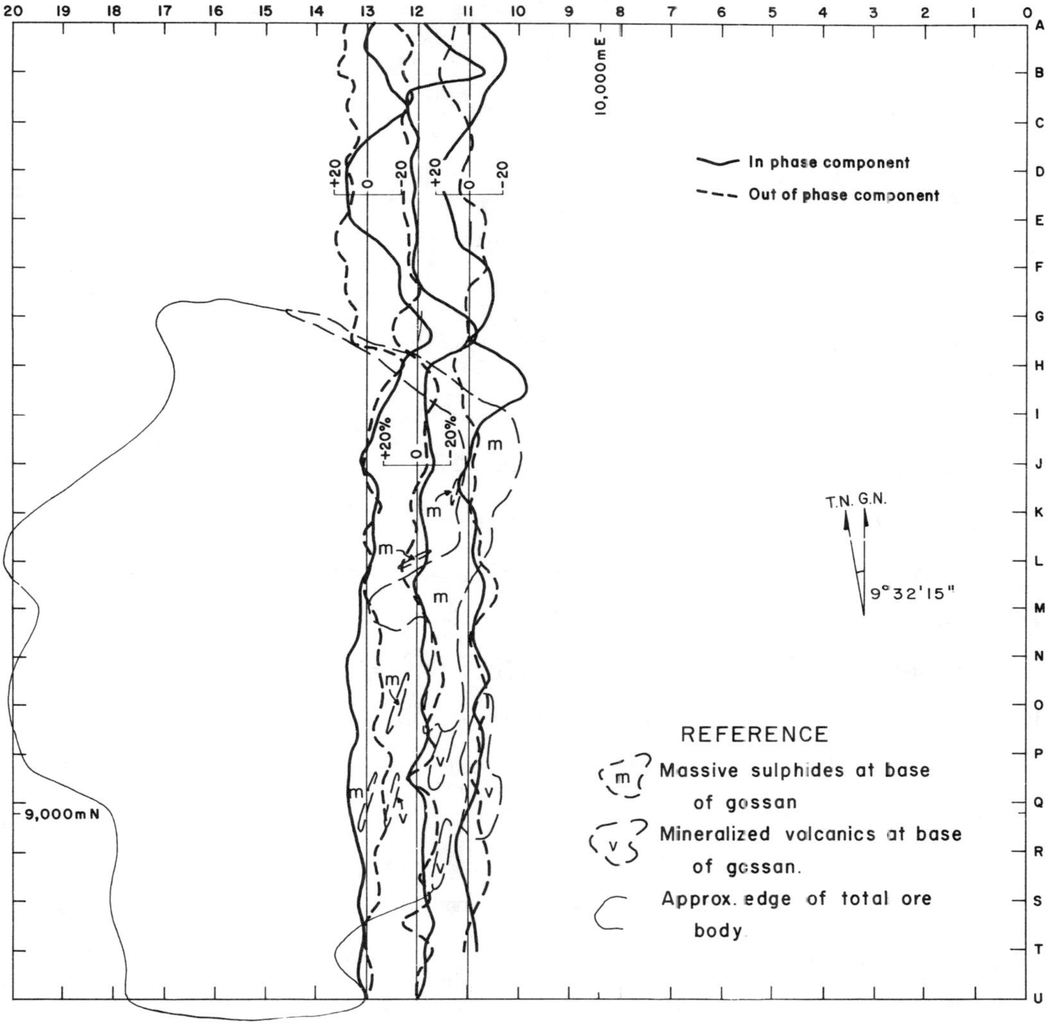

Figure 4 VLF Profiles, Station NWC Lines 11, 12, 13, Woodlawn Test Grid.

The VLF anomalies due to the northern part of the orebody are characterized by asymmetrical anomalies with crossover points near the centre of the body. Major features of the responses are the prominent in-phase lows with amplitudes of as much as -35% beyond the northern edge of the orebody. The in-phase responses to the south of the crossover points are of much lower amplitude and are elongated due to the presence of the bulk of the conductive ore material beneath this region. The out-of-phase response is about one third of the amplitude of the in-phase response and is more obvious beyond the northern edge of the orebody.

Large reverse crossover anomalies were also obtained at the northern end of the grid in the vicinity of Line B (9465mN). These anomalies are comparable in amplitude to those obtained over the Woodlawn orebody. The source of the northern anomalies is believed to be highly conductive black shales or tuffs.

Black Shale Area

A VLF profile was also completed over the Black Shale some 2.5km south of Woodlawn where a prominent self potential anomaly occurs (Cifali and Whiteley, this volume, p 321) which led to the drilling of the sequence. Drilling and geological mapping revealed the presence of a wide, steeply dipping tuffaceous black shale containing stringers and veins of mainly pyrite mineralization with traces of base metal sulphides. This unit strikes approximately northwest.

Station NDT was used and the results of the VLF profile are shown in Fig. 5. An approximately symmetrical reverse crossover anomaly was obtained over the weakly mineralized Black Shale. The location of the crossover was near the centre of the formation, coincident with the centre of a strong resistivity and IP anomaly (Whiteley and Tyne, this volume, p 423). The amplitude of the in-phase anomaly is about half the amplitude of the in-phase anomaly at Woodlawn. The in-phase anomaly also shows a flattening of the maximum and minimum which is the characteristic response of a thick, vertical, tabular conductor. The out-of-phase response is considerably larger than that over the Woodlawn orebody and comparable in magnitude to the in-phase response over the Black Shale, indicating a zone of only moderate conductivity in comparison with the Woodlawn orebody.

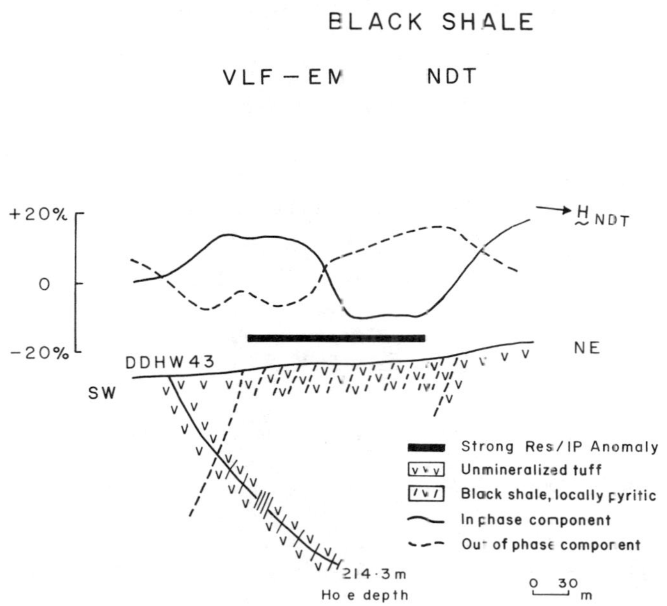

Figure 5 VLF Profiles, Station NDT Black Shale area.

Conclusions

The Woodlawn orebody produces large VLF EM responses expected from a shallow, highly conductive, massive sulphide orebody. The anomalies showed many of the characteristics expected for a conductive target of different strike directions and dips and demonstrate the need to explore with at least two VLF stations whose primary magnetic field directions are approximately orthogonal.

One confusing aspect of the VLF responses over the orebody is the lack of an anomaly near the southern end of the body where the massive mineralization is present although at slightly greater depth. This cannot be easily explained by an attenuation mechanism or a change in conductivity of the orebody along strike.

Large VLF anomalies of comparable magnitude and character to those over the orebody except for an increased out-of-phase response were detected both near the orebody and over the weakly mineralized Black Shale some 2.5km away. Although a relatively larger out-of-phase anomaly over the shale indicates a lower conductivity than the Woodlawn orebody, these anomalies suggest that it is unlikely that the VLF response from the orebody would be sufficiently diagnostic to clearly distinguish it from other conductive rocks in the region.

References

Caven, R.J. 1976. Geophysical Surveying at the Umex Thierry Deposit, Canad. Min. Jour. Oct. 1976 p. 34 to 43

Geonics Ltd., 1969. EM 16 VLF Electromagnetic Unit Operating Instructions. Geonics Ltd. Toronto, Canada.

Heiland, C.A., 1946. Geophysical Exploration. Prentiss Hall, N.Y.

Paal, G., 1965. Ore Prospecting based on VLF - radio signals. Geoexploration, Vol.3, No.3. p. 139 to 147

Paterson, N.R. and Ronka, V. 1969. Five years of Surveying with the VLF EM Method. Presented at the 1969 Am. Int. Meeting Society of Exploration Geophysicists.

Phillips, W.J., and Richards, W.E. 1975. A Study of the Effectiveness of the VLF Method for the Location of Narrow-Mineralized Fault Zones, Geoexploration, Vol. 19, p. 215 to 226.

Ragowsky, T.P. and Bowes, W.A. 1970. EM 16 Model Experiments, Geonics Ltd. Toronto, Canada.

7.2. Three Audio Frequency Magnetic Traverses Across the Woodlawn Orebody

K. VOZOFF AND R. J. G. LEWIS

Contents

Abstract

A limited audio frequency magnetic (AFMAG) survey was completed at Woodlawn during peak primary signal strength. Results obtained showed that the orebody produces an ample AFMAG response although equipment problems did not enable complete interpretation. It is suggested that a more extensive survey with improved AFMAG equipment would provide additional useful information on orebody and host rocks and surrounding geology.

Introduction

The audio frequency magnetic (AFMAG) method was the first serious attempt to free electromagnetic exploration methods from the limitations associated with a transmitter, and was a forerunner of the audio frequency magnetotelluric (AMT) method. By using natural electromagnetic noise fields at audio frequencies it was hoped to do away with the limitations imposed by finite transmitter power and with the complications which arise from transmitter positioning. These natural fields result from electrical discharges in the atmosphere (Ward, et al 1966). In spite of some spectacular results (Sutherland, 1970; Collett and Bell, 1970) AFMAG came into disuse because the signals were often too weak for the equipment which was available, leading to erratic results. Nevertheless, the approach is valid in principle and when conditions are favourable, AFMAG anomalies can be used to find and map contacts between regions of differing conductivities, as well as zones of enhanced conductivity which have limited width.

As implemented, the method is analogous to the receiving function in the vertical loop dip angle method (Telford, et al, 1976). The primary magnetic field is assumed to be horizontal, its orientation varying randomly in time within the horizontal plane. A portion of the primary field penetrates the earth and, if a localized conductive zone is present secondary magnetic fields will accompany the telluric eddy currents which are induced within it. Like the dip angle receiver, the AFMAG operation balances the vertical with a portion of the horizontal. It indicates the ratio of vertical to (phase coherent) horizontal magnetic field in a vertical plane parallel to the traverse. The arctangent of this ratio, the dip angle, will normally be zero, because the field will have no vertical component in a homogeneous or horizontally layered region.

Only when large lateral changes of conductivity occur is a vertical component set up. The change might be a highly conductive, narrow feature or merely a contact between adjacent regions of a large resistivity contrast (Madden and Vozoff, 1971). The strike length and depth extent of these features are as important as their conductivity contrast, so that many economically uninteresting but extensive features give rise to anomalies of regional extent. This result further discouraged widespread use of the method.

Equipment and Field Procedures

The equipment used at Woodlawn, a McPhar Model A652, was possibly the only AFMAG unit available in Australia at the time. This unit operates at two frequencies, 140Hz and 470Hz. The two orthogonal coils are rotated in a vertical plane parallel to the traverse until a "balance" is observed on the output meter. At that stage the shorter coil is aligned in the direction of the (time average) maximum field within that plane. In the absence of an anomaly the coil will be horizontal at balance.

The survey was scheduled for the summer because that is the peak signal strength season and was carried out from the 12th to the 14th of January, 1976 along three traverses. At Woodlawn, signals were well above the instrument noise level at all measurement times (Ginsberg, 1974). A station spacing of 15m was used.

Results

Locations of the three traverses are shown in Fig. 1 and the results in Figs. 2 to 4. In the plotting convention used, a conductor axis is indicated when angles cross from above to below the line as we move to the right. Such "normal" crossovers are indicated on the figures.

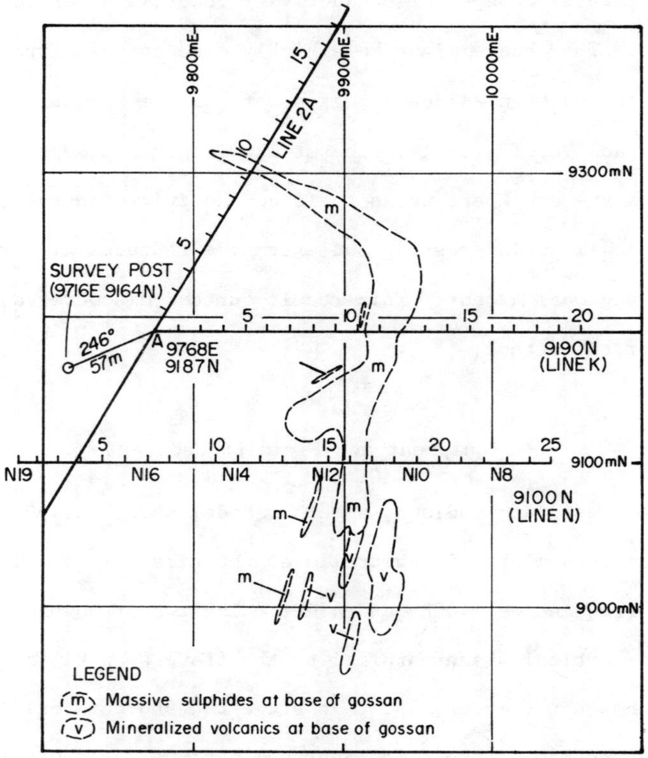

Figure 1 Location of AFMAG traverses.

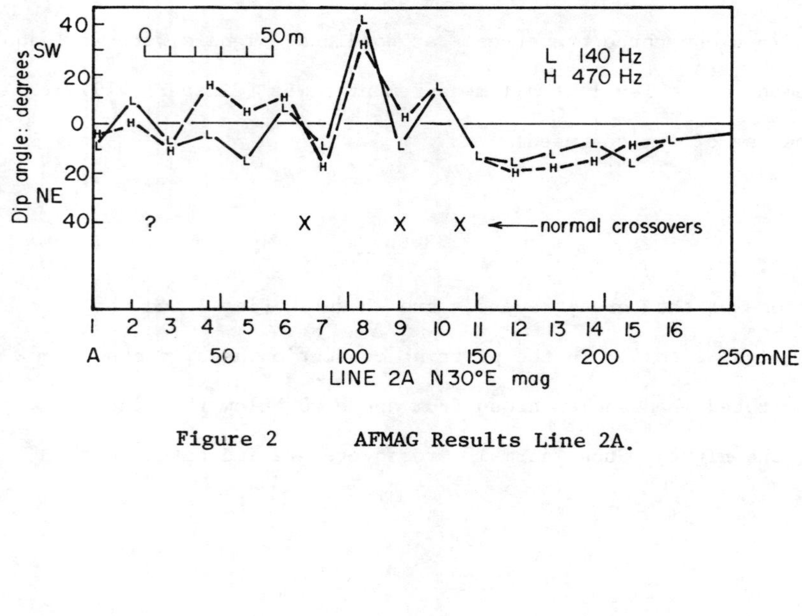

Figure 2 AFMAG Results Line 2A.

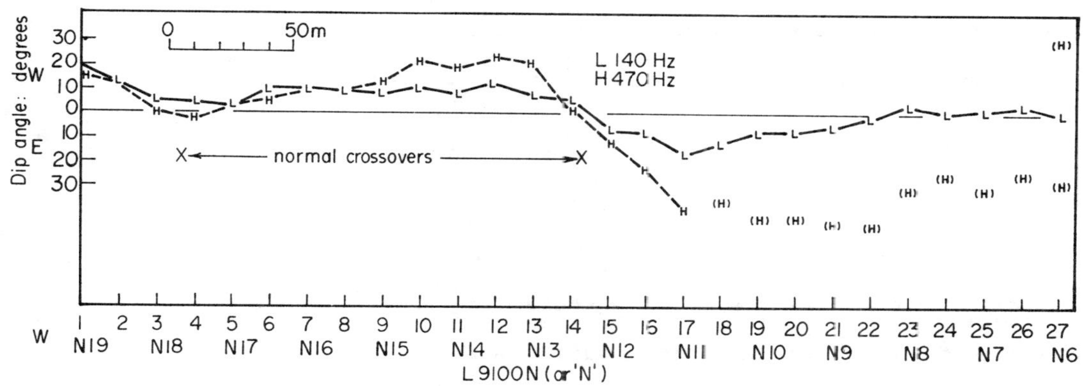

Figure 3 AFMAG Results Line 9100mN (Line N). (Note station numbers 1 to 27 refer to this line only and not the Woodlawn Test Grid - Ed.)

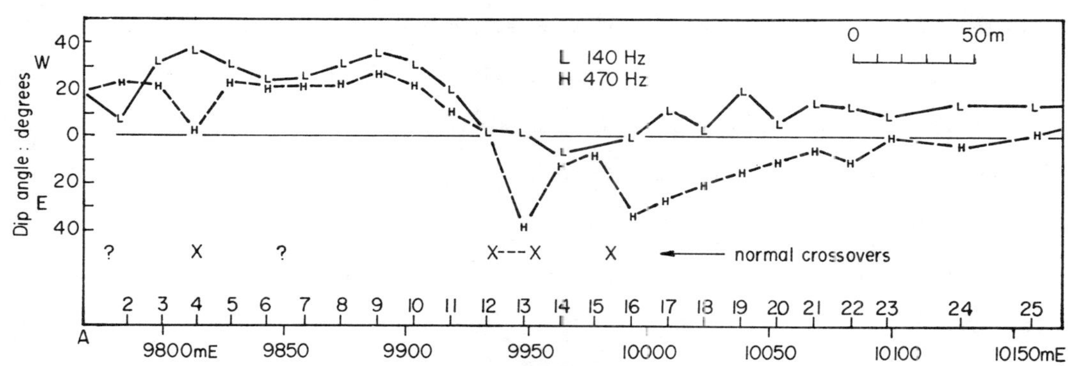

Figure 4 AFMAG Results Line 9190mN. (Line K). (Note station numbers 2 to 25 refer to this line only and not the Woodlawn Test Grid - Ed.)

Line 2A shows erratic dip angles south of the subcrop (stations 9 and 10), becoming more regular to the north. This, plus the fact that the dip angles are relatively small, indicates that the sulphides show no particular alignment in a direction at right angles to that line.

Currents in the north to south directions were more consistent, showing a north to south elongation of the conductor as shown by the results on lines 9190mN (Line K) and 9100mN (Line N). Both show smoother variations, although Line N had to be terminated when rain caused the unit to malfunction, and several of the readings on the eastern half uncertain.

Results onLine N (Fig. 3) show a clear crossover between stations N12 and N13 coincident with the subcropping boundary of the orebody and a poorly defined crossover near station N18. This can be interpreted in two ways: either two discrete conductors occur or a broad conductive zone extends between stations N13 to N18. In the latter case the poorly defined crossover at N18 may reflect the downdip edge of the massive sulphide orebody. (Drill hole information shows the projected downdip limit at station N20. Since the AFMAG traverse on Line N did not extend beyond station N19 it is likely that the actual effect of this boundary was not observed: - Ed.). The AFMAG data indicate a westerly dip of the orebody. The calculated effective conductivity thickness product is moderate due to the short strike length of the orebody.

Line K (Fig. 4) does not extend far enough westward to have crossed the downdip limit of the body. The low frequency dip angles appear to have a regional value of 10° to 15° to the west. This could arise for several reasons. Reading uncertainty was $\pm 5^{o}$ and may be a contributing cause. A gradual drop in resistivity or a conductor beyond the eastern end of the traverse would have this effect, since an anomaly persists for about one skin depth. At Woodlawn one skin depth would be about 1km at the low frequency and half that at the higher frequency. Another, less likely but possible cause, would be an east to west striking conductor to the south of the line. The vertical magnetic field induced by such a feature would be partly coherent with that of a north to south striking conductor and would give rise to an apparent regional dip (Ward et al, 1966).

Discussion and Conclusions

The Woodlawn orebody produces an ample AFMAG response. To be of any real value this result must be augmented with further information on the extent of the anomaly, the anomaly background and particularly the effects of black shales in the footwall. It was not possible to obtain this information at the time of the survey, due to the effects of rain on the equipment.

The anomalies suggest other features in addition to the outline of the massive sulphides. A geological contact further to the east is also suggested.

Some of the reasons for the declining interest in the AFMAG method no longer exist since the advent of numerical (computer) modelling. The effects of contacts and topographic features can now be modelled and interpreted. Other reasons for the loss in popularity of the method do still exist. No improvements in the equipment have been made for perhaps 10 years, and related procedural questions (i.e. the use of a well defined polarization in the horizontal plane) have yet to be satisfactorily settled, although a start has been made in this direction (Ward et al, 1966; Madden and Vozoff, 1971). Certainly, equipment which requires summer time electrical storms but cannot withstand a gentle drizzle requires some modification.

Acknowledgments

Costs of the fieldwork were borne by a Macquarie University Research Grant. We are also grateful to the Director, BMR, for the use of the AFMAG unit.

References

Collett, L.S., and Bell, C.K., 1970, AFMAG use in geological interpretation. Paper presented at Can.Inst. Min. Meeting Toronto, April, 1970 9 pages.

Ginsberg, L.H., 1974, Extremely Low Frequency (ELF) atmospheric noise level statistics for Project Sanguine: IEEE Trans. on Communications, v. COM-22, No.4, P. 555 to 561.

Madden, T.R., and Vozoff, K., 1971, Explanatory booklet to VLF Model Suite (2nd edition): Exploration Aids Inc., Needham, Mass.

Sutherland, Don B., 1970, AFMAG for electromagnetic mapping, in Mining and Groundwater Geophysics 1967: Econ. Geol. Report No.26, Geol. Survey of Canada, (L.W. Morley, Editor), p. 2288 237.

Ward, S.H., O'Connell, J., Rivera, R., Ware, G.H., and Fraser, D.C., 1966, AFMAG Applications and Limitations: Geophysics, v. 31, No. 3, p. 576 to 605.

PART 2

CONTINUOUS WAVE, NEAR SOURCE METHODS

7.3. Audio Magnetotelluric Measurements on the Woodlawn Orebody Using a Grounded Dipole Source

M. J. SMITH

Contents

Abstract

Audio magnetotelluric (AMT) measurements with a grounded electric source dipole were completed at Woodlawn using frequencies from 20Hz to 3.3kHz. A large AMT anomaly, which was strongest at the low frequencies, was obtained over the Woodlawn orebody showing indications of dip on both apparent resistivity pseudo sections and profiles.

Traverses over the nearby Black Shale gave AMT anomalies which were similar in amplitude and resolution to those over the orebody when the remote transmitter dipole straddled the black shale. In that case, the greater strike length of anomaly due to the shale unit would distinguish it from the anomaly due to the short strike length orebody.

The Audio Magnetotelluric Method

The audio magnetotelluric (AMT) method is a well known exploration technique entailing the simultaneous measurement of electrical and magnetic fields over a fairly wide frequency range. The observed values of the electrical field (E) and the magnetic field (H) at one locality yield apparent resistivity values which constitute a vertical sounding of the electrical properties of the subsurface at the locality over a depth range related to the range of frequencies (f) studied. The apparent resistivity (pa) is calculated by the formula

$$\rho a = \frac{1.26 \times 10^5 (E/H)^2}{f}$$

In general, the electrical and magnetic fields which can be measured are the discharges associated with thunderstorms that propagate around the world in the electromagnetic waveguide between the earth and the ionosphere. These natural fields are usually detectable in the frequency range from 10Hz to tens of Khz, although a band centred about 2Khz is attenuated significantly during propagation. Such observations have been shown to be most informative when recorded over horizontally stratified rocks and therefore the technique has been applied widely to investigations of broad sedimentary basin structures.

However, in various geographical localities and at various times of year, natural AMT source fields are too weak to be accurately measured. Also in regions of strong lateral variations in resistivity, the repeatability of observations is strongly dependent on the direction to the thunderstorm source dominant at the time of the observations.

Application of AMT Using Artificial Source Fields to Massive Sulfide Exploration

In the complex geoelectrical halfspace of most massive sulphide deposits, the use of an artificial source field offers some advantages:

a. the amplitudes of the E and H fields are high and constant ensuring accurate measurements;

b. the direction of the source field is constant during any single survey but may be varied to yield new data on the same grid which may assist interpretation;

c. the range of the source field may be varied to optimize field work.

Goldstein and Strangway (1975) have shown that there is a good case for the application of AMT measurements using an artificial source in mapping near vertical geoelectrical boundaries and possibly locating conductive sulphide deposits.

Equipment Used at Woodlawn

The AMT instrumentation employed for the Woodlawn survey consisted of a receiver, transmitter and generator.

The receiver measured electrical fields using a short cable connected to brass stakes of low contact resistance. Magnetic fields were measured with ferrite cored coils. The filtered and amplified signals from both these detectors were determined with a General Radio Type 1568A Wave Analyzer (frequency range 20 Hz to 20 Khz, voltages ranges 100 microvolts full scale to 300 volts full scale, input impedance 1 megohm).

Square wave frequencies of 20Hz, 33Hz, 93Hz, 330Hz, 915Hz and 3.3Khz were transmitted from a Geotronics 5 KVA, EMT 5000 IP/EM transmitter connected to a 150m to 400m length of wire grounded by multiple aluminium foil electrodes. Current output is generally of the order of 1 to 9 amps. The EMT 5000 transmitter is powered by a 120V AC, 400Hz, three phase generator driven by a petrol motor.

Ancillary equipment for the surveying included voltage regulator, dummy load and radio transceivers.

Field Procedures Employed at Woodlawn

The field procedures generally followed the optimum receiver configurations for AMT measurements about a grounded electric dipole proposed by Goldstein and Strangway (1975).

AMT measurements at Woodlawn were conducted on lines G, J and L of the Woodlawn Test Grid utilizing transmitters oriented in various directions and located at various distances from the orebody as shown in Fig. 1. AMT measurements were also performed over the Black Shale, 2.5km southwest of the orebody. The surveys were carried out in November and December 1975.

At each site the transmitter operator switched on the frequencies called for by the receiver operator.

For all AMT measurements, the receiver array comprised a short electric dipole oriented parallel to the transmitter dipole (i.e. measuring the Ex component), and a horizontal coil oriented perpendicular to the transmitter dipole (i.e. measuring the Hy component). Observation points were spaced 30m apart with line lengths of between 200m and 350m.

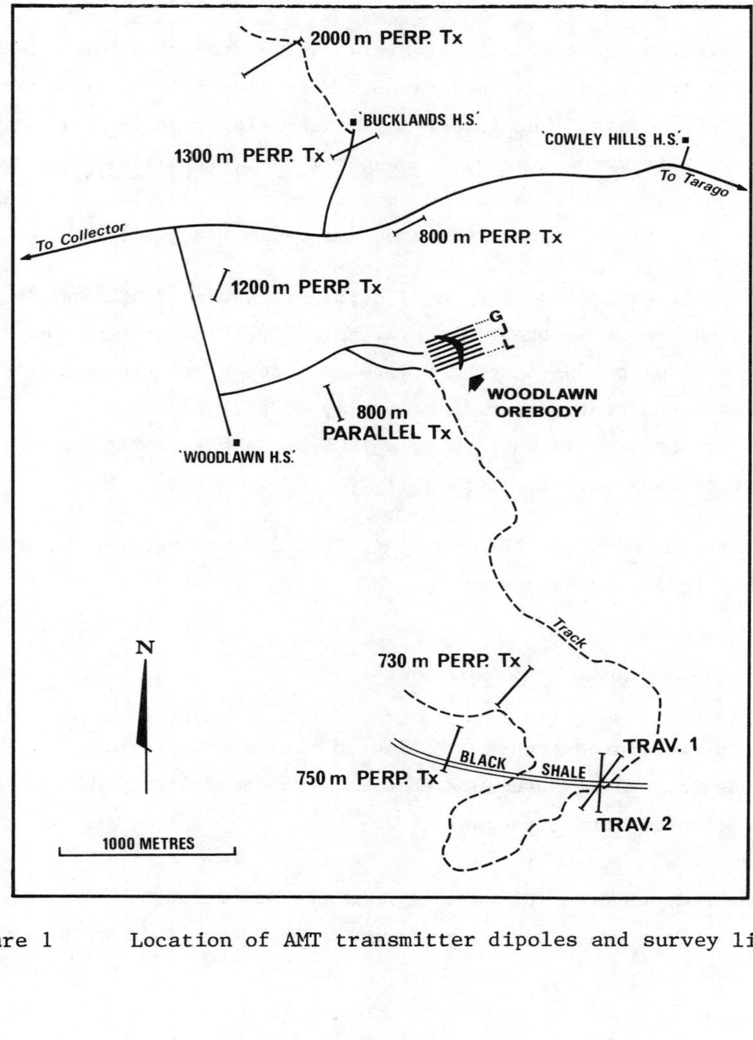

Figure 1 Location of AMT transmitter dipoles and survey lines.

The natural AMT fields were recorded at various times during the work. Comparisons with the Ex and Hy fields observed with the transmitter switched on indicate excellent signal to noise ratios for the H measurements at all frequencies employed, both over the host rocks and over the orebody. For the E field, the signal to noise ratio is excellent at all frequencies over the country rocks. However, over the orebody, the anomalous Ex value dropped to within one order of magnitude of the noise level. In general, the measurements at 20Hz and 33Hz gave the lowest signal levels and required the greatest care.

Discussion of Results

The AMT survey data are presented both as contoured pseudo sections of apparent resistivity and stacked profiles of apparent resistivity. Occasionally, the Ex and Hy values are also plotted. The pseudo sections employ a linear horizontal scale and a logarithmic frequency scale; these plots are contoured at logarithmic intervals of apparent resistivity in order to emphasise low resistivity regions. On the profiles, the horizontal scale is again linear while the vertical scale of apparent resistivity is logarithmic with minimum values towards the top of the plot.

Line J (9220mN)

The apparent resistivity pseudo section obtained using a 30m receiver dipole and a 200m long parallel transmitter dipole located 800m west of the orebody is shown in Fig.2, together with plotted values of measured E field. An excellent anomaly is clearly evident at J10 to J11, coincident with the upper portion of the sulphide body. The contrast between the response of the sulphides and background is strongest at the lowest frequencies, and is approximately five orders of magnitude. The contoured plot is slightly asymmetrical and this reflects the westerly dip of the conductor. The anomaly shown on the pseudo section is further highlighted when presented as profiles. Fig.3 emphasizes the very high contrast observed over the orebody and shows more clearly the asymmetry in the values due to the westerly dip of the orebody. Of some interest is the apparent shoulder on the anomaly at stations J14 and J15 as this feature may reflect the depth extent of the dipping sulphide lens.

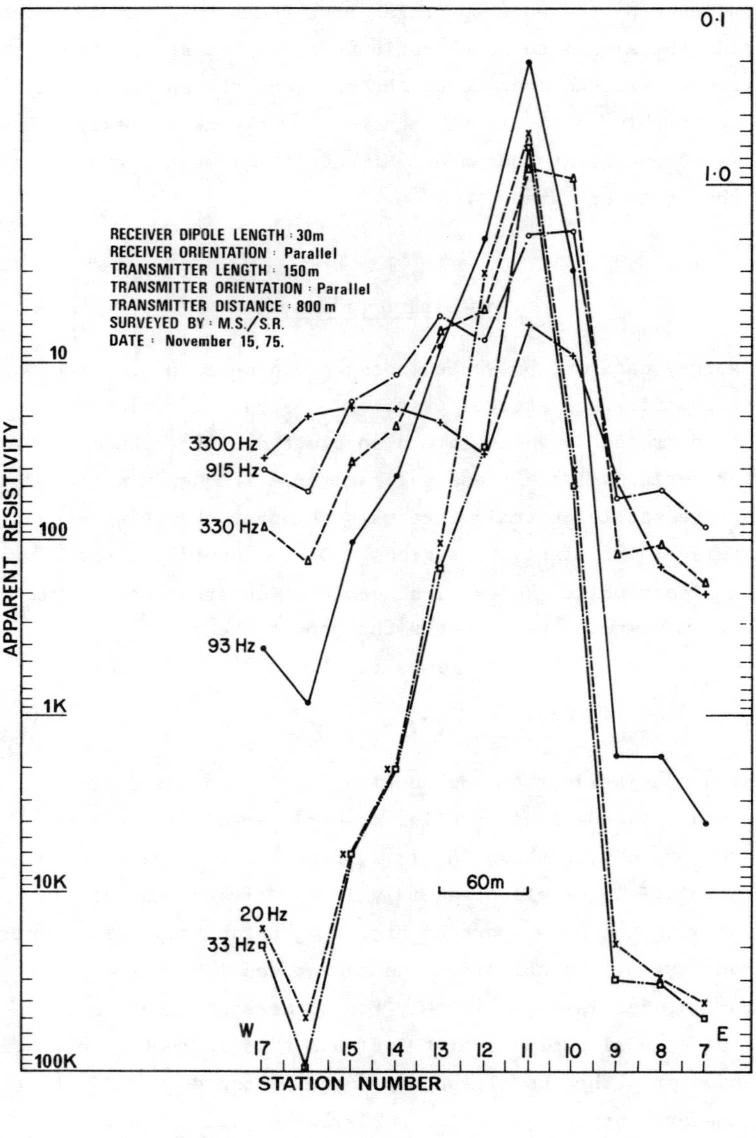

Figure 2 Pseudo sections of electric field (E) and apparent resistivity, Line J, Woodlawn Orebody.

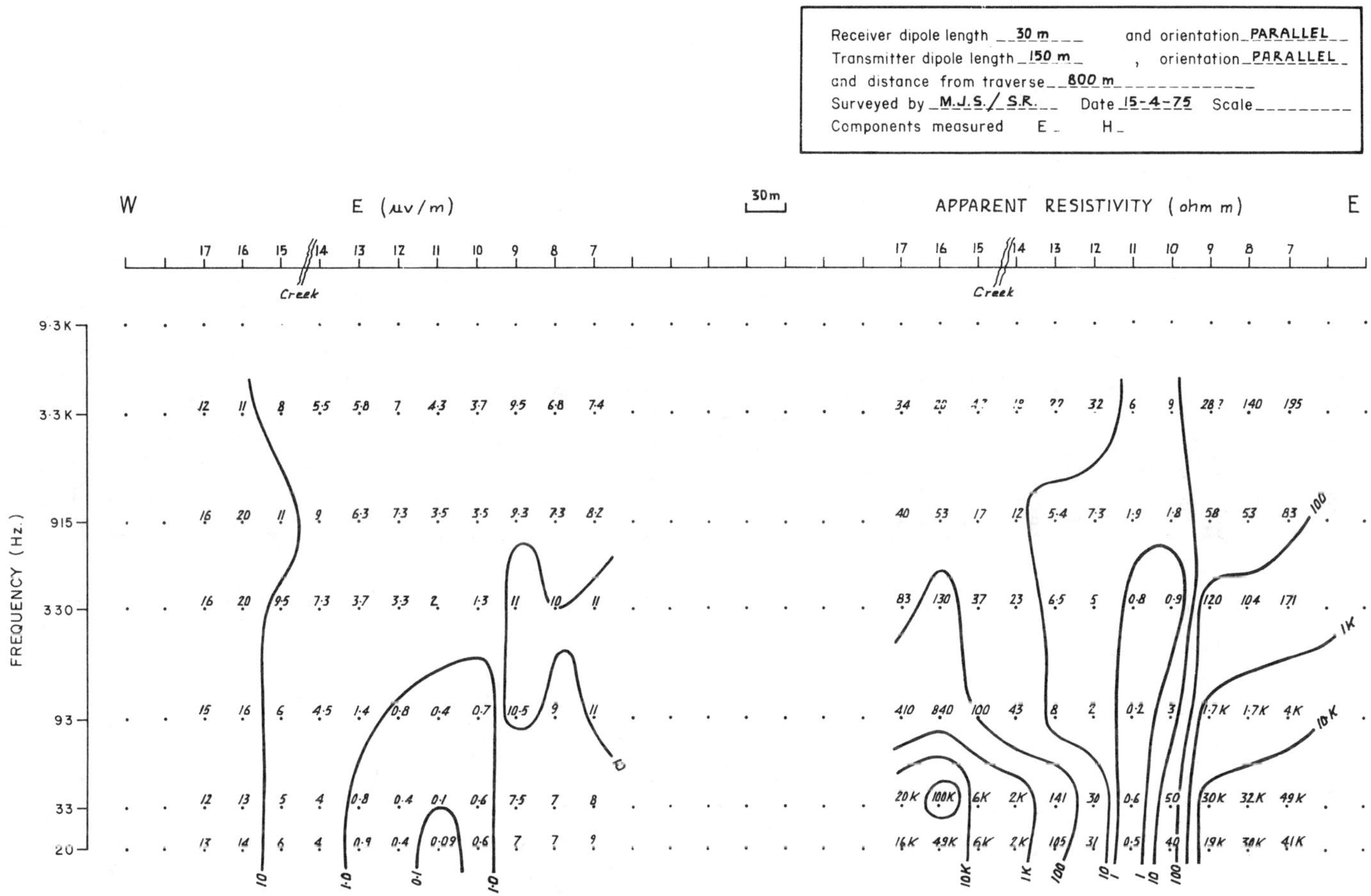

Figure 3 Stacked profiles of apparent resistivity, Line J, Woodlawn orebody.

Black Shale Traverse

This traverse is slightly oblique to the strike of the shale. AMT measurements were conducted on this line using a 30m electric dipole with a 200m long perpendicular transmitter dipole located 730m northwest of the traverse. The transmitter was grounded in very resistive volcanic rocks and the current output remained at only 1 ampere. The resulting apparent resistivity pseudo section (Fig.4) shows a very strong conductor between stations 9.25 and 10.75, corresponding to the black shale outcrop. The anomaly is fairly symmetrical and is strongest at the lower frequencies. The maximum anomaly is four to five orders of magnitude. The profile of apparent resistivities on this line (Fig.5) further demonstrates the strength of the anomaly, (especially on the 33Hz profile) and the approximate symmetry. The dip of the shale is thought to be nearly vertical.

Conclusions

The experimental audio magnetotelluric surveying conducted over the Woodlawn orebody has provided the following information:

1. AMT measurements using a grounded dipole transmitter can indeed be performed efficiently in a massive sulphide environment.
2. The resulting values of apparent resistivity are meaningful and can be interpreted in terms of major geoelectrical discontinuities.
3. The shallow, highly conductive Woodlawn orebody yielded excellent AMT anomalies. The AMT data provided a good qualitative indication of the dip of the conductor.
4. The strength of the anomaly is related to the position of the transmitting dipole with respect to the strike of the conductor. There is a limitation to the range from a transmitting dipole at which the primary E field signal is strong enough to be accurately measured. This limiting range will probably vary from region to region.
5. The shallow, highly conductive Black Shale yielded excellent AMT anomalies. If the transmitter dipole straddles a very long conductor such as a black shale then the resulting anomaly is much stronger in amplitude and sharper in resolution than for other transmitter locations off the Black Shale.

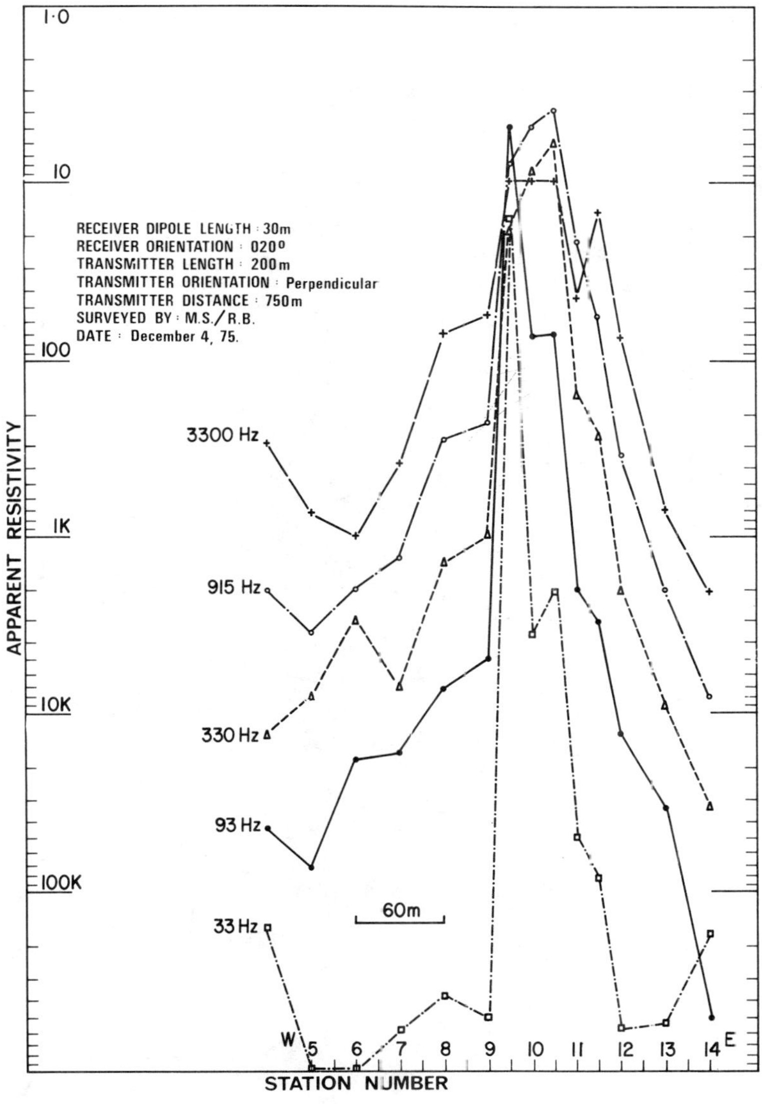

Figure 4 Pseudo section of electric field (E) and apparent resistivity, Black Shale.

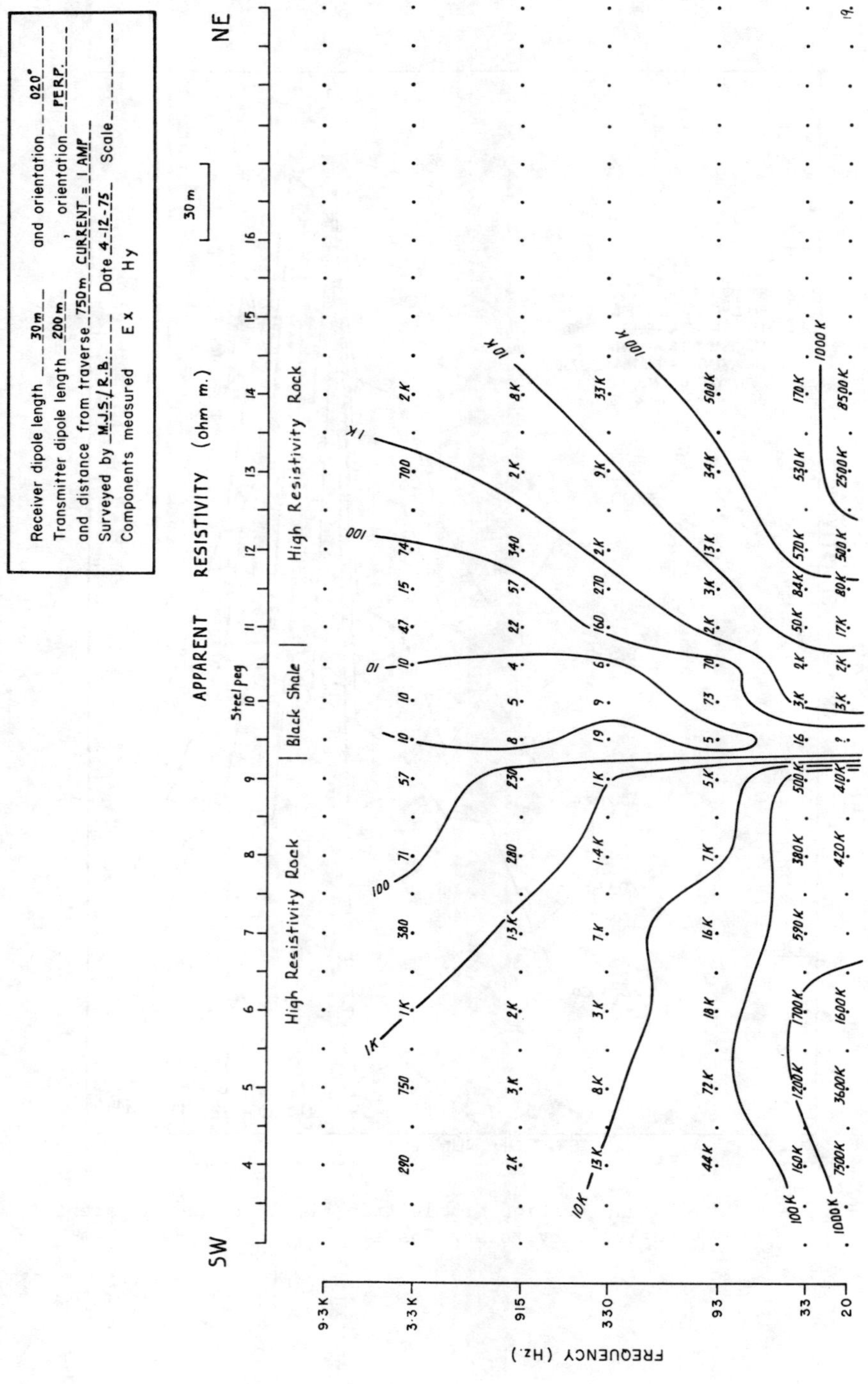

Figure 5 Stacked profiles of apparent resistivity, Black Shale.

6. In general the amplitude and character of the AMT anomaly obtained over the Woodlawn orebody when the transmitter was located to the side of and parallel to the orebody was about the same as the amplitude and character of the AMT anomaly obtained over the Black Shale, when the transmitter straddled and directly engaged the Shale.

7. A sulphide body could possibly be distinguished from a black shale by virtue of the usually much greater strike length of the latter.

Reference

Goldstein, M.A. and Strangway, D.W. 1975 "Audio frequency magnetotellurics with a grounded electric dipole source". Geophysics, Vo.40, No.4, p669.

7.4. Vertical Loop, Dip Angle Electromagnetic Traverses at Woodlawn

R. J. WHITELEY

Contents

Abstract

A single broadside, vertical loop, dip angle electromagnetic traverse over the centre of the Woodlawn orebody produced a weak anomaly. The Black Shale did not produce a broadside anomaly.

Fixed transmitter dip angle profiles on the northern half of the orebody enabled the axis of the highly conductive massive sulphide lens to be traced and the approximate northern limit of the orebody to be determined.

Introduction

The vertical loop, dip angle electromagnetic (EM) method is well known as a rapid and simple exploration technique for shallow, massive, electrically conductive targets (Ward, 1969, p.233).

The method uses a vertical transmitting loop which may be mobile or fixed and a horizontal mobile receiving loop. In exploration it is usual for the centre of the receiving loop to be maintained in the plane of the transmitting loop. The resultant electromagnetic field produced by interaction of the primary field from the transmitter and the secondary field from a buried conductor between the loops is elliptically polarized. The tilt or pitch of the major axis of this ellipse can be determined by rotation of the receiving loop about a horizontal axis in the plane of the transmitting loop, until a minimum signal is obtained.

The amount of rotation or dip angle, in either a positive or negative sense, is measured in degrees from the horizontal.

Reconnaissance dip angle EM surveys are usually carried out in broadside fashion with both transmitting and receiving loops occupying successive stations along adjacent traverses approximately perpendicular to the strike of a suspected buried conductor. The vertical transmitter loop is oriented at rightangles to the traverse direction. The distance between the transmitter and receiver may be up to a few hundred metres with currently available portable power packs.

More detailed definition of a target may be achieved using the fixed transmitter dip angle method. In this method the transmitter and receiver are again located on adjacent traverses normal to the expected strike of the target. The transmitter loop is rotated about a vertical axis so that the plane of the transmitter loop includes the position which the receiver occupies at successive stations along a traverse. Dip angle measurements are made at the receiver as in the broadside method.

In most situations both the broadside and fixed transmitter dip angle methods produce crossover anomalies in the vicinity of a conductive target with a peak and a trough on either side of the conductor and a zero or crossover above the conductor. To assist interpretation of both methods Ward, (1969, p. 315 to 335) has presented numerous anomaly curves for a range of scale models representing simplified subsurface conductors. In general, the simplest anomalies are recorded along profiles normal to the strike of thin steeply dipping conductors of large strike length. Under some circumstances, it is possible to estimate a variety of target parameters including strike length, depth, dip and width by using both broadside and fixed transmitter methods with correctly designed field procedures.

Survey at Woodlawn

A limited dip angle EM survey with both broadside and fixed transmitter methods was undertaken at Woodlawn using a McPhar VHEM system operating at 600 and 2400Hz.

Broadside Results

A single east to west broadside traverse was completed near the centre of the orebody using a loop separation of 30m. The transmitting loop was located on Line J (9220mN) and the receiving loop on Line K (9190mN). Fig.1 shows the results of this traverse at both frequencies. Easterly dips have been plotted as positive angles and westerly dips as negative angles.

As shown in Fig.1 a low amplitude, complex crossover occurs near the centre of the orebody between the peak and the trough. This anomaly has a minimum of about -3 degrees at station 11.5 on the western side of the orebody and a maximum of +5 degrees at station 10. The anomaly at both frequencies is similar and repeated readings at two stations along the traverse indicate errors of about ± 1 degree.

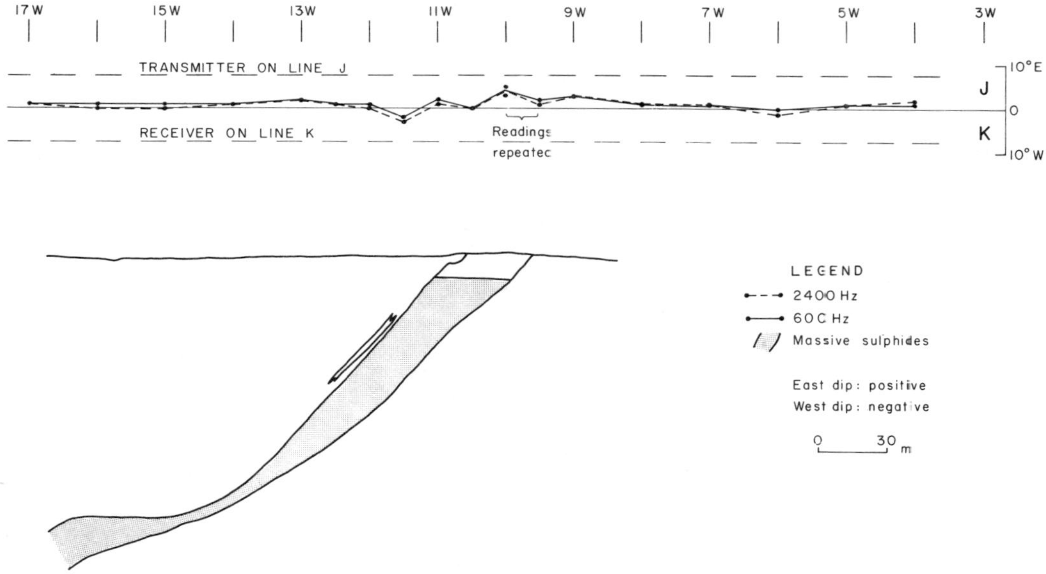

Figure 1 Broadside EM Profile Line J (9220mN)

Since the width of the orebody is comparable to the loop separation a well developed kink on the crossover which occurs near the centre of the orebody between the peak and the trough is probably produced by the edges of the conductor.

Broadside traverses using coil spacings of 30 and 60m were also completed over the Black Shale some 2.5km south of Woodlawn orebody. No discernible anomalies were obtained over this shale and variations in dip angles amounted to less than ± 2 degrees.

Fixed Transmitter Results

The fixed transmitter dip angle method was used in a detailed survey of the northern part of the orebody. The dip angles obtained for both the high and low frequencies are shown in Fig.2 using the same plotting convention as for the broadside method. Crossover points are indicated by arrows, and the outline of the orebody at the base of the gossan is also shown. The results were basically similar for both frequencies.

Initially the transmitter loop was fixed on Line K (9190mN) near the centre of the crossover anomaly obtained with the broadside method. Dip angles were then measured along Line J (9220mN) 30m to the north. A large

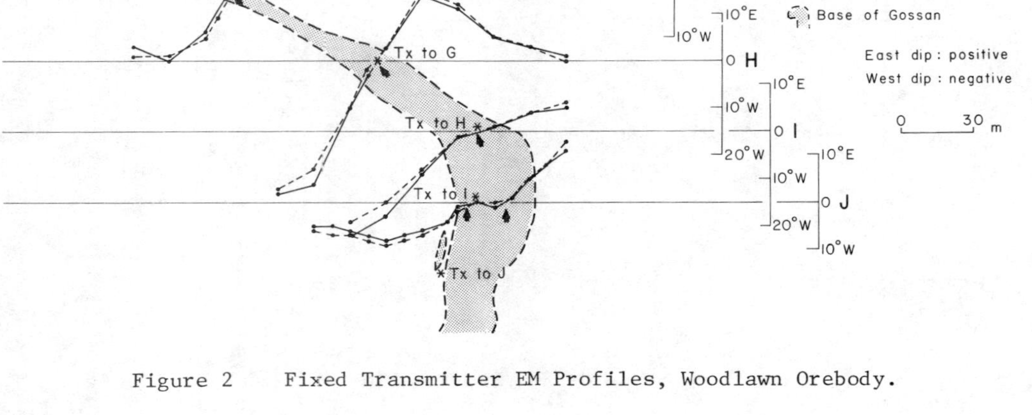

Figure 2 Fixed Transmitter EM Profiles, Woodlawn Orebody.

anomaly, which was not completely defined on the eastern side of the orebody, was obtained. The trough on the western side of the orebody has an amplitude of about -9 degrees. A double crossover is indicated near the centre of the orebody and the anomaly appears to have a steeper slope on the eastern or footwall side.

The transmitting loop was then fixed near the centre of the crossover on Line J and the receiver profiled along Line I (9250mN). The single crossover anomaly obtained on this line was not completely defined although the dip angles were considerably larger on the western side with the crossover point near the centre of the orebody. The anomaly also appears to have a steeper slope on the western or hanging wall side i.e. in the opposite sense to the previous case. This is possibly due to the abrupt change in strike direction of the orebody north of Line I.

Repeating this procedure, the transmitter was fixed near the centre of the crossover on Line I and dip angles were measured on Line H (9280mN). A large amplitude crossover anomalywith a trough of about -25 degrees on the western side and a peak of +15 degrees on the eastern side was obtained. The distortion of this anomaly is probably due to both the obliqueness of the traverse direction to the strike of the orebody and the presence of substantially more conductive material closer to the receiving loop on the western side of the orebody.

The transmitting loop was then fixed at the centre of the crossover on Line H and the receiver moved along Line G (9310mN). The crossover anomaly obtained on this line has a trough of -15 degrees on the western side and a peak +20 degrees on the eastern side. Gradients on either side of the crossover are similar for this anomaly.

The transmitting loop was then fixed near the centre of the crossover on Line G and the receiver moved along Line F (9340mN). A very broad, low amplitude dip angle anomaly was obtained on this line with the suggestion of a crossover in the vicinity of station F17, although the anomaly is not completely defined. A more prominent false on reverse crossover occurs at about station F13.5. The large reduction in amplitude of the suspected crossover anomaly on Line F and the broadness of anomaly indicates that the northern limit of the conductive orebody is between Lines G and F.

The line joining the crossover points closely defines the axis of the conductive orebody.

Conclusions

The single broadside traverse across the centre of the Woodlawn orebody using a loop separation comparable to the width of the orebody produced a low amplitude crossover anomaly with features suggesting a westerly dipping body of substantial width. The results at both the high and low frequencies were basically similar. Broadside traverses across the Black Shale using an identical coil separation and one twice as large failed to produce any anomaly.

Large anomalies were produced using the fixed transmitter method over the orebody. This method enabled the axis of the northern part of the conductive orebody to be accurately traced and the strike extent to the north to be approximately determined. Some of the anomalies were not fully defined, although the features they displayed were typical of those expected from dip angle traverses oblique to the strike of dipping and vertical conductive bodies.

References

Ward, S.H., 1967. The Electromagnetic Method, Mining Geophysics, Chapter 2, part C, Soc. of Expl. Geoph. Tulsa, 1969.

7.5. *Slingram Electromagnetic Surveys of the Woodlawn Orebody and the Black Shale*

R. J. HAREN AND R. J. WHITELEY

Contents

Abstract

An extensive horizontal loop Slingram survey was completed over the Woodlawn Test Grid using frequencies of 880 and 3250Hz. Results indicate that the Woodlawn orebody is a highly conductive sheet dipping at approximately 50 degrees. The Slingram method enables the orebody geometry and its conductivity-thickness product to be determined.

A single Slingram profile was also completed over the Black Shale which shows a response very much less than that recorded over the Woodlawn orebody.

Introduction

Extensive horizontal loop, Slingram electromagnetic (EM) surveys were commenced over the Woodlawn ore deposit in mid 1974. Further surveys over the orebody and across the Black Shale 2.5km to the south of the orebody were conducted at various times up to the commencement of mining activity in 1977. During this extended period ground conditions varied widely, from snow covered or sodden to dry. It is possible that some recorded responses were influenced by differences in surface conductivity resulting from the variable ground conditions.

Field Measurements

Descriptions of the Slingram method and field procedures are given by Ward (1971) and Parasnis (1973). At Woodlawn, a dual frequency ABEM 35/88 EM Gun, operating at either 880 or 3520Hz was employed. A connecting cable between the transmitting and receiving coils on this instrument allows in-phase and out-of-phase measurements to be made alternately at both frequencies.

The connecting cable also acts as a reference length ensuring a constant separation of the transmitter and receiver coils. At Woodlawn a coil separation of 61m (200ft) was used. This separation was expected to give an approximate depth of investigation of about 45m for a vertical or steeply dipping conductive target.

Slingram traverses were carried out over the orebody on the east to west lines F to R of the Woodlawn Test Grid and on lines A and T to the north and south of the orebody respectively. Data from lines A and T provided information on the anomalies associated with a variety of sources which could include sulphide stringers, geological contacts and faults. A station spacing of 15.2m was generally used throughout the survey although this was reduced to 7.6m and locally to 3.8m in the vicinity of the orebody.

Results at Woodlawn

Figs. 1 and 2 show in-phase and out-of-phase stacked profiles at 880 and 3520Hz respectively over the Woodlawn orebody. Data are plotted midway between the transmitter and receiver loop position. The

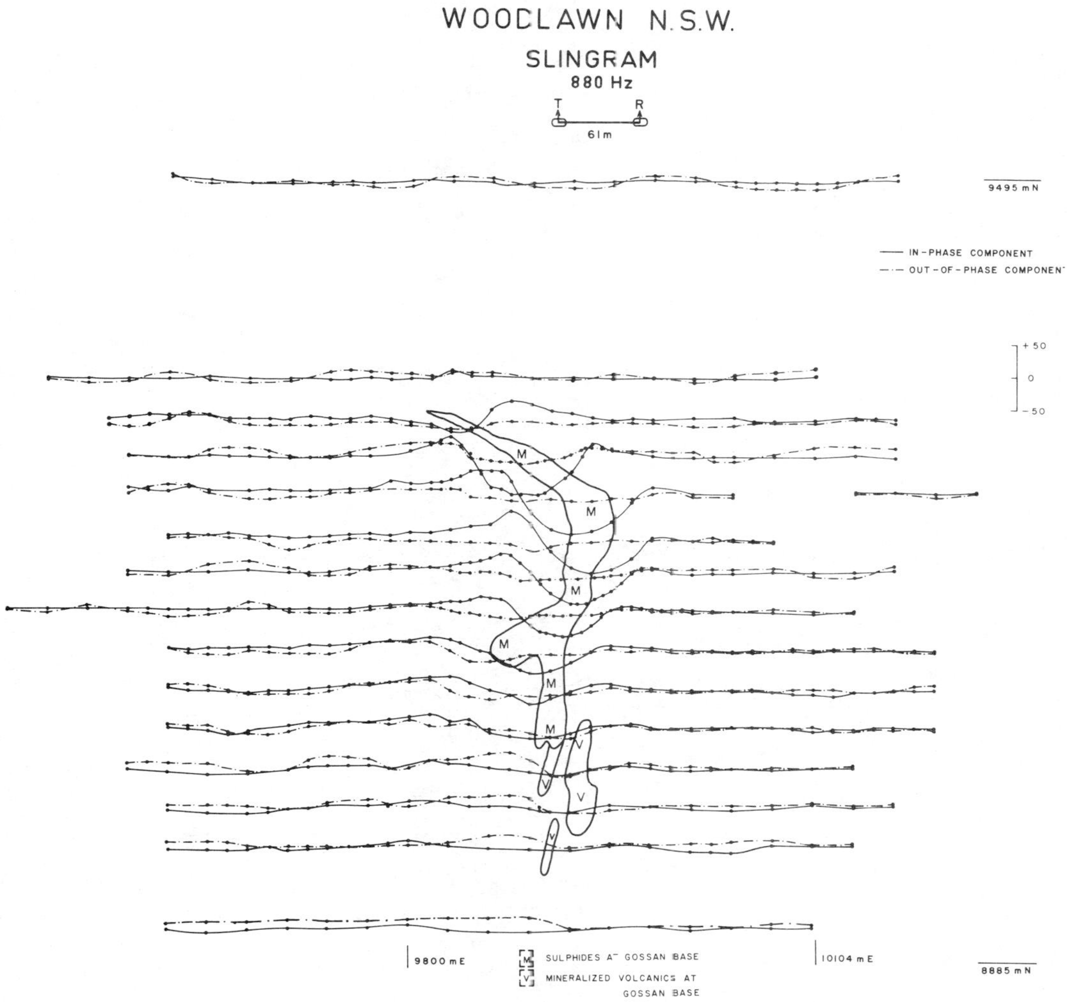

Figure 1 Horizontal loop Slingram EM showing stacked profiles of in-phase and out-of-phase components at 880Hz over the Woodlawn orebody.

location of the massive sulphide orebody at the base of the gossan is also shown.

These figures show the large in-phase and small out-of-phase anomalies associated with the Woodlawn orebody. The in-phase Slingram anomaly accurately defines the strike length and the strike directions of the orebody. Anomalies located over the centre and southern part of the orebody where the traverses are orthogonal to the strike and are

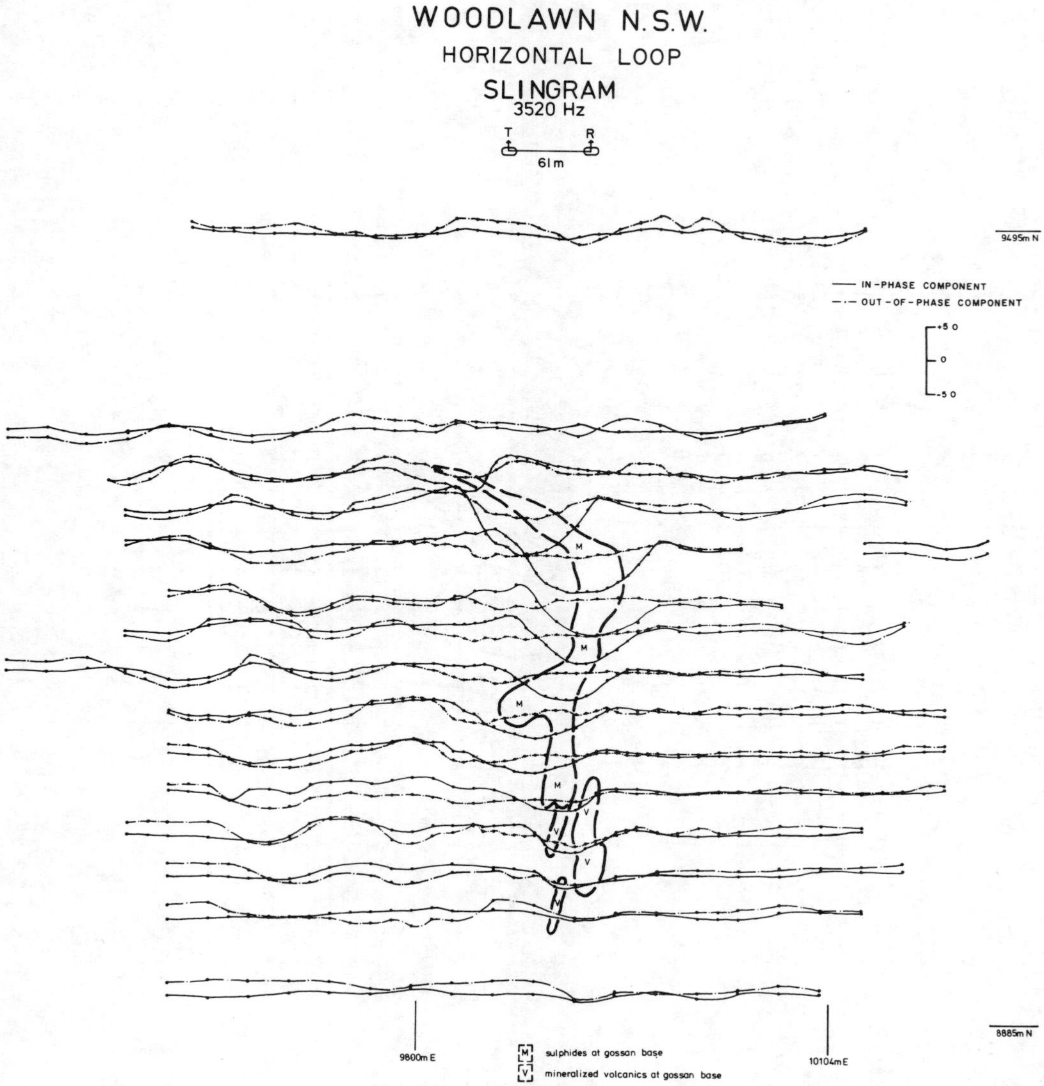

Figure 2 Horizontal loop Slingram EM showing stacked profiles of in-phase and out-of-phase components at 3520Hz over the Woodlawn orebody.

some distance from the change in strike direction (i.e. Lines K to N), show the classic response of an excellent tabular conductor at shallow depth, dipping to the west (Ketola and Puranen, 1967; Nair et al., 1968).

The Slingram anomalies obtained over the orebody permit calculation of a number of parameters by using master curves derived from scale model studies based on thin, dipping, tabular conductive targets (Strangway, 1966; Ketola and Puranen, 1967). Dip, depth to top, and conductivity-thickness products were calculated from these curves. These parameters are listed in Table 1 for lines H, J, K and M. These lines were chosen since the Slingram anomalies obtained were typical of those over a tabular conductor. For comparison actual parameters obtained from drilling are also listed.

Table 1. Woodlawn Orebody Geometric and Parametric Estimates from Slingram Data.

Line	Estimated Dip (degrees)	Actual Dip (degrees)	Estimated Depth (m)	Actual Depth (m)	Estimated Conductivity-Thickness product (S)	Actual Width (m)
H	30-45	85	6	7	50	25
J	<40	50	6	10	40	25
K	45	45	8	14	75	21
M	50	45	8	20	40	32

Data from the northern end of the orebody (lines G to I) reveal the complications introduced when traverse lines are not perpendicular to the strike, thus causing a general broadening of the anomalies. Response amplitudes are also reduced on these lines suggesting a thinning and termination of the conductor to the north. This causes asymmetry of the anomaly profiles leading to erroneous estimates of dip of between 30° to 45° on line G. The orebody is almost vertical at this location. The model curves do not provide accurate estimates of conductor parameters in this situation.

Fig. 3 shows a Slingram profile at 3520Hz normal to the northern end of the Woodlawn orebody. The approximate section width of the near vertical orebody is also shown. Some asymmetry is again evident due to the complex conductor geometry in this region.

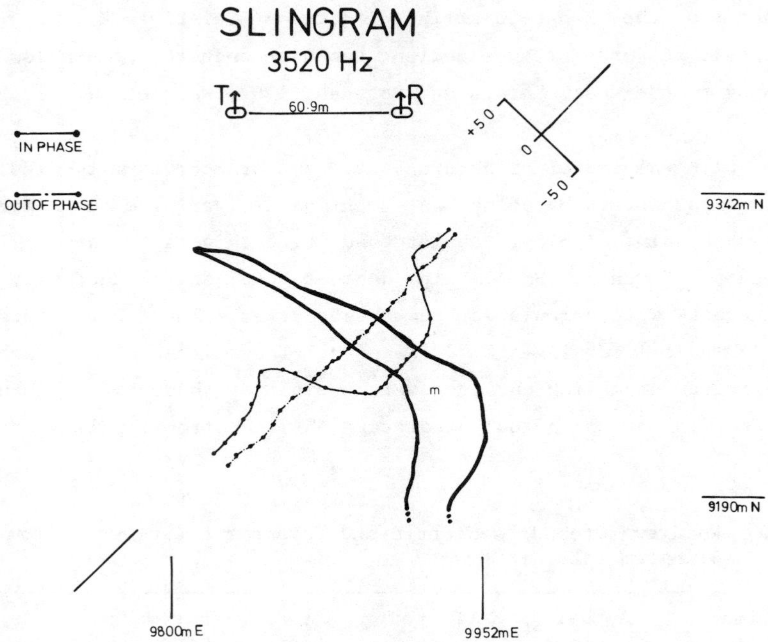

Figure 3 Horizontal loop Slingram EM profile normal to the northwesterly striking part of the Woodlawn orebody.

Fig. 4 shows the Slingram anomaly at two frequencies along traverse line J (9220mN), near the centre of the orebody. The outline of the massive sulphide orebody at the base of the gossan is also shown. Similarities in the anomalies at both frequencies suggest that the conductive orebody is shallow. Dip angle computed from the master curves is less than 40° whereas the actual dip is 50°. This underestimate is probably due to the strike of the orebody changing sharply approximately 30m to the north of line J. This is only one half the dipole spacing and it would have an effect on the electromagnetic coupling. The estimated depth of burial from data on this line is 6m which compares with the depth to the base of the gossan of 10m. The calculated conductivity-thickness parameter for the orebody beneath line J is approximately 40S.

Fig. 5 shows Slingram data from Line M (9131m) near the southern end of the orebody, with the cross-section derived from drilling also shown. Line M is almost perpendicular to the strike of the orebody. Estimates of dip, and width of the conductor beneath this line are reasonably accurate as may be seen from Table 1. (These dips are estimated

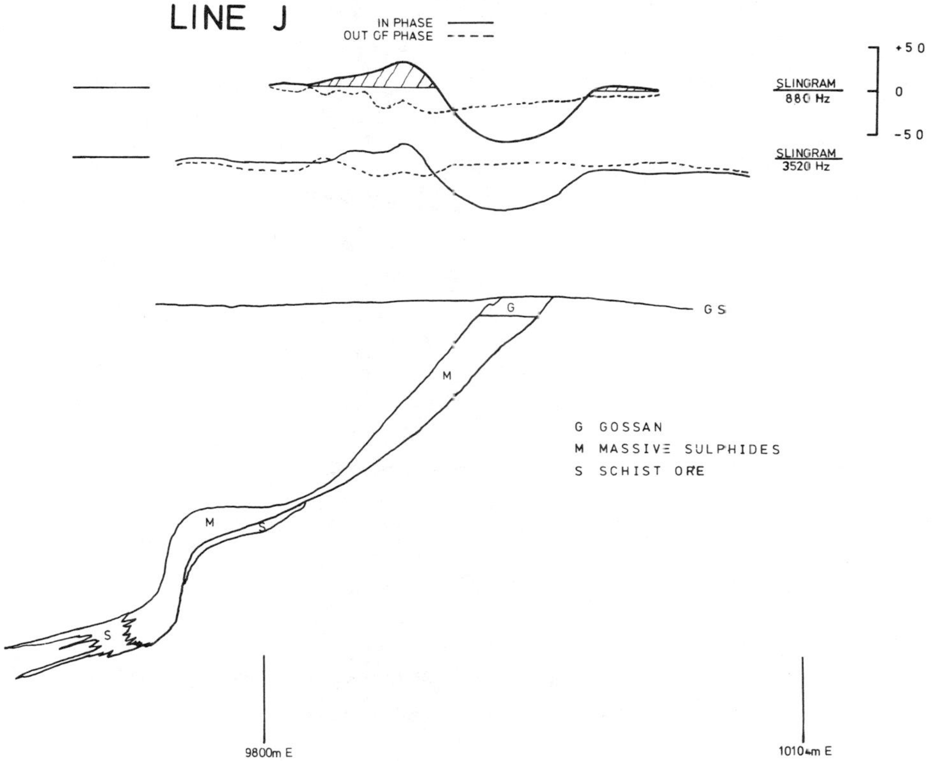

Figure 4 Horizontal loop Slingram EM anomalies along Line J at 880Hz and 3520Hz.

from the asymmetry of individual anomaly profiles. It should be noted that the migration of the anomalies with frequency as shown in Fig. 4 and also on Fig. 3 is the reverse of what would normally be expected ie a migration towards the down-dip direction with decreasing frequency. This may be due to the presence of a highly conductive supergene zone at the base of the gossan - Ed).

The use of in-phase versus out-of-phase response diagrams for qualitative correlation of anomaly type has been a standard practice in EM exploration for some time (Parasnis, 1962). Fig. 6 shows these diagrams for the 880Hz data along line H, J, K and M. The character of these plots is seen to change from line H, through lines J and K, and then to Line M. This variation reflects the changing nature of the structure, depth and conductivity of the orebody. Lines H, J and K exhibit anomalies of large

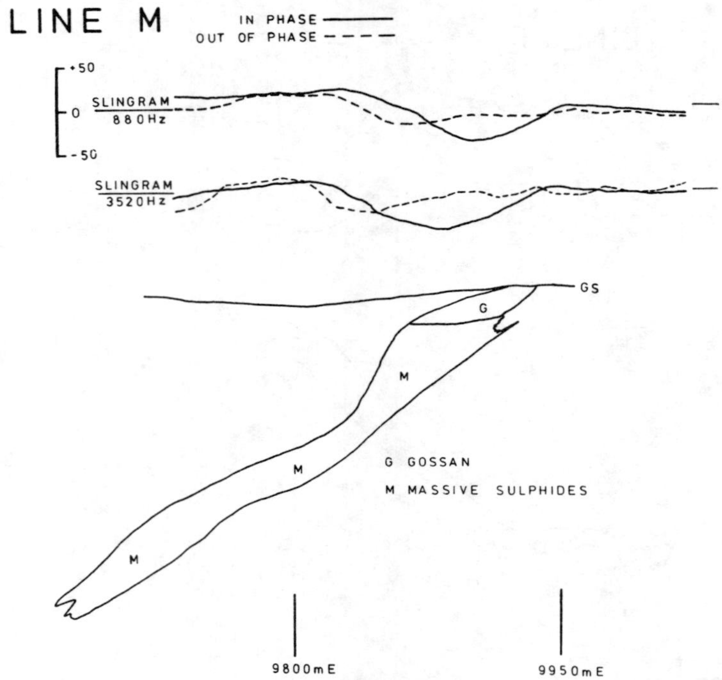

Figure 5 Horizontal loop Slingram EM anomalies along Line M at 880Hz and 3520Hz.

amplitude, whilst line M exhibits a relatively small anomaly which is less complex in shape. Less complex shapes would be expected as the conductive zone deepens.

Scale model experiments were carried out to investigate the Slingram anomalies produced by the Woodlawn orebody and are discussed by Haren (this volume, p 205).

Results at the Black Shale

In the vicinity of Woodlawn, there are a number of black shales containing weak mineralization, mostly pyrite. A large self potential anomaly (Cifali and Whiteley, this volume, p 321) was obtained over the Black Shale, 2.5km south of the Woodlawn orebody. This prompted exploratory drilling of the Shale which showed only trace sulphide mineralization.

Buselli (this volume, p 297) concluded from transient electromagnetic (TEM) measurements that the Black Shale was not a good electrical conductor, with no significant TEM anomaly. A horizontal loop

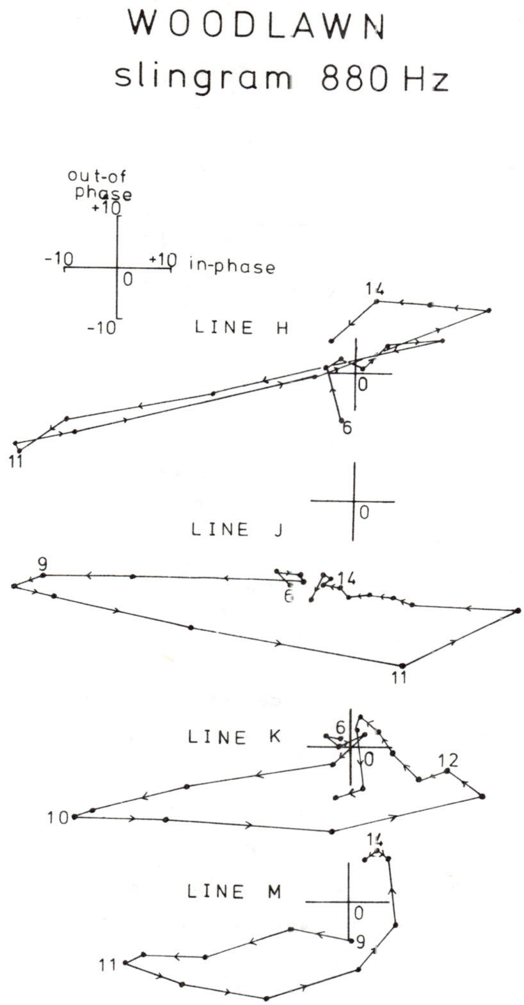

Figure 6 In-phase versus out-cf-phase plots for four lines, from the 880Hz horizontal loop Slingram EM data.

Slingram traverse was completed across the Black Shale using the same transmitter receiver separation (61m) as was used over the Woodlawn orebody. In-phase and out-of-phase results at both 880Hz and 3520Hz are shown in Fig.7. A broad, low amplitude anomaly is associated with the Shale. The amplitude of this anomaly is, however, similar to the amplitude of background variations in this region. From the above results it is clear the massive sulphide orebody at Woodlawn is indicated as a zone of much higher conductivity than the Black Shale and is readily distinguishable.

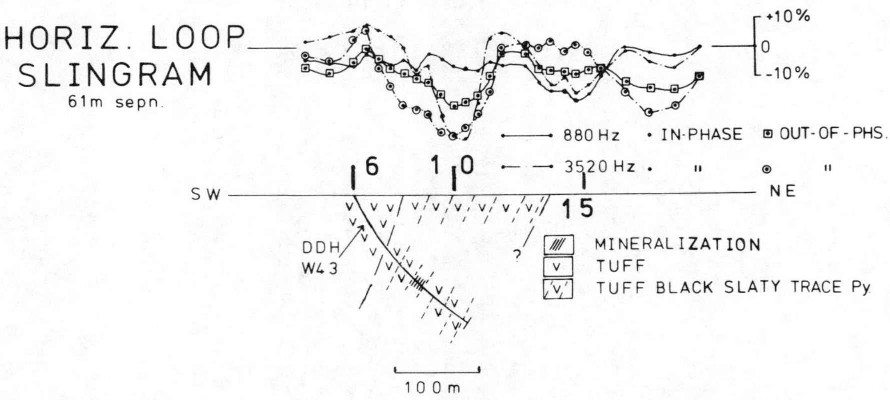

Figure 7 Horizontal loop Slingram EM profile at both frequencies over the Black Shale.

Conclusions

The horizontal loop, Slingram survey has shown that the massive sulphide Woodlawn orebody is an excellent electromagnetic target producing large in-phase and small out-of-phase responses. The character of these responses is typical of a shallow, steeply dipping, highly conductive tabular body. These responses are complicated by variations in strike direction and geometry of the orebody. In spite of this, the actual location and strike of the orebody can be readily determined from the Slingram anomalies. Although direction of dip was indicated from the asymmetry of the anomaly profiles calculated dips were often in error due to the complex shape of the orebody.

A Slingram profile over the Black Shale to the south of the Woodlawn orebody gave a weak anomaly, typical of a poor conductor. This suggests that the Slingram method may be used to distinguish shallow weakly mineralized Black Shale horizons from shallow, highly conductive massive sulphide orebodies in the Woodlawn area.

References

Ketola, M. and Puranen, M., 1967. Type curves for the interpretation of Slingram (horizontal loop) anomalies over tabular bodies. Report of Investigations No.1, Geological Survey of Finland, Otaniemi.

Nair, M.R., Biswas, S.K. and Mazumdar, K., 1968. Experimental studies on the electromagnetic response of tilted conducting half-planes to a horizontal-loop prospecting system. Geoexploration, 6, 207 to 244.

Parasnis, D.S., 1962. Principles of Applied Geophysics. Methuen, London.

Parasnis, D.S., 1973. Mining Geophysics. Elsevier, Amsterdam.

Strangway, D.W., 1966. Electromagnetic parameters of some sulphide orebodies in Mining Geophysics, 1, (S.E.G., Tulsa) 227 to 242.

Ward, S.H., 1967. The electromagnetic method in Mining Geophysics, 2, (S.E.G., Tulsa) 224 to 372.

7.6. Slingram Scale Modelling of the Woodlawn Orebody

R. J. HAREN

Contents

Abstract

Horizontal loop Slingram electromagnetic data collected over the Woodlawn orebody at a frequency of 880Hz are compared with data from simplified and complex, actual scale models of the orebody in both air and saline solution. Small dimension model coils were constructed which enabled model data to be obtained at a scaling factor of 1 to 1000.

The anomalies obtained over the dipping plate model correspond closely to field anomalies near the central section of the Woodlawn orebody on lines J, K and L. Anomalies obtained over the complex, detailed models are almost identical with those collected in the field.

Analogue scale modelling is shown to be an important part of the detailed interpretation of Slingram electromagnetic anomalies.

Introduction

The first adequate description of the basic theory of electromagnetic (EM) scale modelling was by Sinclair (1948). Since that time, many researchers have simulated geological conditions with conductive models that were easy to manipulate in the laboratory environment. Many of the early investigators used models in which metallic sheets were suspended in air (e.g. Clark and Mungal, 1951; Bosschart, 1964; Fraser and Ward, 1967). Since naturally occurring ore zones are surrounded by a conducting medium, wide discrepancies were often noted between EM survey data and data obtained from models surrounded by an insulating medium such as air.

In attempts to make more realistic scale models Hedstrom and Parasnis (1958) and Lowrie and West (1965) used thin metallic sheets to simulate conductive overburden. Researchers also investigated the EM response of conductive model bodies in galvanic contact with the thin overburden and a conductive host medium in the form of conductive solutions (e.g. Gaur et al., 1972; Verma, 1972; Gaur and Verma, 1973).

It has been shown by Ward (1967) and Frischknecht, (1971) from Maxwell's equations that the results of EM field measurements in the presence of a conductive environment are identical with those due to an equivalent physical model if the equation

$$\sigma\mu\ell^2 f = \text{constant},$$

is satisfied throughout the whole domain. σ is the electrical conductivity in Sm^{-1}; μ is the magnetic permeability in Hm^{-1}; ℓ is a linear body dimension in m and f is frequency of the EM field in Hz. (The principle of similitude, Stratton (1941)).

This equation is satisfied provided that displacement currents are neglected or the dielectric constant is taken as zero. The parameter $\sigma\mu f\ell^2$ is defined as the response or induction parameter which determines the response of inductive EM systems to simplified conductive targets. To achieve an equivalent response from a small scale model of a conductive target, this response must be held constant by increasing the frequency of the applied EM field and/or by increasing the conductivity of the model to compensate for the change in the length parameter. When the field instrument is used in the laboratory modelling and the length scale chosen, the only parameter that can be varied is the conductivity since the frequency is set by the instrumentation.

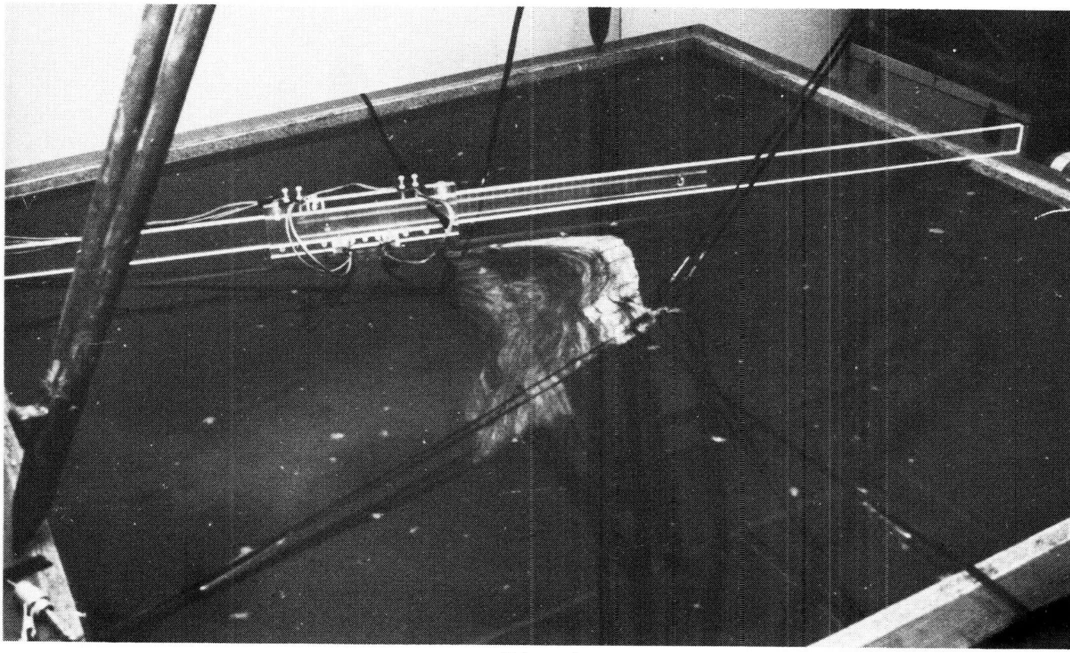

Fig. 1.

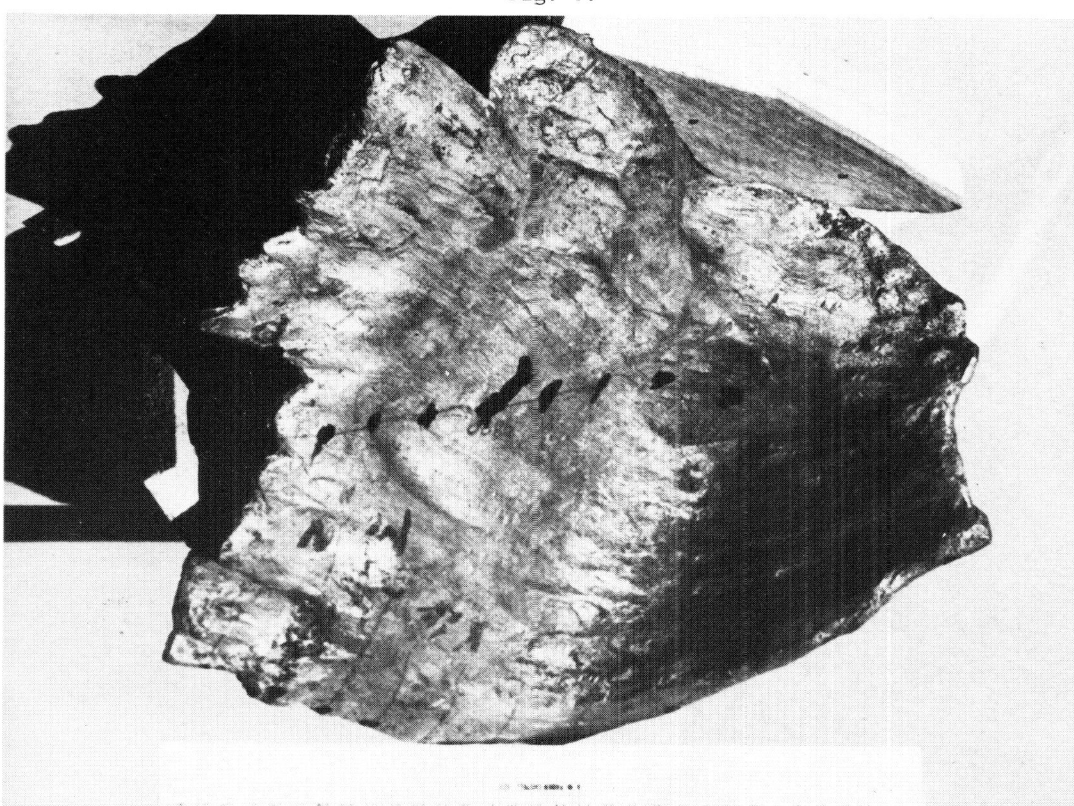

Fig. 2.

The Woodlawn Scale Models

A length scale factor of 1 to 1000 was used for modelling the Woodlawn orebody. Scale model data were collected in air and in the tank shown in Fig. 1. The tank, approximately 1.4m square and 0.8m deep was filled with water containing 50kg of dissolved sodium chloride to simulate the host rock environment. The conductivity of this solution was 5.0 Sm^{-1}.

The shape of the Woodlawn massive sulphide orebody has been defined by a large number of drill holes (Malone, this volume, p 49). Five metre level plans based on drilling data were produced by Jododex at a scale of 1 to 1000. The shape of the massive sulphide body was traced off each level plan and then cut out of 5mm thick wooden templates. The templates were nailed and glued together, filed down and covered in automotive body putty to provide a continuously smooth outer surface.

Two models of the massive sulphide orebody were cast, one in an aluminium alloy and the other in lead so that they differed slightly in bulk conductivity. Fig.2 is a photograph of the lead model of the Woodlawn orebody. Accurate casting in relation to the coordinate reference system enabled precise positioning of the traverse lines across the model, for comparison with the actual field traverses.

A simplified aluminium sheet model of the Woodlawn orebody was also constructed. This had a width of 25mm and was milled so the top was horizontal when the sheet dipped at 50 degrees. It was thus possible to compare the Slingram response of the accurate scale model with that of the simple sheet model.

The alloys used in the manufacture of the scale models and their conductivities are listed in Table 1. Equivalent bulk conductivity and resistivity values of the massive sulphide orebody being modelled are also listed.

Equipment and Modelling Procedure

Equipment used for the Slingram experiments was modified ABEM EMIMOD scale model system. The EMIMOD utilized the ABEM EM Gun EM35/88 compensator, as was used in the Slingram survey of the Woodlawn orebody (Haren and Whiteley, this volume, p 193). The transmitter was, however, replaced by a signal generator and a gantry was constructed to enable measurement over the model tank (Fig.1).

TABLE 1. Metallic alloys used in scale modelling

	Woodlawn 1st Casting	Woodlawn 2nd Casting	Woodlawn Sheet Model
Composition of alloy	Al 90% Mg 10%	Pb 100%	Aℓ 100%
Conductivity of alloy (Sm^{-1})	12.5×10^6	4.6×10^6	20.0×10^6
Equivalent bulk conductivity of ore (Sm^{-1})	12.5	4.6	20.0
Approximate equivalent resistivity of ore (Ωm)	0.08	0.22	0.05

Model data were collected at a frequency of 880Hz, identical to that used in the field survey. Coil spacing was set at 61mm, which at a linear scale factor of 1000, was equivalent to the coil separation of 61m used in the field survey.

In the field, Slingram coils are usually held between 0.5 and 1.0m above the earth's surface. With most modelling gantries the model coils are up to 15mm above the surface, which means that the equivalent field coils would be 15m above the ground when the scale factor is 1000. This problem was overcome by using a first order approximation, and graphing the change in response for different coil height. This allows extrapolation back to zero height assuming the change to be continuous and smooth. Fig.3 shows the change in response at one point for different coil heights. In the model data presented, correction for coil height has been included.

Model Results and Interpretation

Slingram Traverse over Thick Plate

Fig.4 shows the horizontal loop Slingram anomaly at 880Hz over a 25mm wide vertical aluminium sheet model in air. The symmetrical nature of the anomaly

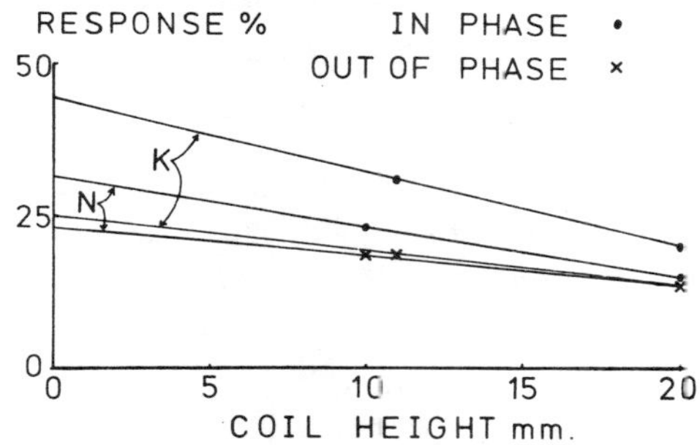

Figure 3 Variation in response on two traverse lines over the dipping Woodlawn orebody model at different coil heights

profile is expected because of reciprocity and the symmetry of the conductive inhomogeneity. The response is large in amplitude as would be expected from a thick conductive target near the surface.

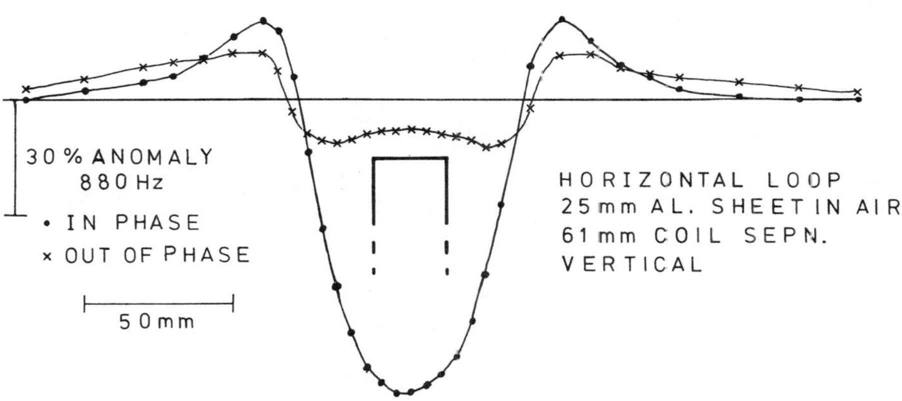

Figure 4 Horizontal loop Slingram anomaly at 880Hz over a vertically dipping sheet model.

Fig.5 shows the Slingram anomaly at 880Hz across the 25mm thick aluminium sheet model dipping at 50 degrees in both air and the salt water solution. The anomaly is now asymmetric with the in-phase response showing a large positive anomaly above the upper surface of the sheet, and a large negative anomaly over the top of the sheet. The extent of the asymmetry is a good indication of the dip of the conductor (Ketola and Puranen, 1967). A minor increase of 8% in the peak to peak response is seen to occur when the sheet is immersed in the salt water solution. The in-phase anomaly when compared with the master curves of Ketola and Puranen (1967) agrees closely with the anomaly produced by a thick highly conductive body dipping at 45 degrees at a depth of 8m.

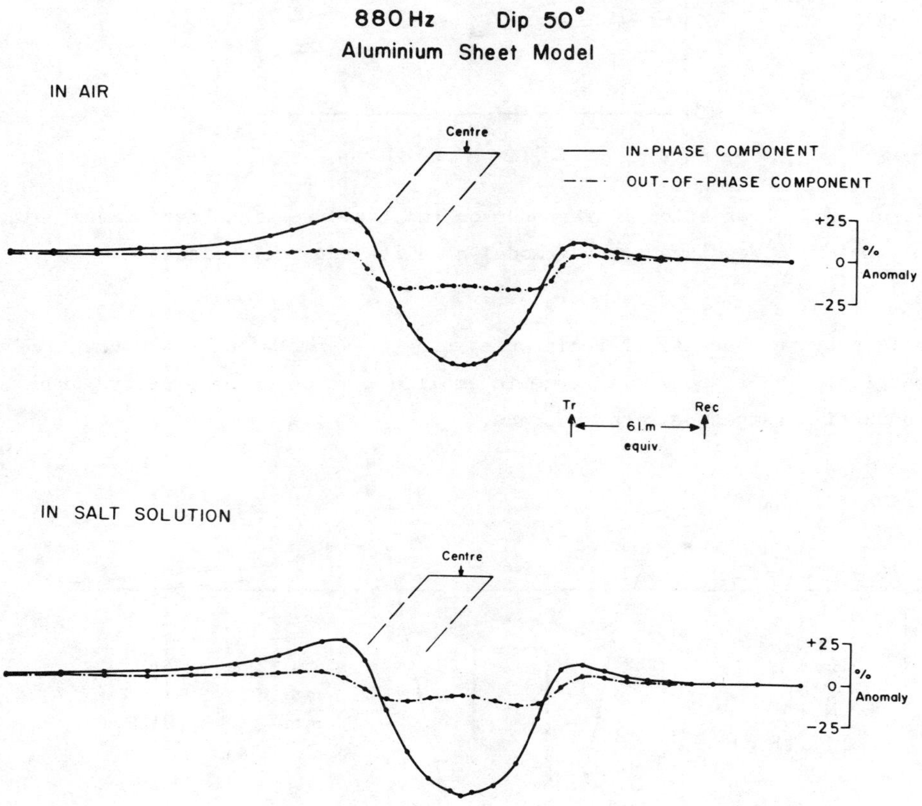

Figure 5 Horizontal loop Slingram anomalies at 880Hz over a dipping aluminium sheet model in air and salt solution.

Slingram Traverse over Woodlawn Model

Horizontal loop Slingram EM modelling was completed over the accurately scaled lead and aluminium alloy models of the Woodlawn orebody. Fig.6 shows stacked profiles of the Slingram data at a frequency of 880Hz for the lead model in air. The complex shape of the model is reflected in the Slingram anomalies by the shift in the anomaly peak position along the different traverses. This positional variation follows exactly the top of the metal model. The asymmetry of the anomalies in the centre of the model indicates that the model

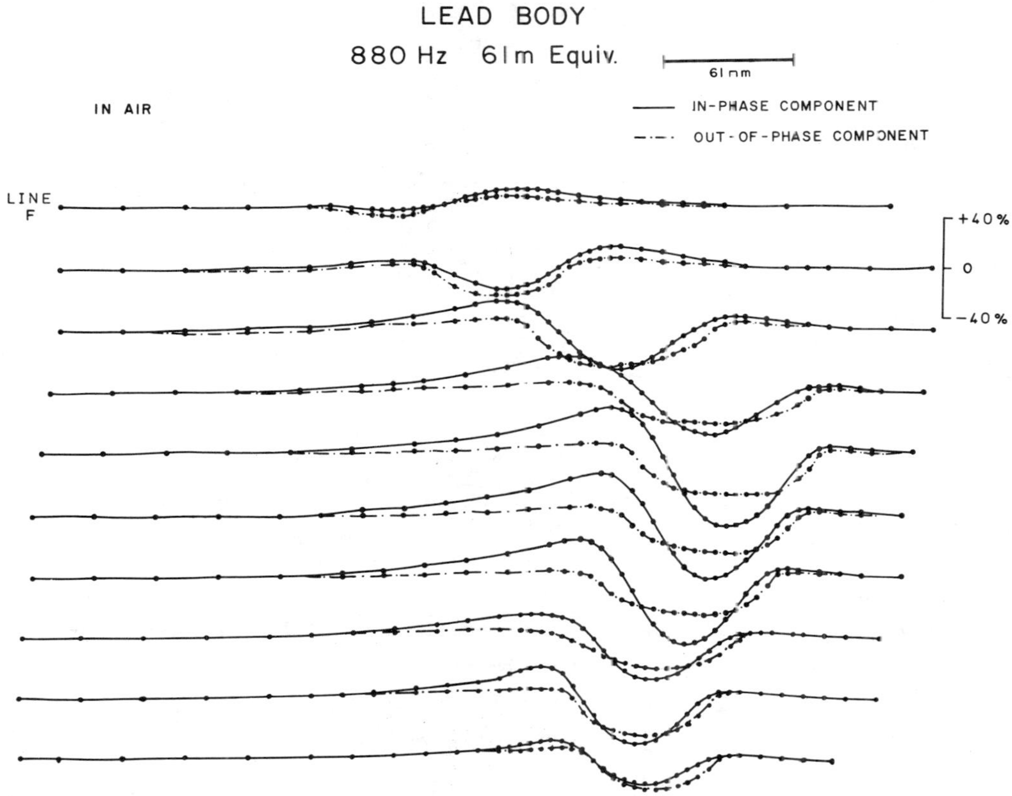

Figure 6 Stacked profiles of horizontal loop Slingram anomalies over the lead Woodlawn model at 880Hz.

has a steep dip, towards the larger peak. Comparison with the sheet model suggests that the dip is approximately 50 degrees.

Figs. 7 and 8 show the stacked profiles over the aluminium model at a frequency of 880Hz in both air and the salt solution, respectively. This model again produces Slingram anomaly that reflects the complex shape of the scale model. The asymmetry in anomaly shape again suggests a steep dip for the model.

A comparison of the anomalies over the lead and aluminium models shows that the lead model produces an out-of-phase anomaly of greater amplitude, as would be expected since it is less conductive.

Fig.9 shows stacked profiles from field survey over the Woodlawn orebody (Haren and Whiteley, this volume, Fig.2, p 196). More lines of data are

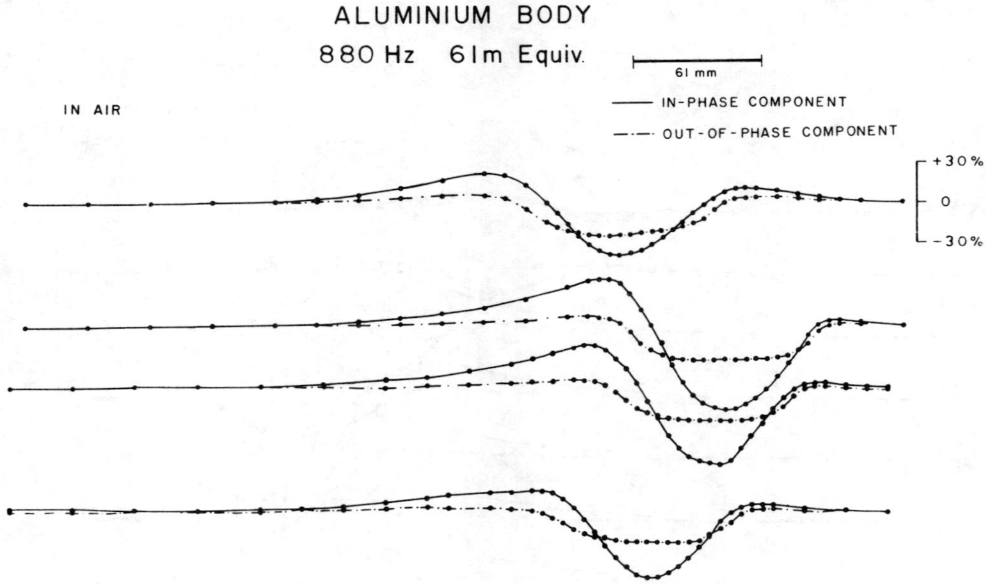

Figure 7 Horizontal loop Slingram anomalies recorded along four traverse lines at 880Hz for the aluminium alloy model in air.

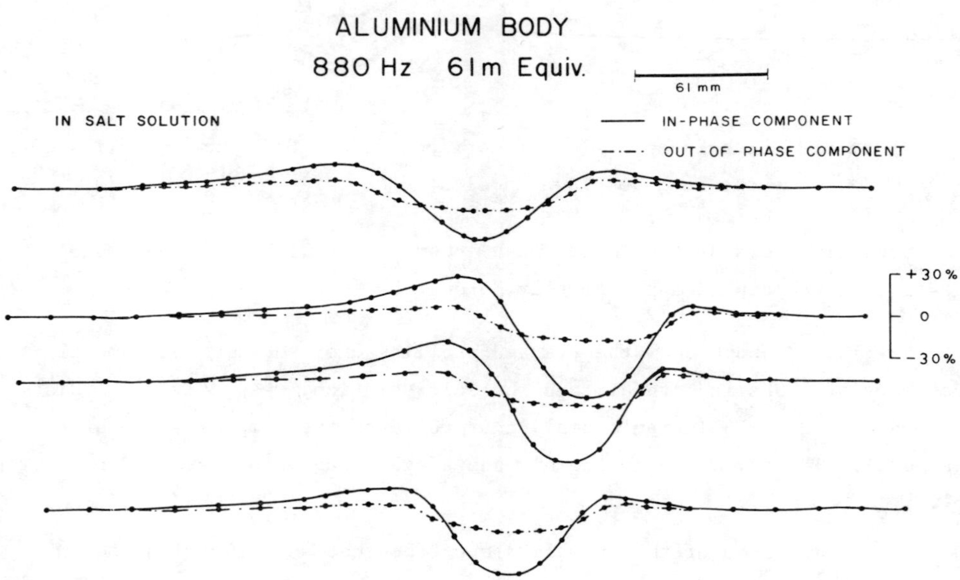

Figure 8 Horizontal loop Slingram anomalies recorded along four traverse lines at 880Hz for the aluminium alloy model in a salt solution.

shown in Fig.9 than were measured over the model. Comparison of the field data and model data in Figs. 6, 7 and 8 show that the anomalies obtained over the scale models are remarkably similar to the field anomalies over the Woodlawn orebody. Anomalies produced by the aluminium models show a better agreement with the field data, and would suggest that the bulk conductivity of the shallow part massive sulphide orebody is approximately 10 Sm^{-1}, which is close to the measured values (Tyne et al, this volume, p99). This indicates that the upper part of the massive sulphide orebody is contributing the bulk of the response seen in the Slingram field anomalies. Model data from the thick dipping aluminium sheet of Fig.5 also compare favourably with the field data from Woodlawn. These results suggest that even for conductive orebodies of complex shape simple models can be a useful aid to the interpretation of Slingram field data.

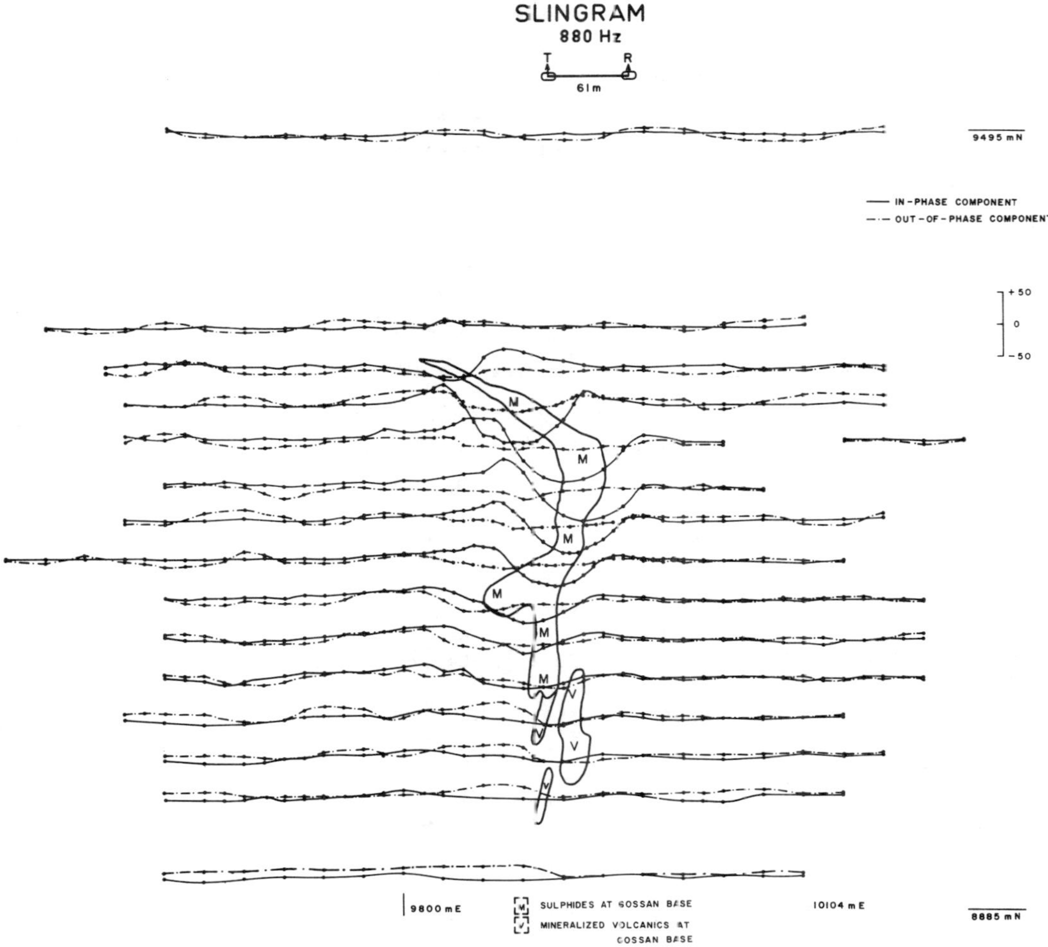

Figure 9 Stacked horizontal loop Slingram anomalies recorded at 880Hz over the Woodlawn orebody.

Table 2 lists the actual parameters of the orebody and those calculated using the interpretational methods applied to the field curves and to the aluminium model data. (Strangway, 1966, Ketola and Puranen, 1967). Comparison of these parameters shows that the model Slingram anomalies are close to those actually measured over the Woodlawn orebody in both shape and magnitude.

Conclusions

The 1 to 1000 scale model of the Woodlawn orebody produced horizontal loop Slingram anomalies which closely resembled the anomalies obtained in the field.

Useful comparative data were provided with a simple, thick, dipping conductive sheet model for the less complex parts of the orebody, away from strike variations and on profiles approximately orthogonal to strike.

The accurately scaled model of the Woodlawn orebody based on the results of extensive exploration drilling and using a saline solution to simulate finite conductivity host rocks, produced Slingram anomalies almost identical to field anomalies in both magnitude and shape. These results indicate that the shallow part of the conductive orebody contributes the dominant EM response at the frequencies used and that the likely bulk conductivity of the orebody is 10 Sm^{-1}.

Horizontal loop Slingram data over highly conductive orebodies such as Woodlawn provide valuable information on strike length, shape, dip and conductivity thickness products. Scale modelling, although time consuming, has a place in detailed EM interpretation.

REFERENCES

Bosschart, R.A., 1964. Analytical interpretation of fixed source electromagnetic prospecting data. Ph.D. Thesis (unpubl.), Delft, The Netherlands.

Clark, A.R., and Mungal, A.G., 1951. Scale model experiments in electromagnetic methods of geophysical exploration. *Can. J. Phys.*, 29, 285-293.

Fraser, D.C., and Ward, S.H., 1967. Analytical and model studies of a rotatable field electromagnetic prospecting system. *Geophysics*, 32(5), 899-917.

Frischknecht, F.C., 1971. Electromagnetic scale modelling. In *Electromagnetic Probing in Geophysics* (Wait, J.R. ed.) Golem Press, Boulder.

Gaur, V.K., Verma, O.P. and Gupta, C.P., 1972. Enhancement of electromagnetic anomalies by a conducting overburden. *Geophysical Prospecting*, 20(3), 580-604.

Gaur, V.K., and Verma, O.P., 1973. Enhancement of electromagnetic anomalies by a conducting overburden, Part II. *Geophysical Prospecting*, 21(1), 159-184.

Hedstrom, E.H., and Parasnis, D.S., 1958. Some model experiments relating to electromagnetic prospecting with special reference to airborne work. *Geophysical Prospecting*, 4, 322-341.

Ketola, M. and Puranen, M., 1967. Type curves for the interpretation of Slingram (horizontal loop), anomalies over tabular bodies. Report of investigations NO.1 Geological Survey of Finland, Otaniemi.

Lowrie, W., and West, G.F., 1965. The effect of conducting overburden on electrical prospecting measurements. *Geophysics*, 30, 624-632.

Sinclair, G., 1948. Theory of models of electromagnetic systems. *Proc. I.R.E.*, 36, 1364-1370.

Strangway, D.W., 1966. Electromagnetic parameters of some sulphide orebodies. In *Mining Geophysics*, 1, S.E.G., Tulsa, 227-242.

Stratton, J.A., 1941. *Electromagnetic Theory*. McGraw-Hill, New York.

Verma, O.P., 1972. Electromagnetic model experiments simulating conditions encountered in geophysical prospecting. Ph.D. Thesis (unpubl.), Univ. of Roorkee, India.

Ward, S.H., 1967. The electromagnetic method. In *Mining Geophysics*, 2, S.E.G., Tulsa, 224-372.

Ward, S.H., Ryu, J., Glenn, W.E., Hohmann, G.W., Dey, A., and Smith, B.I., 1974. Electromagnetic methods in conductive terranes. *Geoexploration*, 12, 121-183.

TABLE 2. Slingram: model and field estimates for Woodlawn

Line	Dip (degrees)			Depth (m)			Conductivity-thickness		Actual width of orebody
	Actual	Field estimate	Model estimate	Actual	Field estimate	Model estimate	Field estimate	Model estimate	
H	85	30-45	60	7	6	8	50	70	25
J	<40	50	45	10	6	6	40	70	25
K	45	45	50	14	8	6	75	70	21
M	50	45	50	20	8	10	40	46	32

7.7. *Large Loop Turam Electromagnetic Survey of the Woodlawn Orebody*

R. J. HAREN

Contents

Abstract

A detailed Turam electromagnetic (EM) survey was carried out over the Woodlawn Orebody using a dual frequency ABEM Turam 2S system operating at 220Hz and 660Hz, respectively. Large horizontal transmitting loops were alternately laid out on either side of the orebody and vertical field anomalies were measured across the orebody.

Results of the Turam survey suggest that the conductivity and geometry of the Woodlawn orebody are such that it is an excellent Turam target.

Introduction

The Turam method as used in base metal exploration is described by Bosschart (1964). Usually a primary electromagnetic (EM) field is created around a large, single turn rectangular loop which is laid on the ground with its long edge approximately parallel to the geological strike. Measurement traverses are then made perpendicular to the long edge of this loop using a receiving system comprising two small, closely spaced vertical dipole coils. Amplitude ratios and phase differences between the received signals at each coil are measured at successive stations.

The large area of the transmitting loop creates large primary flux densities within the ground. In comparison with most moving transmitter EM methods a much greater volume of ground is energized and primary field strengths are much greater.

Apart from the operational difficulties of laying and maintaining large transmitting loops in the field (which led to the use of small multiturn loops; Haren and Whiteley, this volume, p. 233) there are several problems which can adversely affect the performance of the fixed transmitter method in exploration. Firstly, many of the observed anomalies are due to small conductive irregularities in the overburden and near surface weathered layers. These tend to mask the effects of large conductors at depth. This problem is also due, in part, to the use of small receiver coils. Secondly, a large, steeply dipping conductor near the transmitter loop may shield significant conductors at depth or further from the loop. This is due to large close conductors dissipating the energizing magnetic field, leaving only low flux densities in deeper subsurface regions. Conductive overburden causes similar problems which are enhanced because of the excellent coupling of the Turam primary field with horizontal layers.

Survey at Woodlawn

The survey at Woodlawn was carried out with a dual frequency ABEM Turam 2S system. This system operates at frequencies of 220 and 660Hz with currents of up to 3A.

Transmitting loops were alternately laid on either side of the orebody and were oriented with sides paralleling the Woodlawn Test Grid On the eastern, updip side of the orebody the transmitting loop was approximately 427m long, extending from 9403mN to 8976mN and 153m wide extending from 10257mE to 10104mE. The western, downdip transmitter loop was also 427m long and 153m wide extending from 9403mN to 8976mN and from 9739mE to 9586mE. Turam ratio and phase measurements were made over seven separate traverses (Lines H to N inclusive) with a receiver coil spacing of 15.2m.

Transmitter loop size is important in determining Turam anomaly amplitudes. Lajoie and West (1978) have shown that, for tabular conductor models beneath overburden, the largest ratio anomalies are to be expected when the length of the long edge of the loop approximates the strike length of the conductor. At Woodlawn the leading edge of both loops used in this survey was approximately 150m from the nearest edge of the orebody and the length of the loop was somewhat longer than the strike length of the orebody which is approximately 300m.

Data Reduction

In homogeneous ground the amplitude of the primary field decays with increasing distance from the loop. It is usual to remove this decay from the amplitude ratio data and to calculate reduced ratios. Phase differences are unaffected and do not require reduction. Reduced ratios were calculated using a computer program by Williams (1971) and both reduced ratio and phase difference data were plotted at the centre of the receiver coils.

A further reduction of the Turam data to the real and imaginary components of the vertical magnetic field was also carried out (Bosschart, 1964). This reduction facilitates interpretational procedures.

For interpretation, it is usual to assume that a free space approximation applies for the normal magnetic field produced by a horizontal transmitter and that anomalies are not affected by conductive overburden or host material (Bosschart, 1964). However, Ward et al. (1974) have shown that Turam anomalies are greatly modified when the host rock and overburden conductivity are varied and therefore free space assumptions do not always apply in real situations and, if not applicable, inaccurate interpretations may result.

Bosschart (1964) has shown that, under favourable conditions, a shift in the location of the secondary current axis when using different transmitter locations allows estimates to be made of the width of a conductor. The calculation of width allows conductivity estimates to be made from the response parameter.

Discussion of Results

Vertical Field Reduced Ratios and Phase Differences

Fig. 1 shows stacked profiles of reduced ratio and phase difference at 220Hz for the transmitter loop on the eastern, updip side of the orebody. The top of the massive sulphide ore lens beneath the gossan is also shown. The position of the maxima along successive profiles correlates well with the top of the ore body. Drilling has shown that, for the profiles close to Line H (9280mN), the overburden is approximately 7m thick. This means that the receiver coils in this region are well coupled with the secondary magnetic fields induced in the orebody. The reduced ratio anomaly decreases in amplitude toward the southern end of the body, possibly due to a thickening of overburden cover to approximately 20m over the massive sulphide lens in this region. This reduction in anomaly amplitude could also be caused by a reduction in bulk conductivity near the southern end or by a lack of continuity of the conductor due to interfingering of massive sulphides with volcanics.

Fig. 2 shows stacked profiles of reduced ratio and phase difference at 660Hz for the eastern transmitter loop. The anomaly shape and the progressive shift in the position of the maxima along strike again give an accurate picture of the complex shape of the massive sulphide ore lens.

The anomaly shape and amplitude at both frequencies (Figs. 1 and 2) are similar and both data sets enable the conductive orebody to be accurately positioned.

Figs. 3 and 4 show stacked profiles of reduced ratio and phase difference over the Woodlawn orebody for the transmitter loop on the western, downdip side of the orebody at frequencies of 220 and 660Hz, respectively.

Again, anomaly amplitude and shape at both frequencies are very similar, and the position of the orebody is accurately determined.

Comparison of Figs. 3 and 4 with Figs. 1 and 2 shows the orebody to be a dipping highly conductive body of complex shape. Differences in

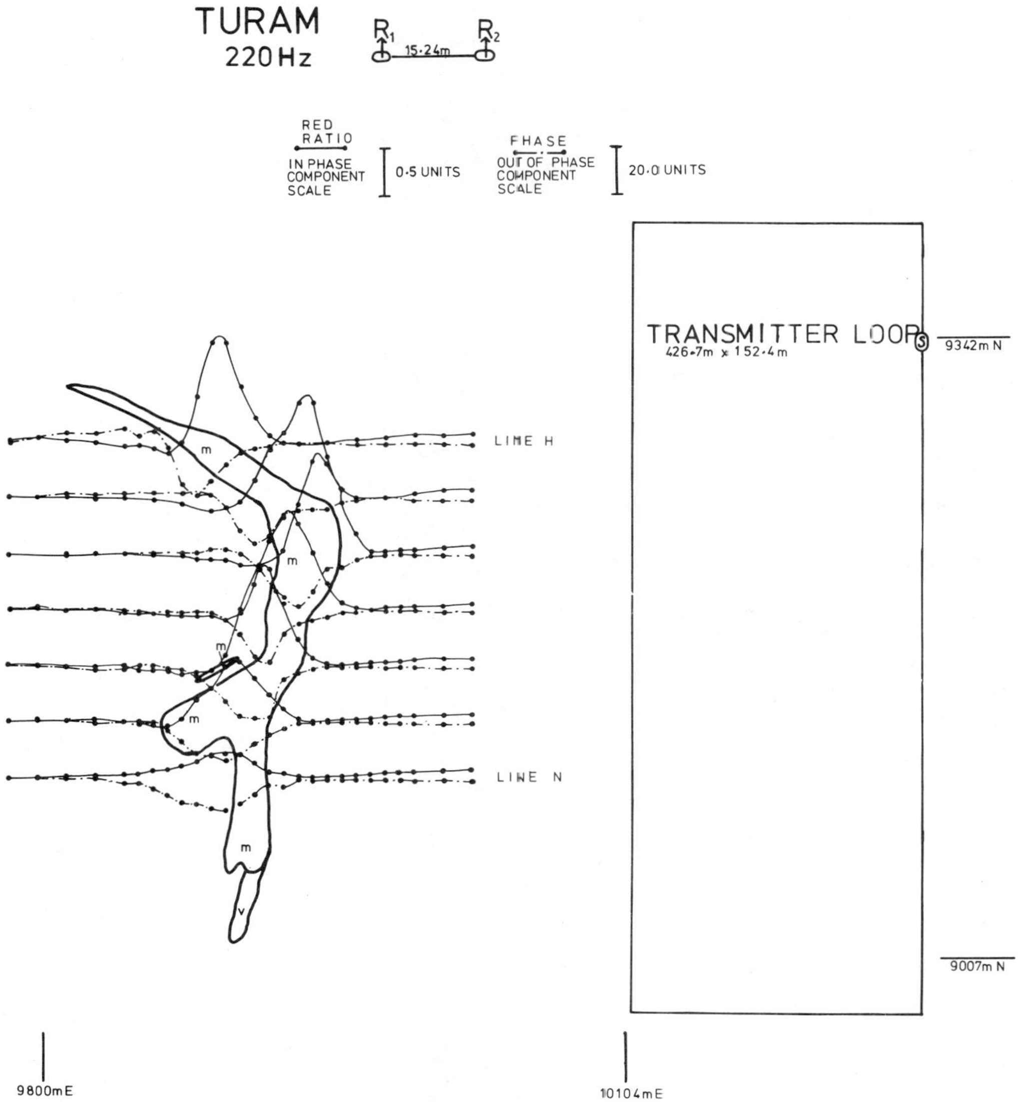

Figure 1 Stacked profiles of reduced ratio and phase difference at 220Hz. Eastern, updip transmitter loop.

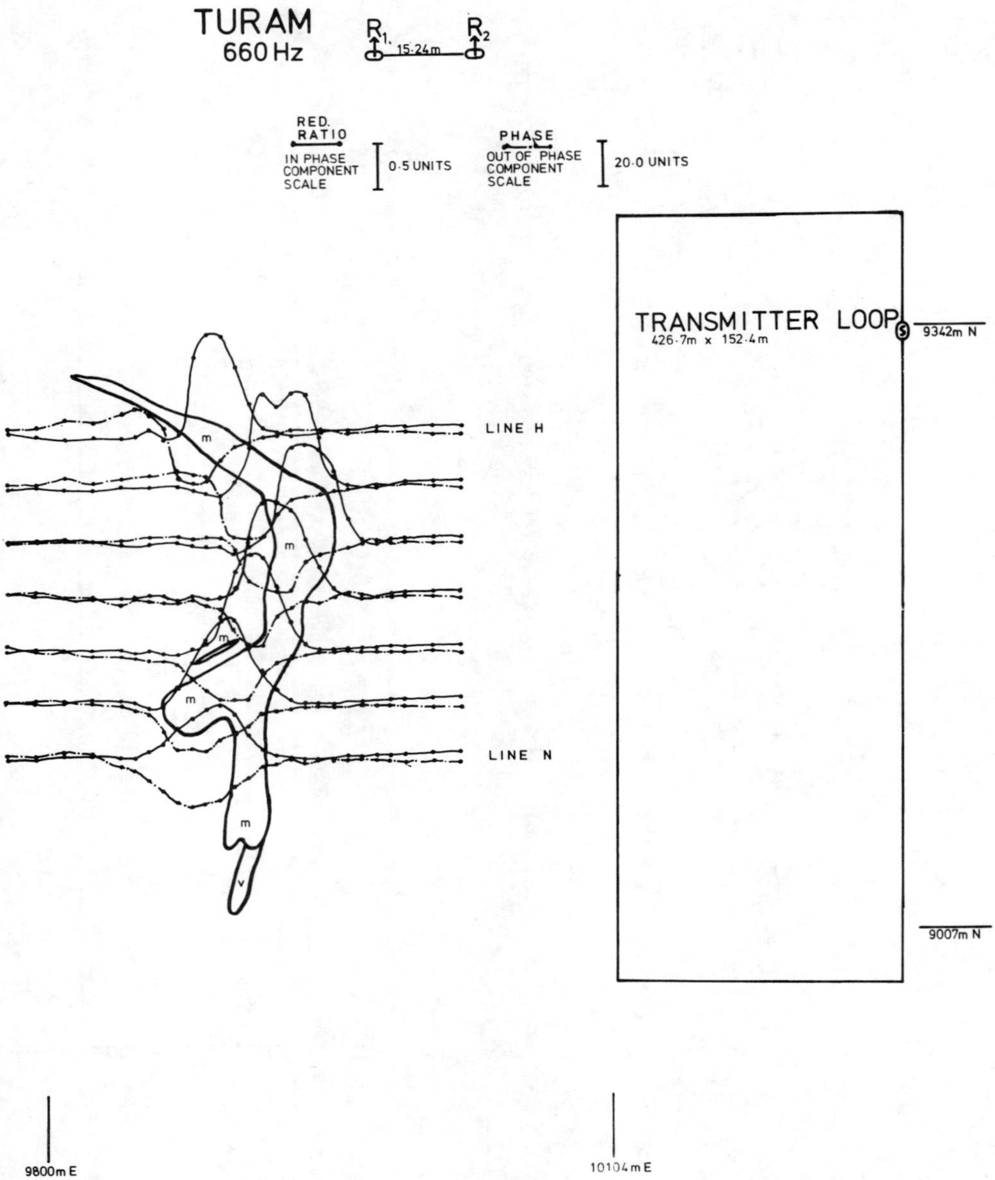

Figure 2 Stacked profiles of reduced ratio and phase difference at 660 Hz. Eastern, updip transmitter loop.

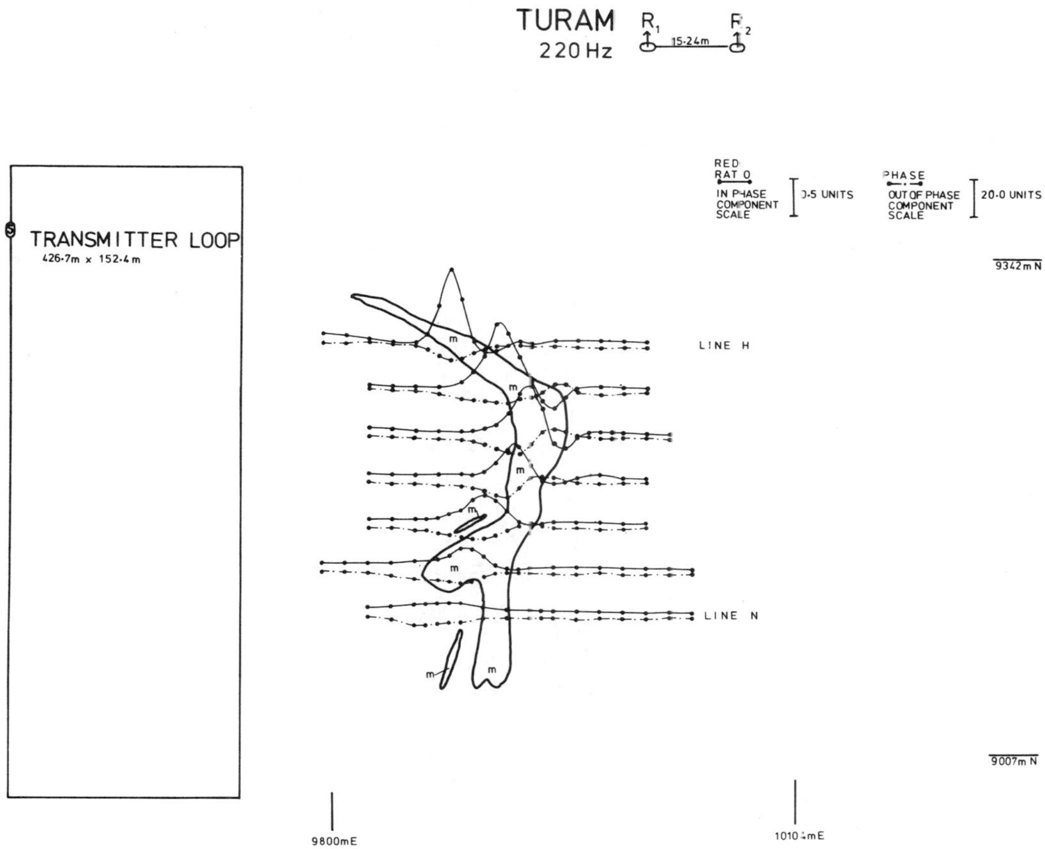

Figure 3 Stacked profiles of reduced ratio and phase difference at 220 Hz. Western, downdip transmitter loop.

anomaly amplitude or shape which occur when the transmitter positions are on the opposite sides of the conductive orebody show that it is not vertical (Bosschart, 1964).

Fig. 5 shows all the vertical field Turam data collected along line J (9220mN) with the drilled cross section. Differences are more clearly seen in the anomaly shape and amplitude for transmitter loops on opposite sides of the orebody. The displacement of the maxima for the transmitter, locations on opposite sides, indicate a width for the conductive orebody beneath line J of about 40M.

Fig. 6 shows a similar plot for the vertical field anomaly along line M (9130mN). Anomalies obtained on this traverse are of lower amplitude

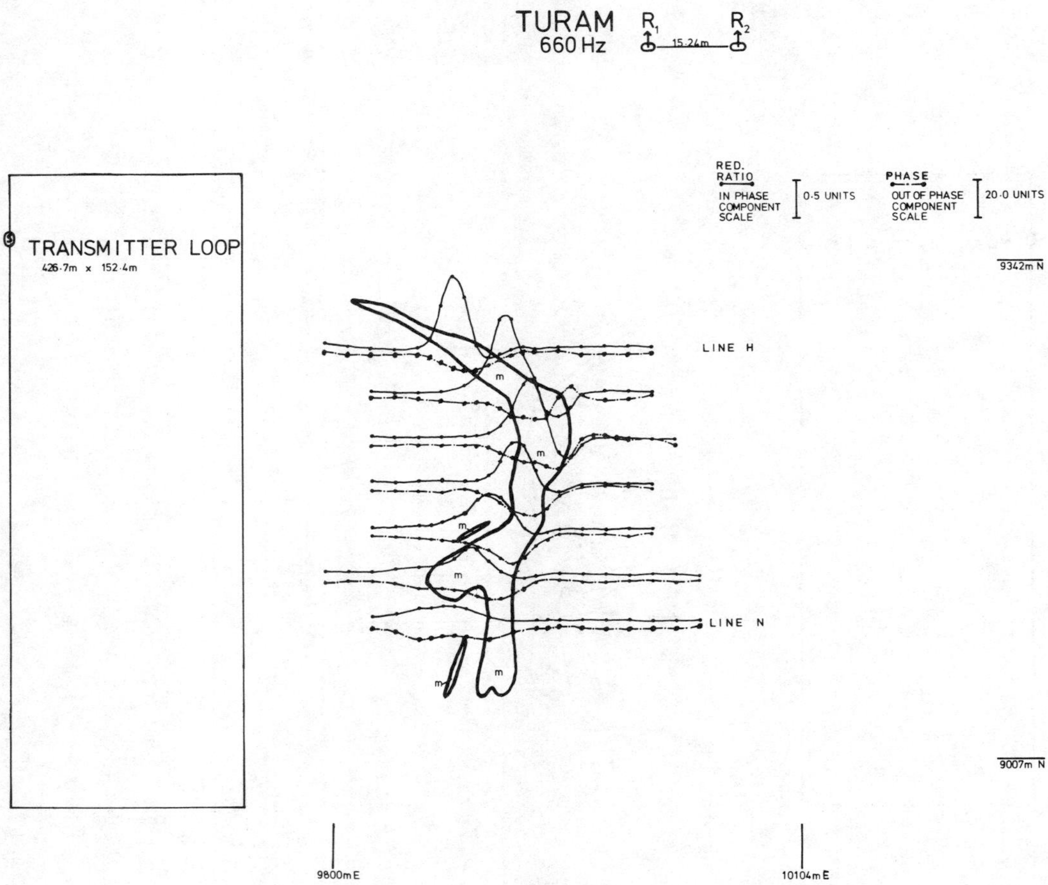

Figure 4 Stacked profiles of reduced ratio and phase difference for the Turam method at 660 Hz. Western, downdip transmitter loop.

and very broad, suggesting that the orebody is deeper or effectively less conductive than beneath line J. The massive sulphide lens changes to stringers approximately 45m to the south of line M.

Real and Imaginary Components of the Vertical Field

Fig. 7 shows stacked profiles of the real (in-phase) and imaginary (out-of-phase) vertical field components for lines H,J,K and M at both 220Hz and 660Hz with the transmitter on the eastern side of the orebody. Since the secondary current centres are taken to be under the inflection points of the curves any migration for different frequencies can give an indication of dip. These plots show the amount of westerly migration of the

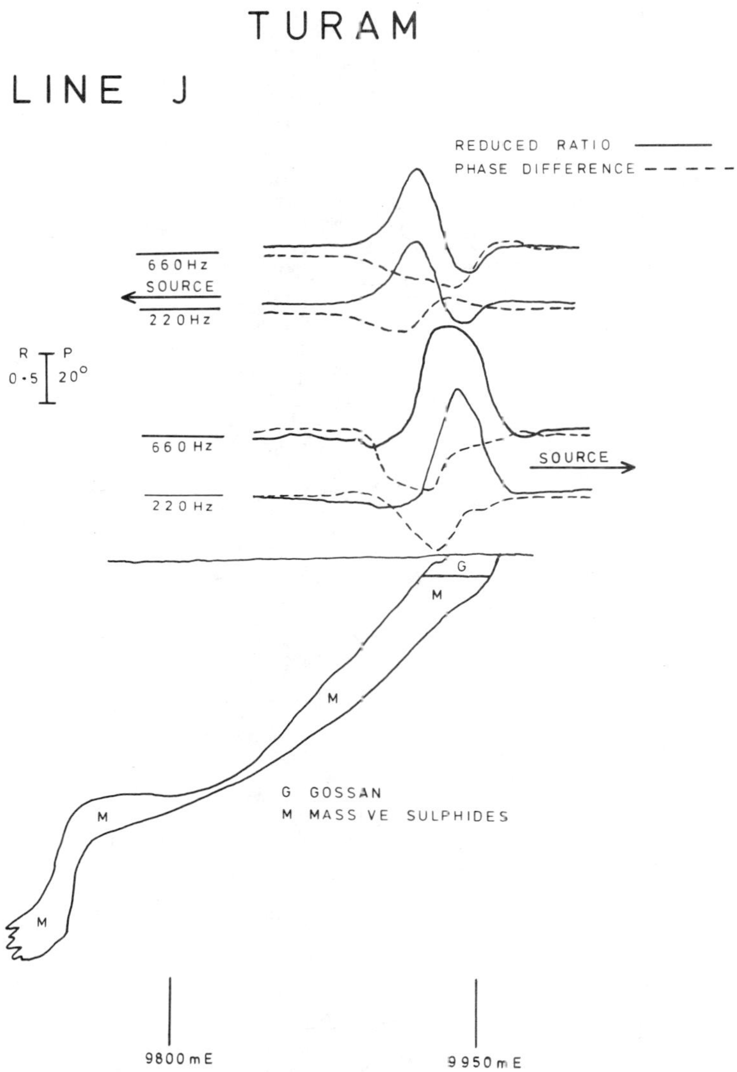

Figure 5 Reduced ratio and phase difference anomalies along Line J over Woodlawn for the two transmitter locations. The cross-section of the orebody under the traverse line is shown.

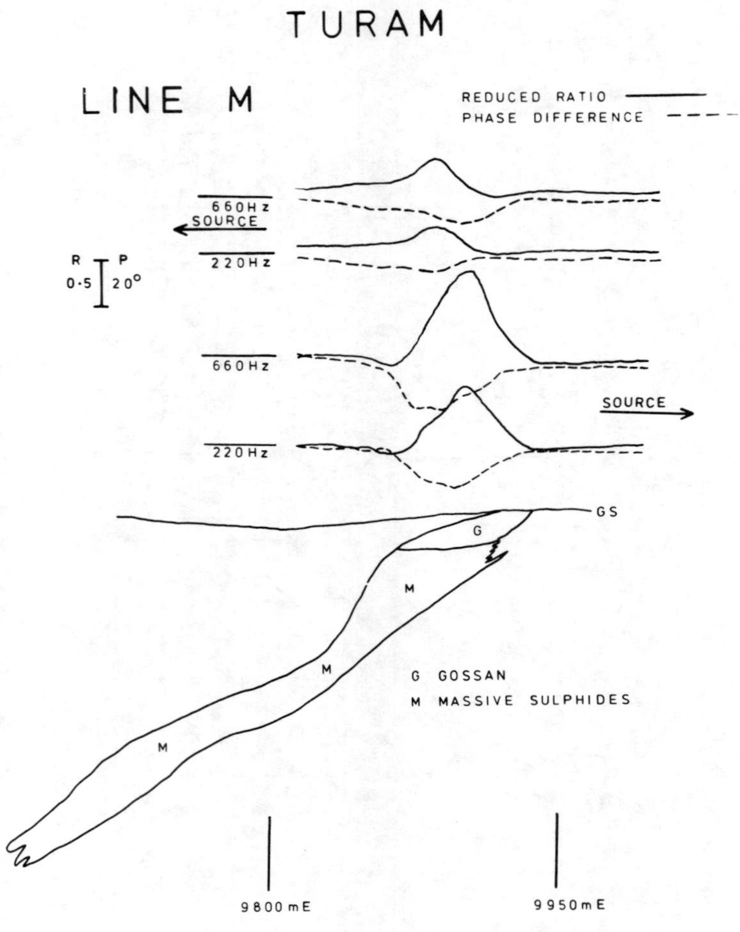

Figure 6 Reduced ratio and phase difference anomalies along Line M over Woodlawn for the two transmitter locations. The cross-section of the orebody under the traverse line is shown.

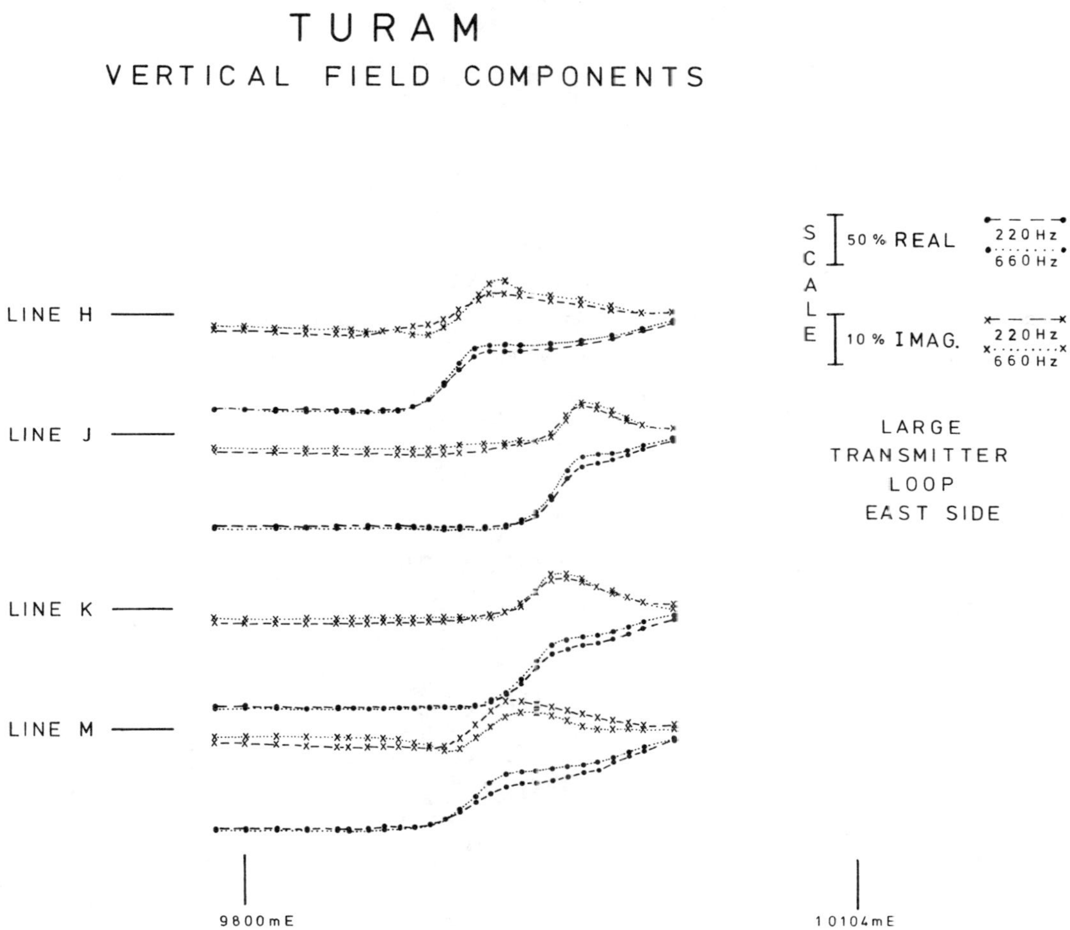

Figure 7 **Real and imaginary vertical field components for the Turam data collected using the eastern transmitter.**

inflection points at the lower frequency is smaller than would be expected for the westerly dipping orebody. This may be due to the extremely high conductivity of the massive sulphide ore causing the secondary current centres for both frequencies to be at practically the same location near the upper surface or the orebody. It would appear that much lower frequencies would be required at Woodlawn to provide a clear indication of actual dip of the orebody, since the data from the frequencies used suggest a very steep dip.

The real and imaginary vertical field components for the same four traverses using the transmitter on the western side of the orebody are shown in Fig.8. These curves are more complicated than for the easterly transmitter location, possibly due to the slowly increasing response as the receiving coils move updip towards the orebody. During the survey measured field strengths dropped sharply once the edge of the conductor had been crossed. The 220Hz and 660Hz plots are very similar and give no clear indication of dip.

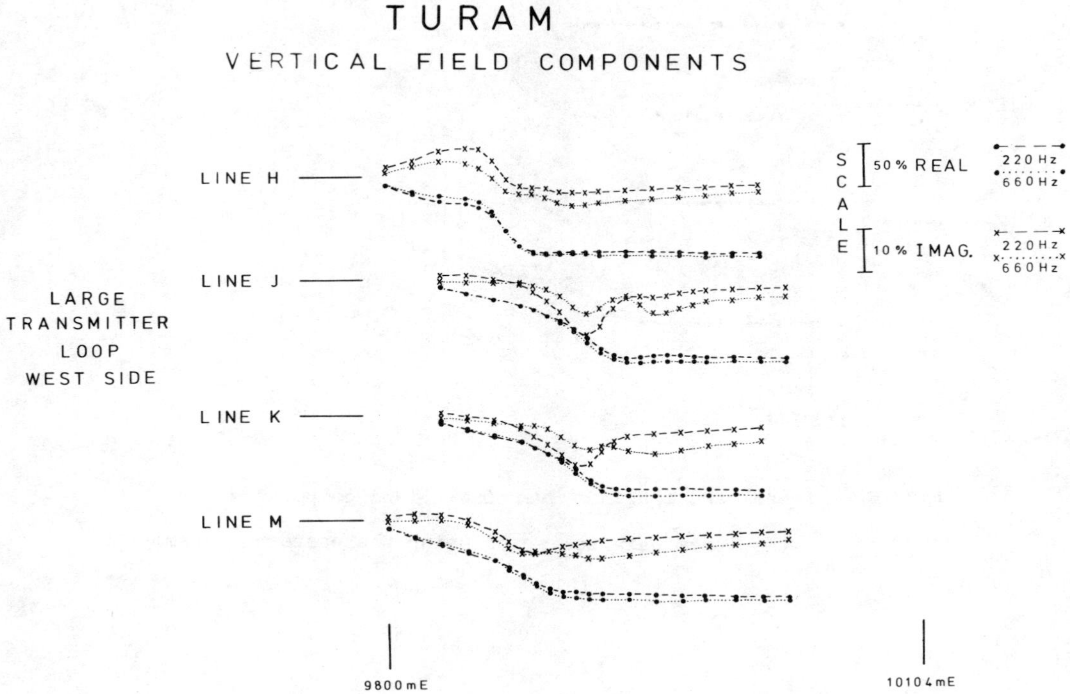

Figure 8 Real and imaginary vertical field components for the Turam data collected using the western transmitter.

Table 1 gives details of the actual body parameters and those estimated from reduced ratio and phase difference plots for Lines H,J,K and M. The depth estimate is taken as the half width of the real and/or imaginary field component curve. The conductor width is determined by the differences in the positions of the inflection points of the response curve for transmitters on either side of the orebody. Dips are not clearly indicated.

Table 1.

Line	Orebody Dip	Est. Depth to orebody (m)	Actual depth to orebody (m)	Conductor quality	Est. width of orebody (m)	Actual width of orebody (m)
H	85°W	15 to 20	7	very good	35	25
J	50°W	10 to 20	10	very good	40	25
K	45°W	15 to 25	14	very good	40	21
M	45°W	18 to 25	20	good	45	32

Conclusions

The Woodlawn orebody is an excellent EM target. The Turam method accurately defines the subcrop of the conductive orebody. Differences in anomalies for the two transmitter positions on opposite sides of the orebody show that it is complex in shape although its width may be estimated. At the high frequencies used, the Turam data indicate a much steeper dip for the orebody than is the case. This is probably due to the very high conductivity and relatively large width of the orebody.

Acknowledgements

Equipment used for the survey was kindly loaned by the Geological Survey of N.S.W. and the Bureau of Mineral Resources.

References

Bosschart, R.A. 1964. Analytical interpretation of fixed source electro-magnetic prospecting data. Ph.D. thesis (unpubl.), Delft, The Netherlands, pp. 103.

Lajoie, J.J. and West, G.F., 1976. The electromagnetic response of a conductive inhomogeneity in a layered earth. Geophysics, 41 (6A), 1133-1156.

Ward, S.H., Ryu, J., Glenn, W.E., Hohmann, G.W., Dey, A. and Smith, B.D., 1974. Electromagnetic methods in conductive terrains. Geoexploration, 12, 121-183

Williams, J.P., 1971. Computer programs for Turam data reduction. Bull. Aust. Soc. Explor. Geophys., 2(1), 29-34

7.8. Experimental Small Loop (Turam) Electromagnetic Surveys at Woodlawn

R. J. HAREN AND R. J. WHITELEY

Contents

Abstract

One of the drawbacks of the conventional Turam method is the positioning and maintenance of the large transmitter loop. For comparison with the large loop method an experimental Turam survey over the Woodlawn orebody was carried out using a square multiturn transmitter loop with 30.5m sides. This small loop acts as an effective dipole source which may be quickly set up and is simple to maintain.

Both vertical and horizontal fields were measured. Results showed that small loop Turam is capable of delineating orebody characteristics with an accuracy equal to the conventional large loop Turam method.

Small loop Turam data collected over the nearby Black Shale, showed it to be a poor conductor with little possibility of it being mistaken for a highly conductive massive sulphide deposit.

Introduction

A large loop Turam survey of the Woodlawn orebody (Haren, this volume, p 219) showed it to be an excellent electromagnetic (EM) target. In common with most Turam surveys employing large single turn transmitter loops problems were experienced at Woodlawn due to breakages caused by wandering stock and wildlife. In addition, although the topography around Woodlawn is gently undulating, there are many prospective areas in the Eastern Lachlan Fold Belt which are characterized by heavy natural scrub and rugged terrain which makes it difficult to deploy and maintain a large loop.

A small transmitter loop requires the clearing of only a small area and can be laid out in a comparatively short time and can be quickly repositioned if required.

Generally, it can be said that small loop fixed transmitter methods produce anomalies that have a reduced amplitude relative to large loops due to a reduction in the flux density applied to the target. Lajoie and West (1976) have shown that where conductive overburden is present, small loops produce larger anomalies than large loops; where conductive overburden is absent the reverse holds true. From the cases studied it appears that there may be an optimum source dimension relative to a particular orebody which maximizes anomalies. This optimum size of loop is probably of similar dimensions to that of the target orebody. However, considerations of loop size must be balanced against operational convenience for a particular survey.

Survey at Woodlawn

In this experimental Turam survey at Woodlawn small transmitter loops were placed alternately at a distance of about 300m on either side of the orebody. The loops were 30.5m square, containing 8 turns of wire. This increased flux density of the primary field which had been reduced because of the smaller area of the loops. The loops were sited between lines J and K, near the centre of the orebody, and oriented with sides along the Woodlawn Test Grid. At these distances and square loop dimensions, the primary field in the vicinity of the orebody may be assumed to have been that due to a vertical axis, dipole source.

An ABEM Turam 2S system operating at frequencies of 220 and 660 Hz was used for these measurements. This system was previously employed for the large loop Turam survey of the Woodlawn orebody (Haren, this volume, p 219).

Both vertical and horizontal field anomalies were recorded along four profiles (H, J, K and M) in the vicinity of the orebody using a receiver coil separation of 6.1m and a station spacing of either 6.1m or 15.2m.

The vertical field amplitude ratios were converted to reduced ratios using a modified computer program for reduction of large loop data (Williams, 1971).

Discussion of Results

Vertical and horizontal field data were plotted at the centre of the pair of the receiver loops.

Vertical Field Reduced Ratios and Phase Differences

Fig. 1 shows stacked profiles of the vertical field reduced ratio and phase differences at 220 Hz for the western source. The positions of the anomaly maxima along successive profiles correlate well with the top of the massive sulphide orebody. The reduced ratio anomaly decreases in amplitude and increases in width towards the southern part of the orebody. This is consistent with the known geometry of the massive sulphides which increase in depth from about 8m at the northern end to over 20m at the southern end.

Fig. 2 shows the stacked profiles of vertical field reduced ratio and phase differences at 660 Hz for the westerly source. Again there is good correlation between the position of the anomaly maxima and the massive sulphides. Only minor differences are evident in the vertical field anomalies for the two frequencies, the most significant variations being in the phase differences anomalies on lines J (9220mN) and K (9190mN). The higher frequency profiles show an increased width for both the ratio and phase anomalies.

Figs. 3 and 4 show the stacked profiles of vertical field reduced ratio and phase differences of the transmitter on the eastern side of the orebody at frequencies of 220 and

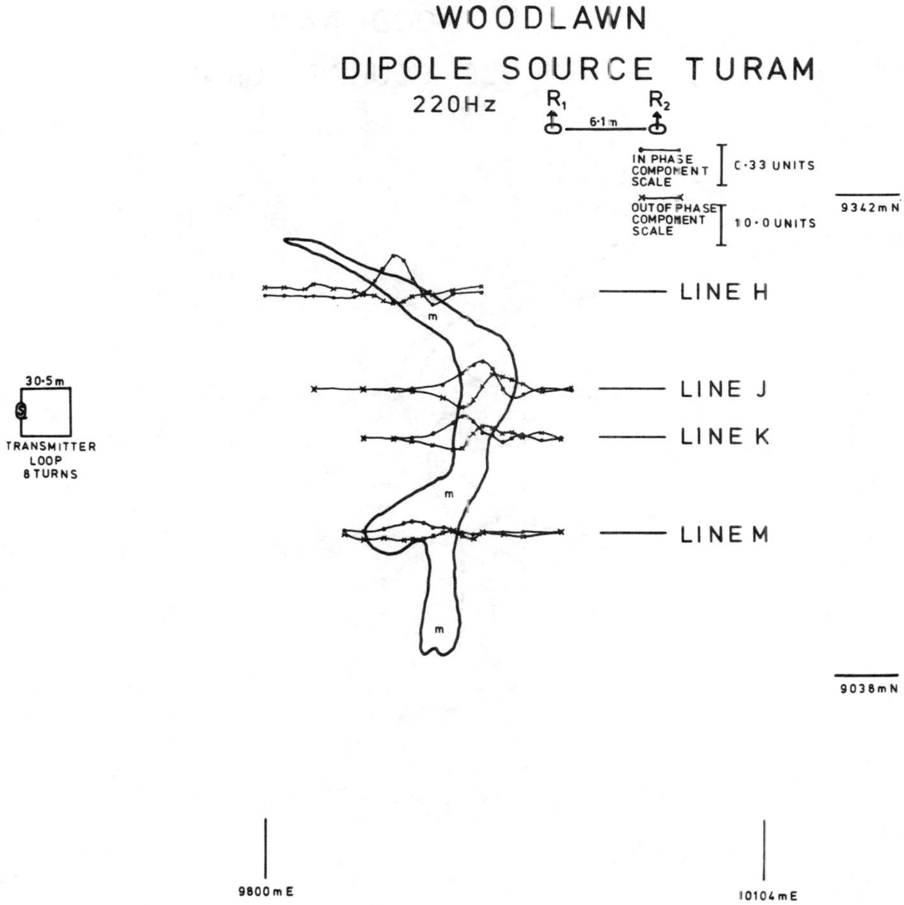

Fig. 1 Vertical field reduced ratio and phase difference anomalies over Woodlawn using a small loop transmitter at 220 Hz on the western side of the orebody.

660 Hz, respectively. Both the reduced ratio and phase anomalies exhibit more complex shapes than do those for the westerly transmitter. The magnitude of the ratio anomaly is approximately equal to that obtained with the westerly transmitter. The phase anomaly exhibits a rapid gradient over the trailing edge of the orebody. The anomalies associated with both frequencies are very similar for the eastern transmitter. The differences between anomaly gradients recorded at the two transmitter sites are likely to be due to the different coupling with the orebody for

GCSWO - I

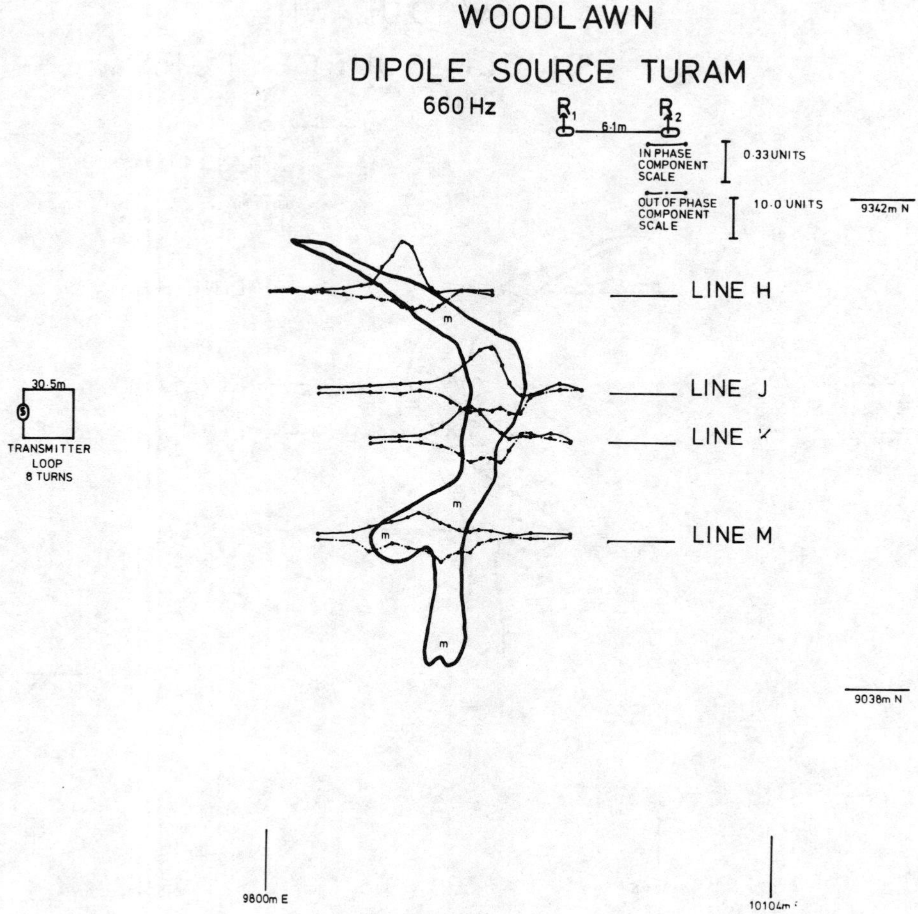

Fig. 2 Vertical field reduced ratio and phase difference anomalies over Woodlawn using a small loop transmitter at 660 Hz on the western side of the orebody.

the up dip and down dip transmitter locations and indicate that the conductor is not vertical. They also reflect the complex shape of the Woodlawn orebody.

Horizontal Field, Reduced Ratios and Phase Differences

The stacked profiles of horizontal field anomalies are shown in Figs. 5 to 8 for the different frequencies and transmitter

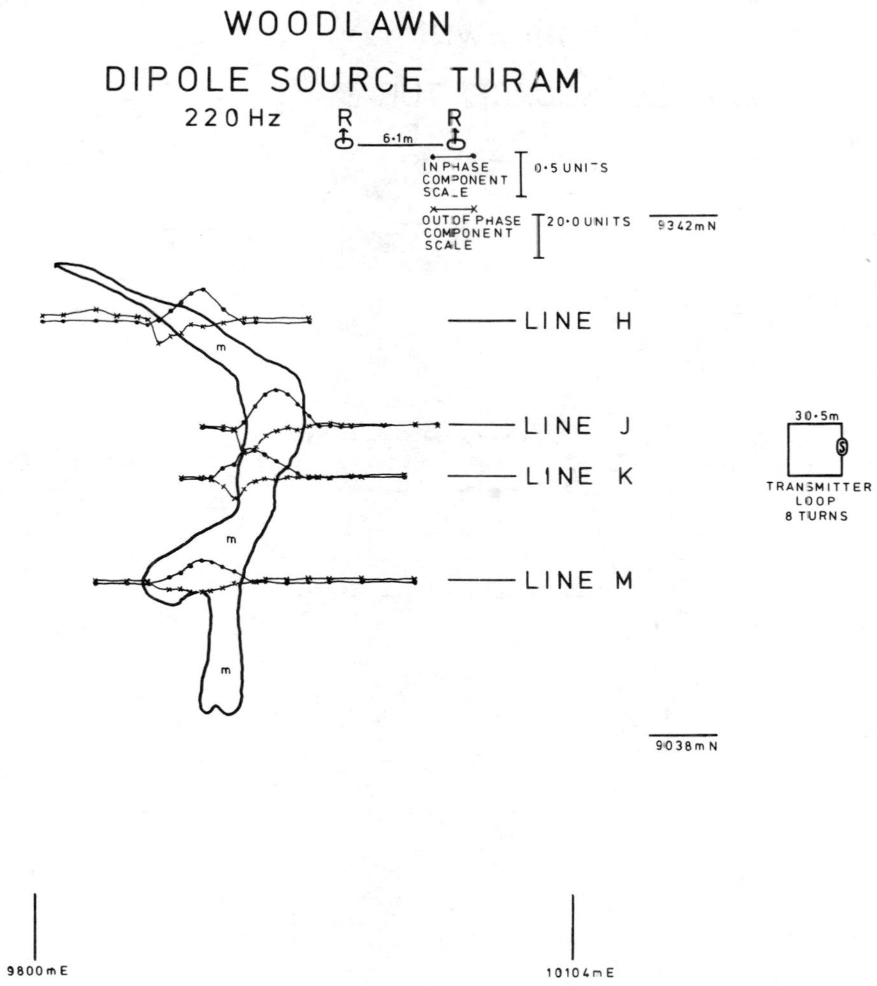

Fig. 3 Vertical field reduced ratio and phase difference anomalies using a small loop transmitter at 220 Hz on the eastern side of the orebody.

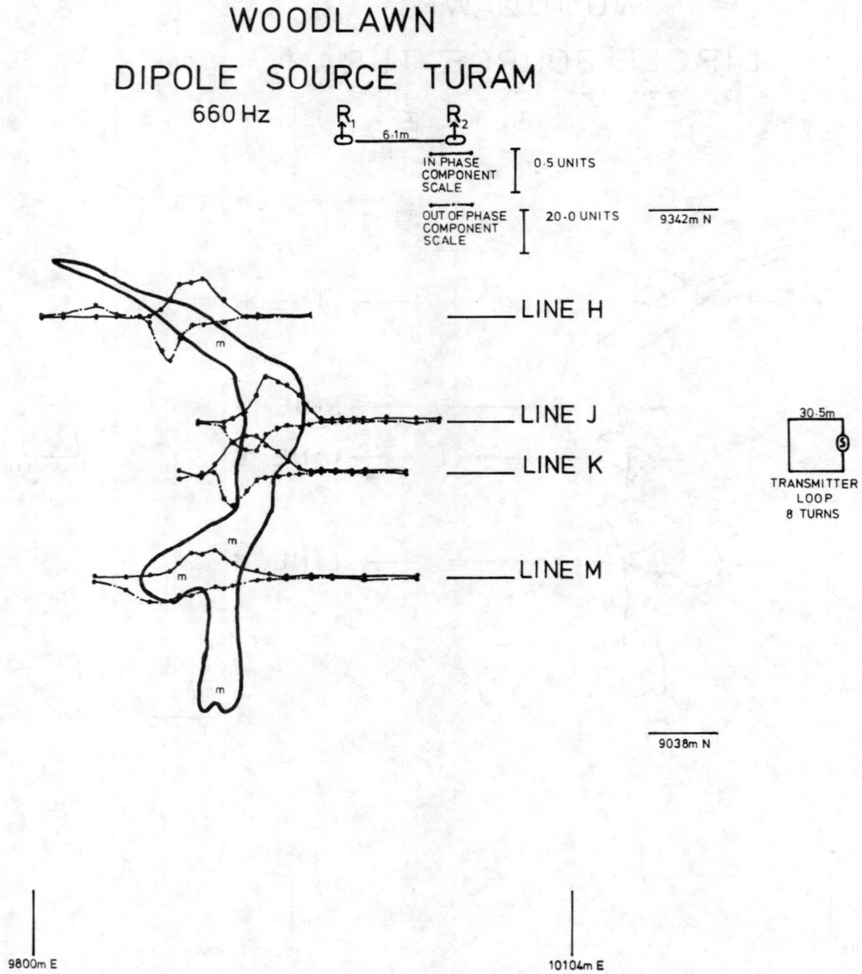

Fig. 4 Vertical field reduced ratio and phase difference anomalies over Woodlawn using a small loop transmitter at 660 Hz on the eastern side of the orebody.

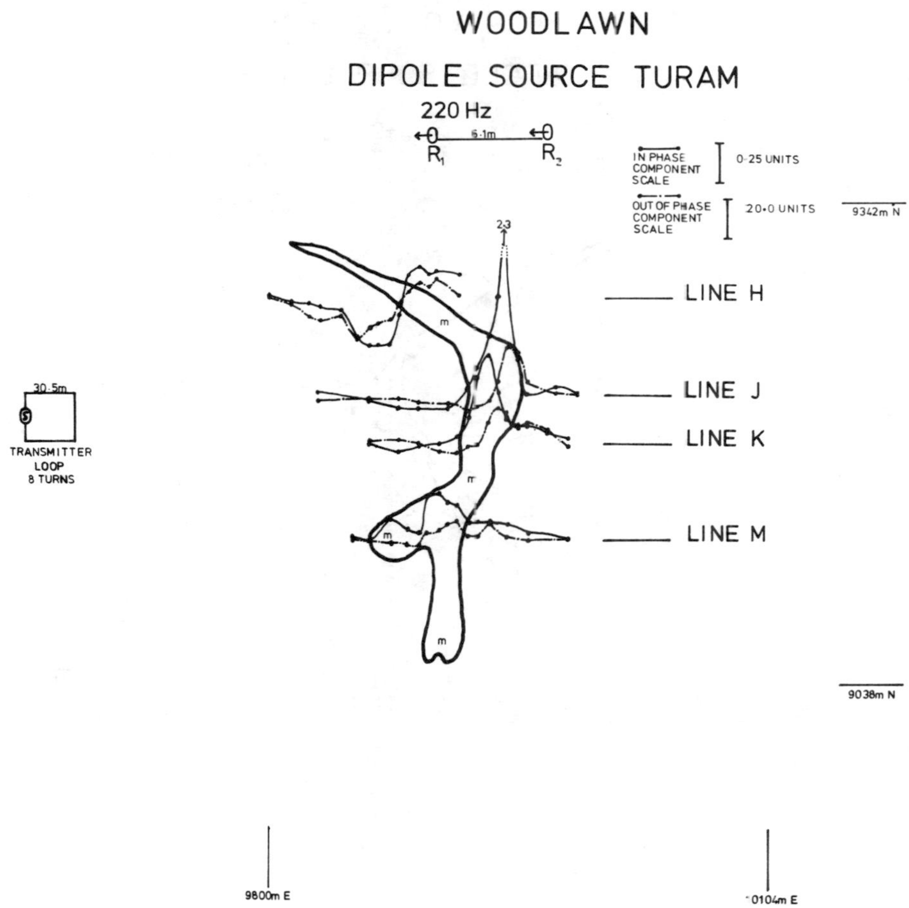

Fig. 5 Horizontal field ratio and phase difference anomalies over Woodlawn using a small loop transmitter at 220 Hz on the western side of the orebody.

locations. The complex nature of the anomalies in both ratio and phase can be seen in all of these figures. The anomalies in the horizontal field strength ratio when using the down dip transmitter location are similar to those obtained by Ward et al., (1974), in a model tank for dipping thin sheets, and are positioned over the centre of the conductive target. The broad anomalies associated with the up dip transmitter location occur at the leading edge of the dipping orebody. Anomalies in the field strength ratio for the up dip and down dip transmitter locations

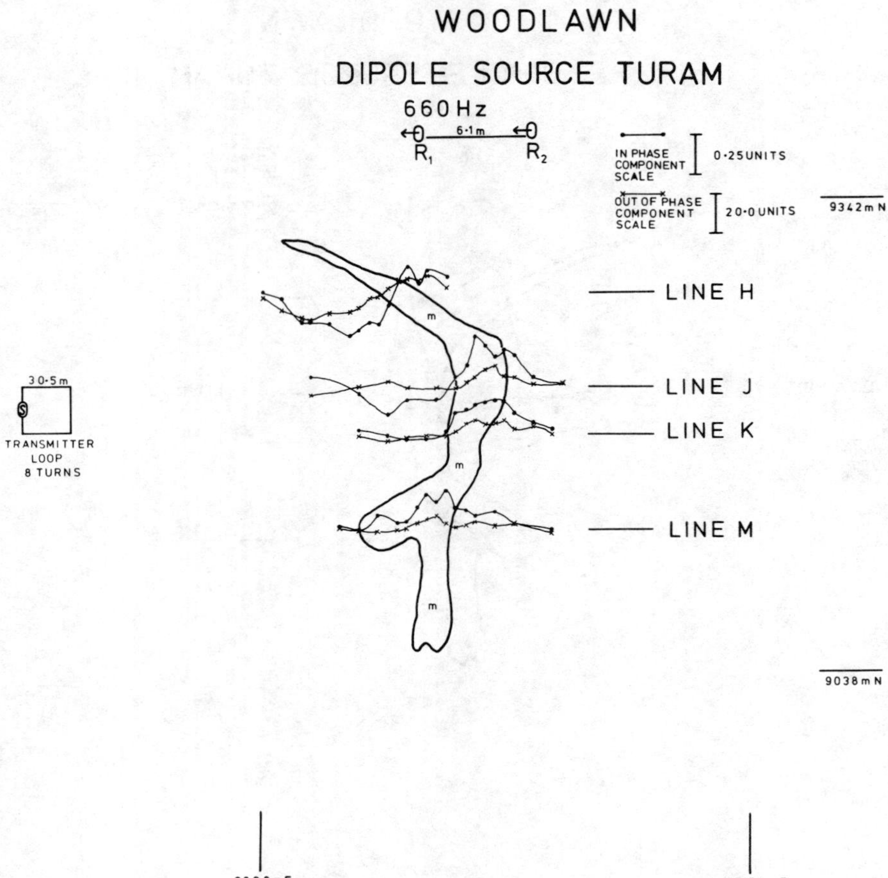

Fig. 6 Horizontal field ratios and phase difference anomalies over Woodlawn using a small loop transmitter at 660 Hz on the western side of the orebody.

are opposite in sign, as was found by Ward et al., ibid. Sharp variations in conductivity at shallow depth are clearly indicated from the horizontal field anomalies. When the transmitter loop is down dip, large horizontal field gradients are associated with the steeply dipping Woodlawn orebody. Large variations in anomaly magnitude at the different frequencies are noted for the down dip transmitter. For both transmitter positions, the 220 Hz ratio anomalies are larger than the 660 Hz ratio anomalies, suggesting that with greater depths of penetration, the response from the orebody is increased.

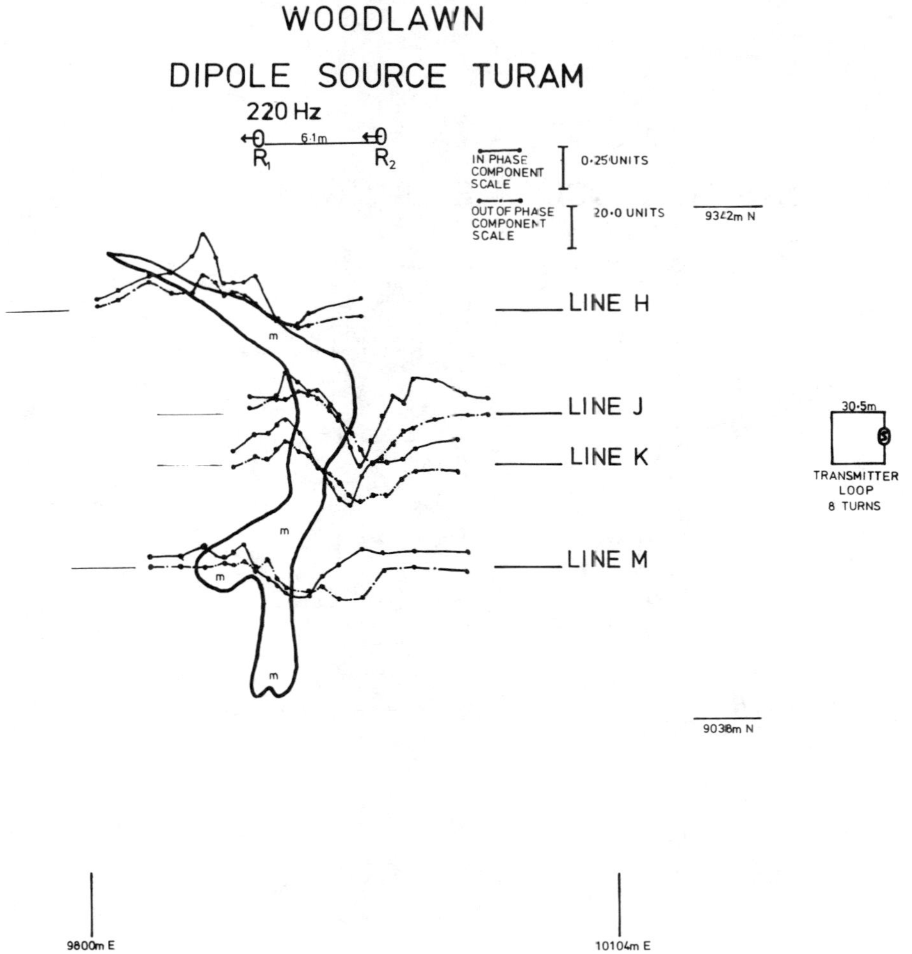

Fig. 7 Horizontal field ratio and phase difference anomalies over Woodlawn using a small loop transmitter at 220 Hz on the eastern side of the orebody.

To facilitate further interpretation, reduced ratio and phase difference profiles were converted to real and imaginary field profiles using the procedure described by Bosschart (1964). This method, which utilizes a cosine and sine transform, combines reduced ratios and phase differences into a data set which enables real (in-phase) and imaginary (out-of-phase) field component variations to be plotted as a percentage of the normal primary field.

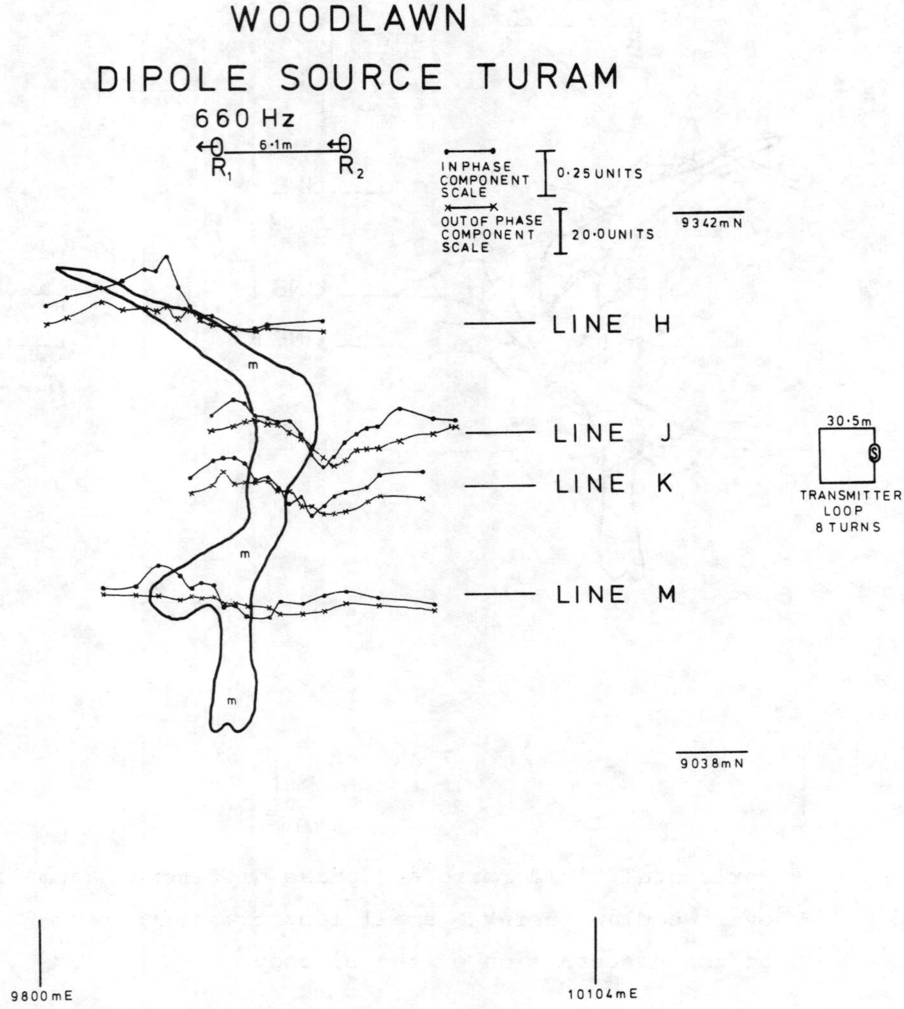

Fig. 8 Horizontal field ratio and phase difference anomalies over Woodlawn using a small loop transmitter at 660 Hz on the eastern side of the orebody.

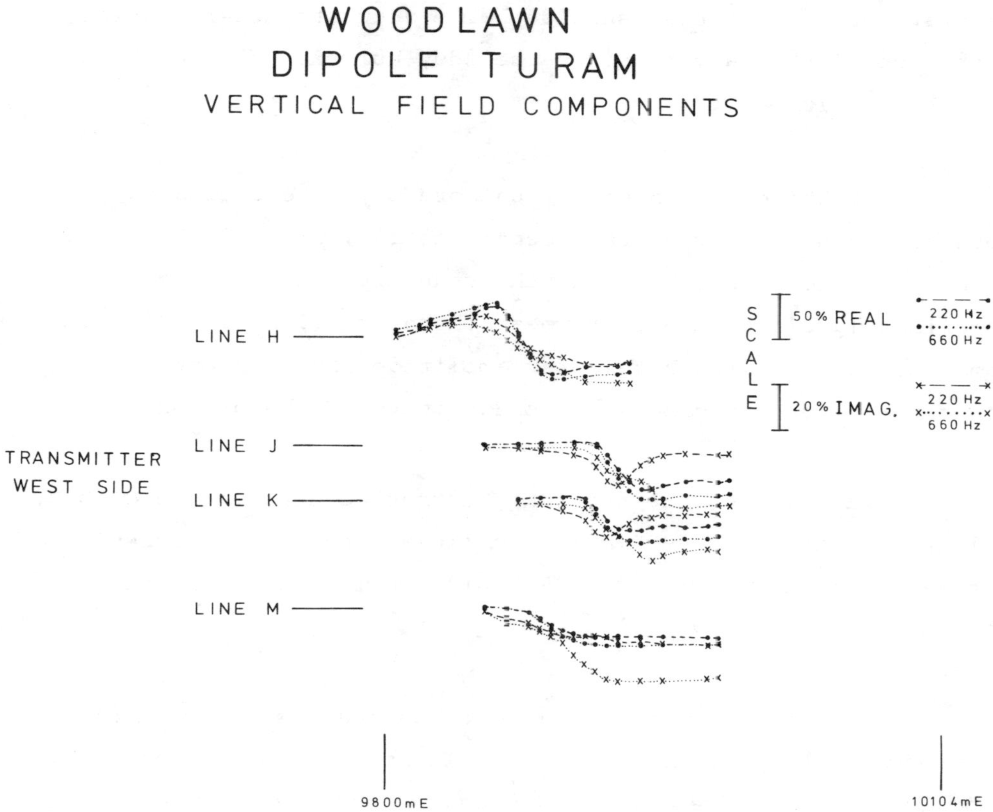

Fig. 9 Real and imaginary vertical field components over Woodlawn using a small loop transmitter at 220 Hz and 660 Hz on the western side of the orebody.

Real and Imaginary Components of the Vertical Field

The real and imaginary components of the vertical field for the two frequencies are shown in Figs. 9 and 10 for the small loop transmitter to the west and east of the orebody respectively. The anomalies associated with the western transmitter (Fig. 9) show that strong eddy currents are induced in the orebody at both frequencies. The anomalies are smoothly varying and amenable to quantitative interpretation. The depth to secondary current sources is taken as half the width of the peak to trough distance. For highly conductive targets, the depth and position of the secondary current centres should enable accurate determination of the top of the target, since the eddy currents flow close to its upper surface.

At the northern end of the orebody, the estimated depth to the secondary current centre is 15m, which is 7m below the base of the gossan. The depth gradually increases to approximately 24m on the extreme southern profile (line M). The anomalies on this line have lower amplitudes and suggest either deeper burial or a decrease in conductivity of the orebody.

The axis of the secondary current centre in the orebody on line H is at H12.5 (9877mE). On lines J and K the current centres occur at J11 and K11 (9922mE). On line M the current centre is at M11.5 (9905mE).

With the transmitter located on the eastern side of the orebody, the anomalies in the real and imaginary field components at both frequencies are again smoothly varying (see Fig. 10), and give an estimated depth to the secondary current centre of about 20m on line H and about 25m on line M to the south. There is no significant migration of the position of the inferred secondary current centres from the higher to the lower frequency which can occur in the direction of dip because of the usually greater penetration at the lower frequency. The lack of migration in this case may be due to the very high conductivity of the orebody resulting in little effective increase in penetration between the two fairly closely spaced frequencies used (220 and 660 Hz).

For the eastern transmitter location (see Fig. 10), the inferred positions of the secondary current centres were at H12 (9891m) for line H, J12.5 (9937m) for line J, K10.7 (9930m) for line K, and M11.7 (9900m) for line M.

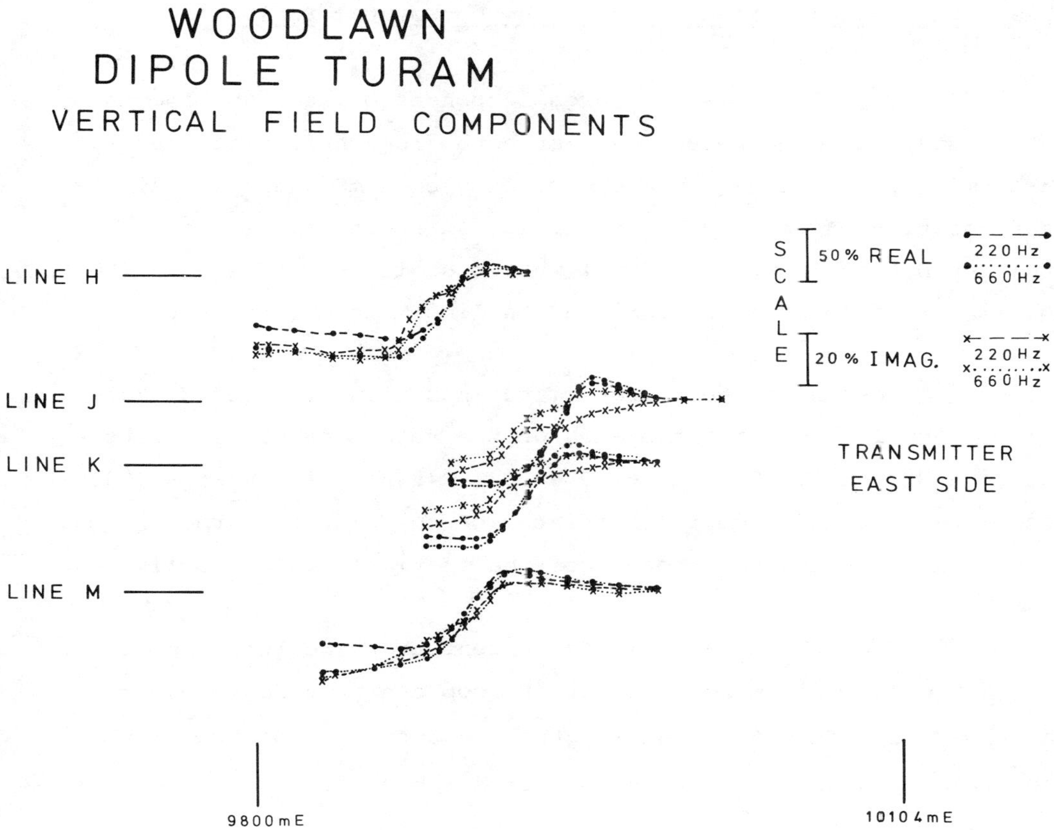

Fig. 10 Real and imaginary vertical field components over Woodlawn using a small loop transmitter at 220 Hz and 660 Hz on the eastern side of the orebody.

If the inflection points on the anomaly curves are taken to indicate the approximate positions of the current centres, the difference in the position of these from opposite sides of the orebody can give an estimate of the width of the body. This is because the induced currents tend to flow near the side surface of the edges of a highly conductive body nearest to the transmitter. Estimations of width of the orebody by this means gives very considerable underestimates in this case. While the actual width of the orebody on lines J, K and M are about 30m, the above method gives estimates of 15m for lines H and J, 6m for line K and very thin for line M.

Real and Imaginary Components of the Horizontal Field

Fig. 11 shows the large responses of real and imaginary horizontal field component data at both frequencies with the two transmitter positions. The centre of the anomaly should define the position of the secondary current axis within the conductor. On line H this is at H (9880mE) whilst on line J it is at J (9930mE), on line K it is at K (9925mE) and on line M at M (9891mE).

For the eastern transmitter position, the horizontal field anomaly indicates the axis of the secondary current centre on line H at H (9900mE), line J at J (9943mE), line K at K (9930mE) and line M at M (9900mE). Estimates of the approximate widths of the orebody give similar values to those from the vertical field anomaly.

From these data, it is evident that the Turam method using a small multiturn transmitter loop can provide information on the strike direction and length, the depth to the top of the conductor and to some extent the direction of dip.

Comparison of Small and Large Loop Vertical Turam Data

A comparison of the 660 Hz small loop Turam data and the normal Turam data (Haren, this volume, p 219) on line J, shown in Fig. 12 for the transmitters located on both the east and west sides of the orebody. The large loop transmitter had dimensions of 427m by 153m with the leading edge in each case approximately 165m from the centre of the orebody and approximately parallel to its strike. The shape of the reduced ratio and phase difference anomalies from the two types of transmitter suggest that data from either source can be used for interpretation. The difference in anomaly amplitude is due to the smaller receiver separation of 6.1m for the small loop source compared to a receiver separation of 15.2m for the large loop.

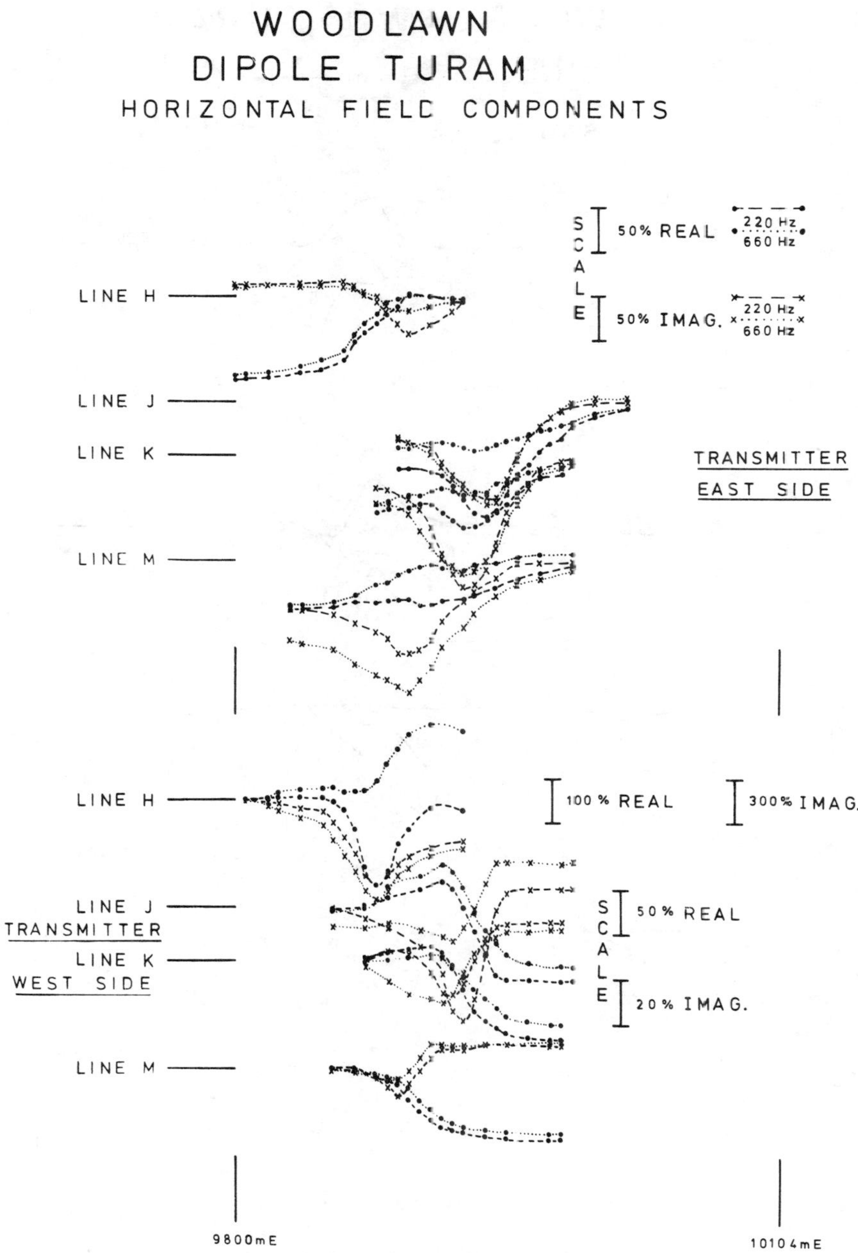

Fig. 11 Real and imaginary horizontal field components over Woodlawn using a small loop transmitter at 220 Hz and 660 Hz on both sides of the orebody.

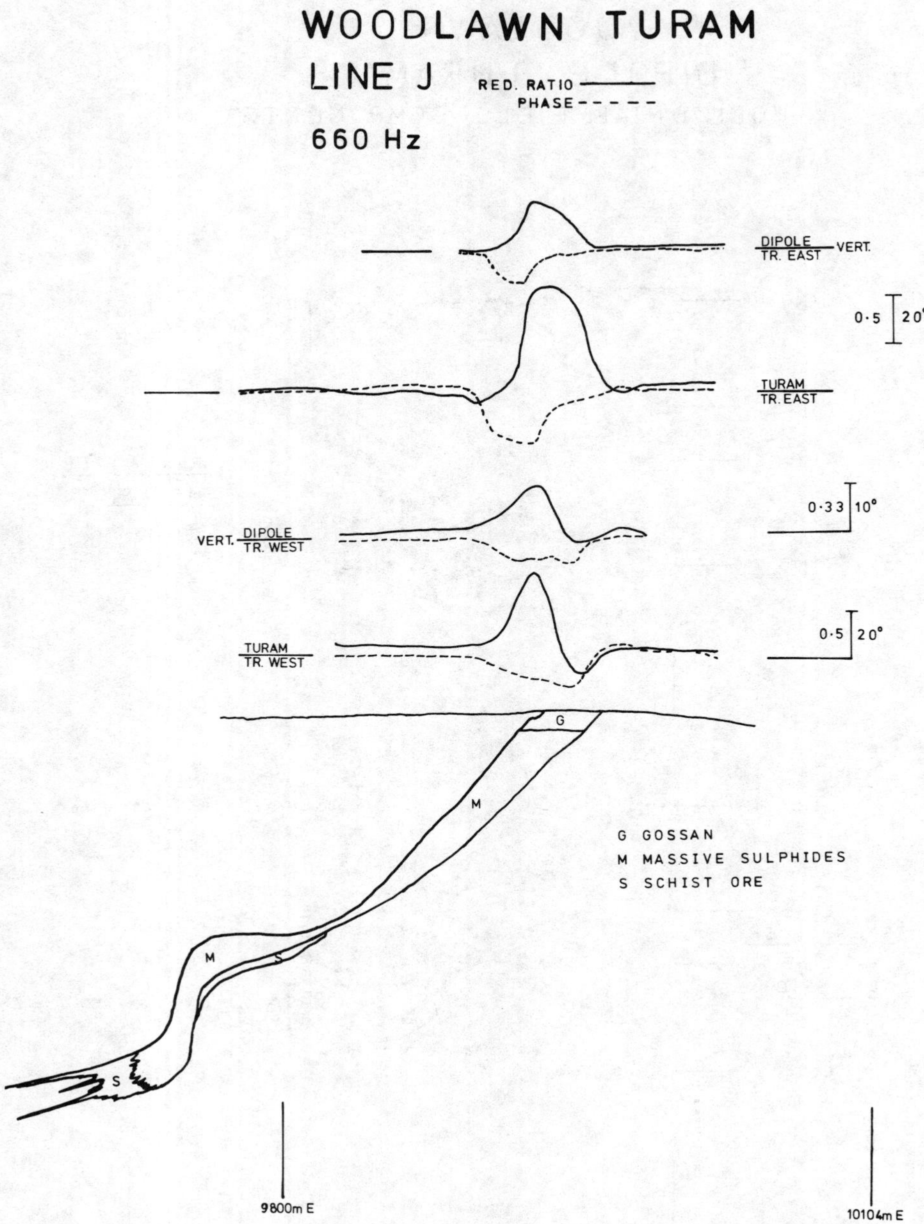

Fig. 12 Comparison of the reduced ratio and phase difference vertical field anomalies from small loop and large loop Turam surveys at 660 Hz.

It is possible to convert the 6.1m receiver separation data for the small loop data to an equivalent 15.2m receiver separation. This has been done on data from line J as shown in Fig. 13 at 220 Hz. For both positions the small loop data shows a larger response. Similar data for 220 Hz for line K is shown in Fig. 14 for transmitters to the east and west of the orebody. Along this line, the amplitudes of the responses are almost identical. It should be noted, however, that the eastern small loop transmitter was located near the back edge of the large loop, while it was at the centre of the large loop on the western side. Thus the small loop transmitter, placed in a position similar to the Lajoie and West (1976) model, would give a larger response in each of these cases.

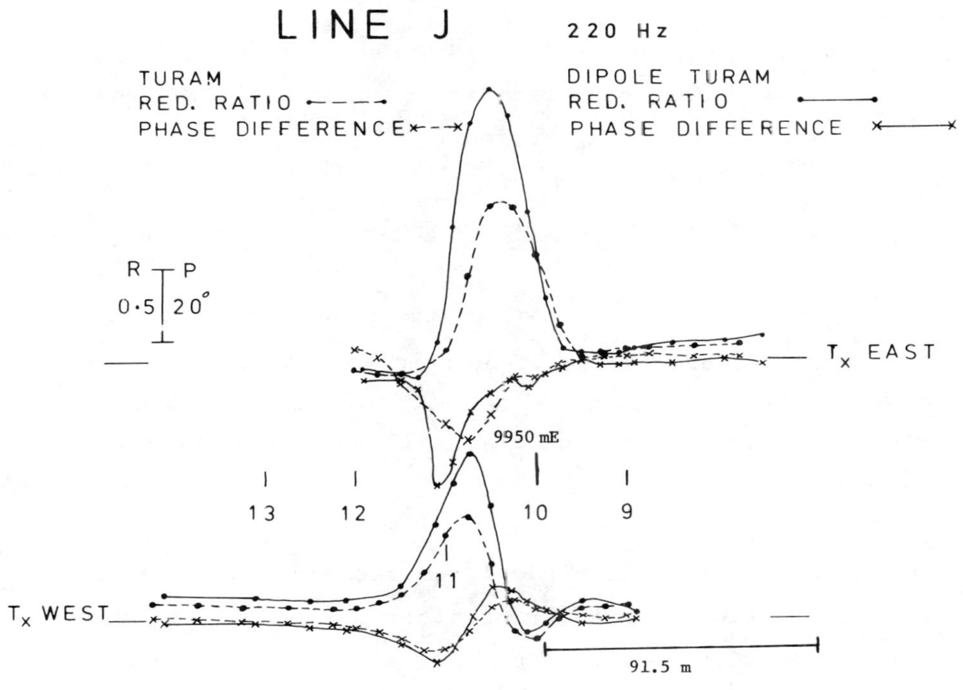

Fig. 13 Vertical field Turam data on line J, for both transmitter types at 220 Hz (equivalent receiver array separation).

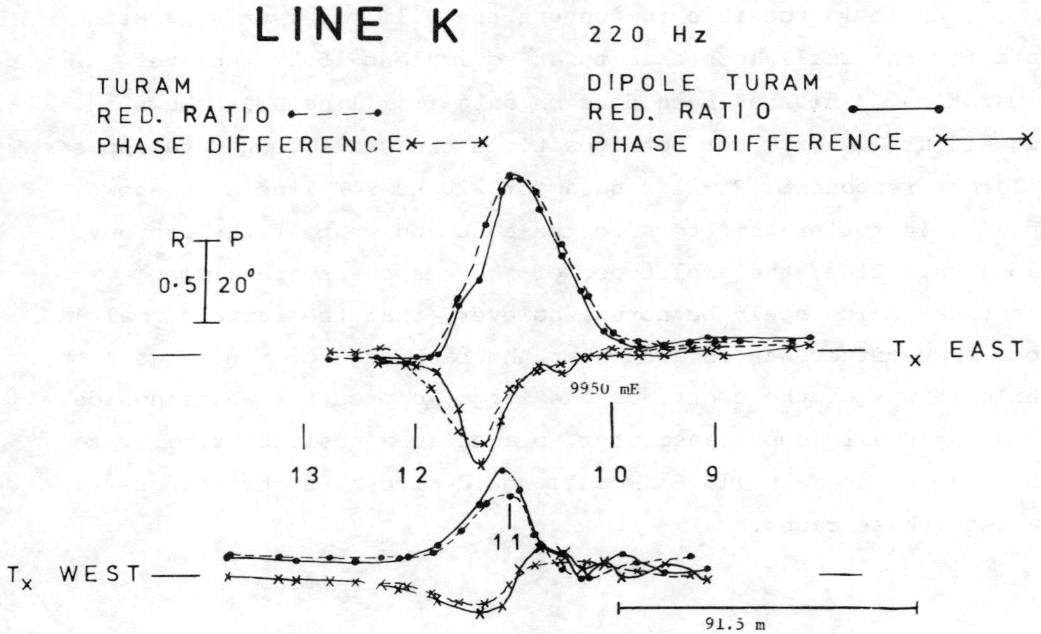

Fig. 14 Transformed vertical field Turam data on Line K, Woodlawn for both transmitter types at 220 Hz (equivalent receiver array separation).

Results at the Black Shale near Woodlawn

In the vicinity of Woodlawn the Black Shale some 2.5km to the south of the orebody (Malone, this volume, p 49) was drilled on a large self potential anomaly (Cifali and Whiteley, this volume, p 321). Results of a small loop survey are shown in Fig. 15.

A weak, but discernible Turam anomaly, is associated with the Black Shale for both transmitter positions. In this survey, the transmitter loop used was 12m a side with 26 turns and was placed approximately 300m on either side of the Black Shale. Data were collected with a 15.2m receiver coil separation. The amplitudes of the anomalies from the Black Shale are much less than from the Woodlawn orebody.

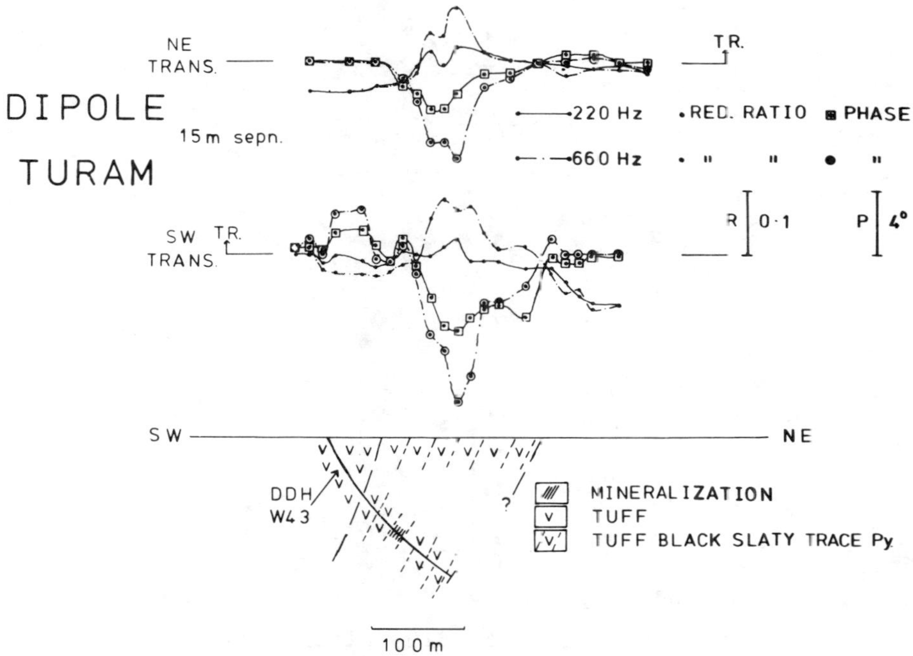

Fig. 15 Vertical field reduced ratio and phase difference anomalies for two different small loop transmitter positions at 220 Hz and 660 Hz, Black Shale.

From the real and imaginary field components shown in Fig. 16, a depth to the top of the conductive zone was estimated to be between 30 and 40m. This together with the low amplitude of the anomalies indicates that the conductive zone is a fairly shallow weak conductor.

Conclusions

The Woodlawn orebody is an excellent target for ground based electromagnetic methods of exploration. Small loop Turam surveys measuring both vertical and horizontal fields, have shown that valuable information can be obtained regarding the strike length and nature of the conductive body. The data from

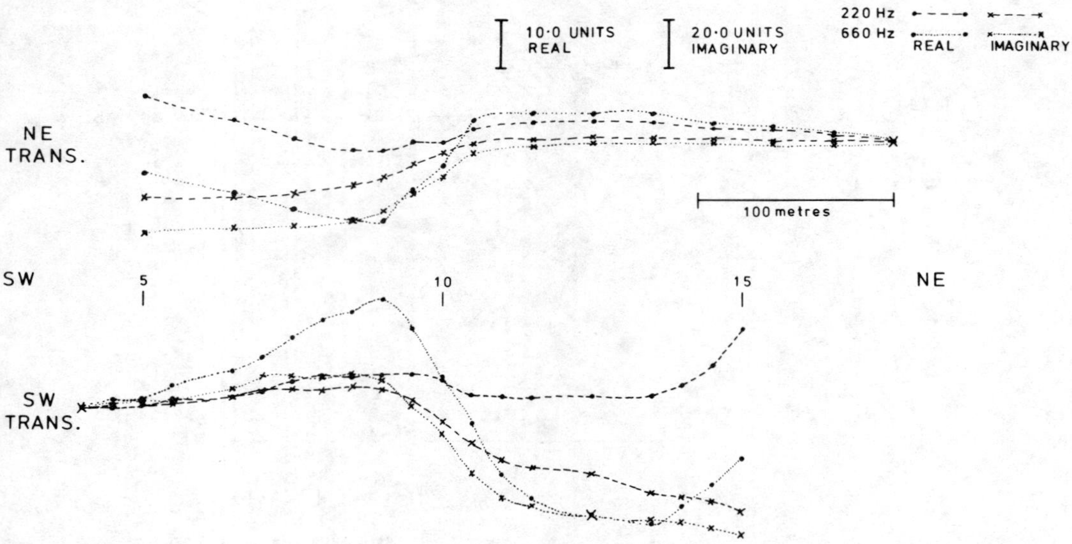

Fig. 16 Real and imaginary vertical field components for the data displayed in Fig.15, Black Shale.

positions on either side of the orebody confirm that it is dipping and of a complex arcuate shape.

Vertical field anomalies obtained with the small loop transmitters have similar amplitudes and shape to those obtained with the large loop transmitters. This similarity becomes more evident when the data are corrected so as to be from equivalent receiver separations.

For massive sulphide targets close to the surface, a small dimension, multiturn loop transmitter survey is an effective method of exploration, particularly where difficult terrain could cause problems in the deployment of large loops.

Vertical field anomalies from the Black Shale near Woodlawn are of low amplitude and suggest that the source there is shallow and of poor conductivity unlike the highly conductive Woodlawn orebody.

Acknowledgements

The Bureau of Mineral Resources and the Geological Survey of New South Wales kindly loaned the ABEM Turam system.

References

Bosschart, R.A. 1964. Analytical interpretation of fixed source electromagnetic prospecting data. Ph. D. Thesis (unpubl.), Delft, The Netherlands.

Lajoie, J.J. and West, G.F. 1976. The electromagnetic response of a conductive inhomogeneity in a layered earth. Geophysics, 41, (6A), 1133-1156.

Ward, S.H., Ryu, J., Glenn, W.E., Hohmann, G.W., Dey, A. and Smith, B.D., 1974. Electromagnetic methods in conductive terrain. Geoexploration, 12, 121-183.

Williams, J.P., 1971. Computer Programs for Turam data reduction. Bull. Aust. Soc. Explor. Geophys., 2, (1), 29-34.

PART 3

PULSE TRANSIENT METHODS

7.9. Ground Pulse Electromagnetic Survey, Woodlawn

D. CRONE

Contents

Abstract

A Crone pulse electromagnetic (PEM) survey using dual horizontal loops measured very large anomalies of up to 500 times background over the Woodlawn orebody. A conductivity-thickness product for the orebody of about 90 siemens was calculated from the magnitudes of late time samples. Early sample values indicated a higher conductivity near the footwall of the orebody. Dip calculations proved inaccurate where the profiles were not orthogonal to the strike of the orebody or were near an abrupt change in strike direction.

Introduction

The pulse electromagnetic (PEM) equipment manufactured by Crone Geophysics Ltd., was developed in cooperation with Newmont Exploration Ltd., with the aim of providing a deep penetration ground EM method for use in areas of high surficial conductivity.

The PEM method is very flexible and may be used with horizontal, vertical or Turam type survey methods, with the further advantage that exact alignment or separation of the transmitter and receiver coils is not required since the method operates in the time domain.

The primary current pulse through the transmitter loop is a double side ramp of 12.2 ms duration. An initial current on period of 10.8 ms is terminated by a 1.4 ms downward sloping ramp to the current off point. The current off period is 9.4 ms during which 8 samples of the EM transient decay curve are measured at the receiver loop.

The positions of these samples after the current off point are listed on Table 1. They provide an approximately equivalent frequency range of 2000 to 20 Hz.

Field Procedure

At Woodlawn, the horizontal loop (Slingram) configuration was used with a constant separation (a) of 61m (200 ft) between the transmitter and receiver loops.

A transmitter loop approximately 4m (13 ft) in diameter, was laid in a rough circle on the ground.

Table 1 Position of Samples on transient decay curve

Sample Number	Sample Window (usecs)	Position (usecs)
1	100 - 200	150
2	200 - 400	300
3	400 - 700	550
4	700 - 1100	900
5	1100 - 1800	1450
6	1800 - 3000	2400
7	3000 - 5000	4000
8	5000 - 7800	7200

In all, three profiles were surveyed with the PEM method, Line G (9310mN), J (9220mN) and N (9100mN) using a station spacing of 30.5m (100 ft).

The field crew consisted of one observer for the receiver and two assistants for the transmitter. A total of 48 stations were occupied during the survey which required about 2.5 hours in rainy weather.

Results and Discussion

The three lines surveyed all produced PEM anomalies of very large magnitude up to 500 times background in the vicinity of the Woodlawn orebody. These are plotted in Figs. 2 to 4 as profiles at the centre point between the transmitter and receiver loop positions. A logarithmic vertical scale has been used on these profiles to contain the large response amplitudes.

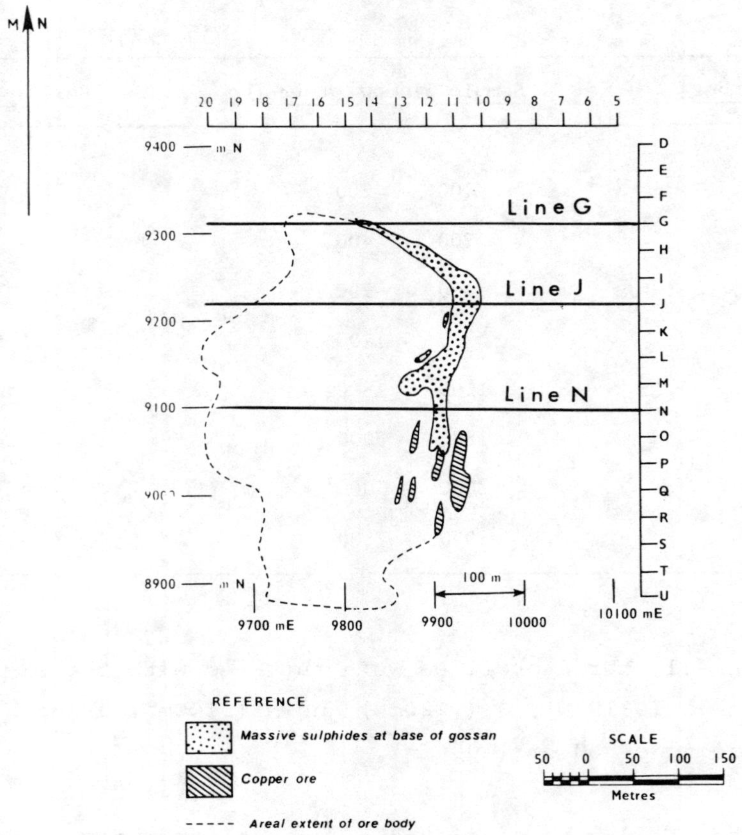

Figure 1 Location of PEM Lines

Standard horizontal loop theoretical curves may be used for interpretation of these profiles (e.g. , Ketola and Puranen, 1967) but these must be inverted.

On line N, 9100mN, (Fig. 2), transient decay sample 1 (high frequency) indicates a conductive source with a horizontal width of about 43m (140 ft) centred at N12. Samples 3, 4 and 5 (middle frequency samples) show a zone of higher conductivity about 12m (40 ft) in horizontal width occurring on the footwall side of the orebody. The higher conductivity is probably due to the footwall copper zone which could be expected to be more

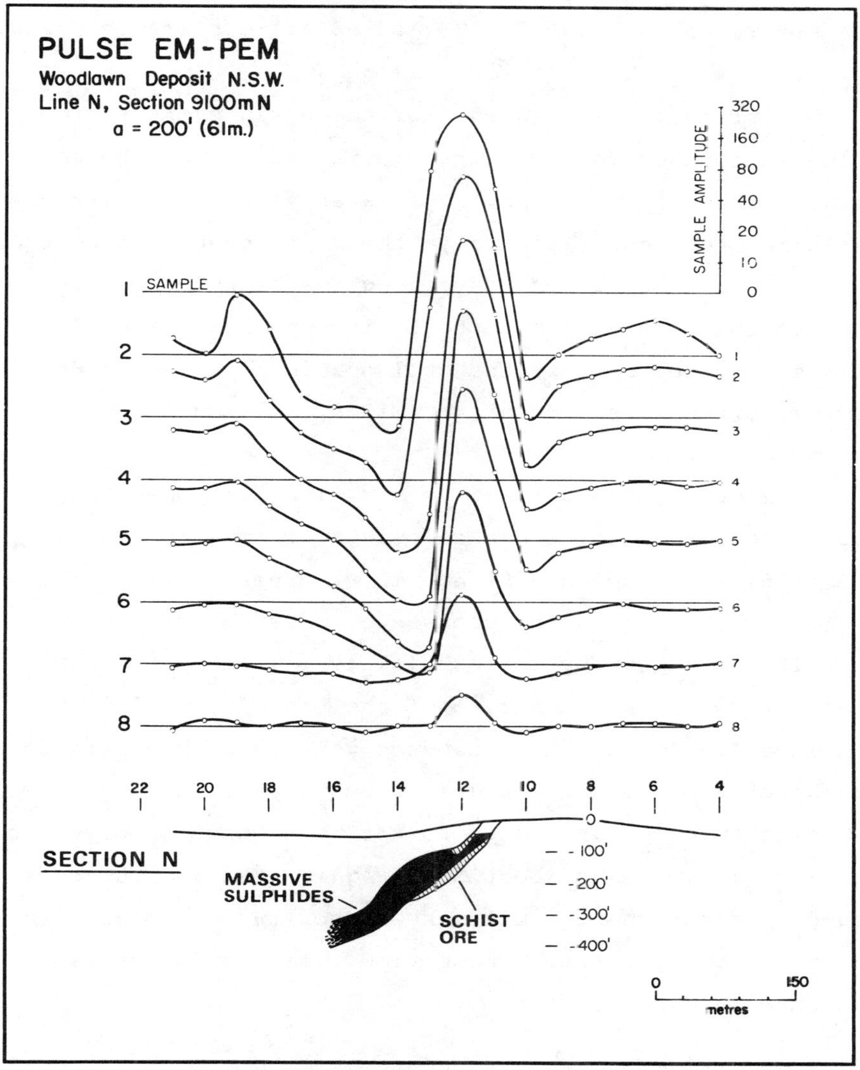

Figure 2 PEM Profile Line N (9100mN)

conductive than the lead and zinc sulphides. Conductivity-thickness products were calculated from all 8 samples at station N14. Starting from sample pairs 1 and 2 to 7 and 8, these products are 9, 25, 35, 55, 88, 93 and 88 siemens respectively.

The products calculated from the early samples are always low since they include currents flowing along the edges as well as within the ore zone. The later values of 88 to 93 siemens reflect currents flowing in the most conductive portion of the ore zone. The dip of the conductive orebody was predicted to be 30° to the west with a depth to its upper surface of approximately 10 per cent of the coil spacing i.e. about 6m (20 ft). A narrow weaker conductor occurs at N18.5.

On line J, 9220mN (Fig. 3) a similar response was obtained. The horizontal width calculated from sample 1 is about 55m (180 ft) between J10 and J11.8. This is wider than the 35m (115 ft) shown on the cross-section from drilling results. This discrepancy is most likely caused by the sharp northwesterly swing in the strike of the body 30.5m (100 ft) north of line J. Again a more conductive zone with a horizontal width of about 15m (50 ft) appears to be present on the footwall side. The calculated dip is 30° to the west which appears too low, the actual dip being about 50°. Again, the inaccuracy is believed due to the change in strike direction influencing the shape of the anomaly curve. The narrow, weak conductor is again detected at J18.

The response on line G, 9310mN (Fig. 4) illustrates the difficulty of interpreting from profiles which do not cross the strike at right angles. These responses indicate a dipping conductor whereas the geological section shows the orebody to be almost vertical near the surface. In addition, the direction of dip as indicated by these responses is opposite to that indicated on the other lines. The shape of the responses on this line is

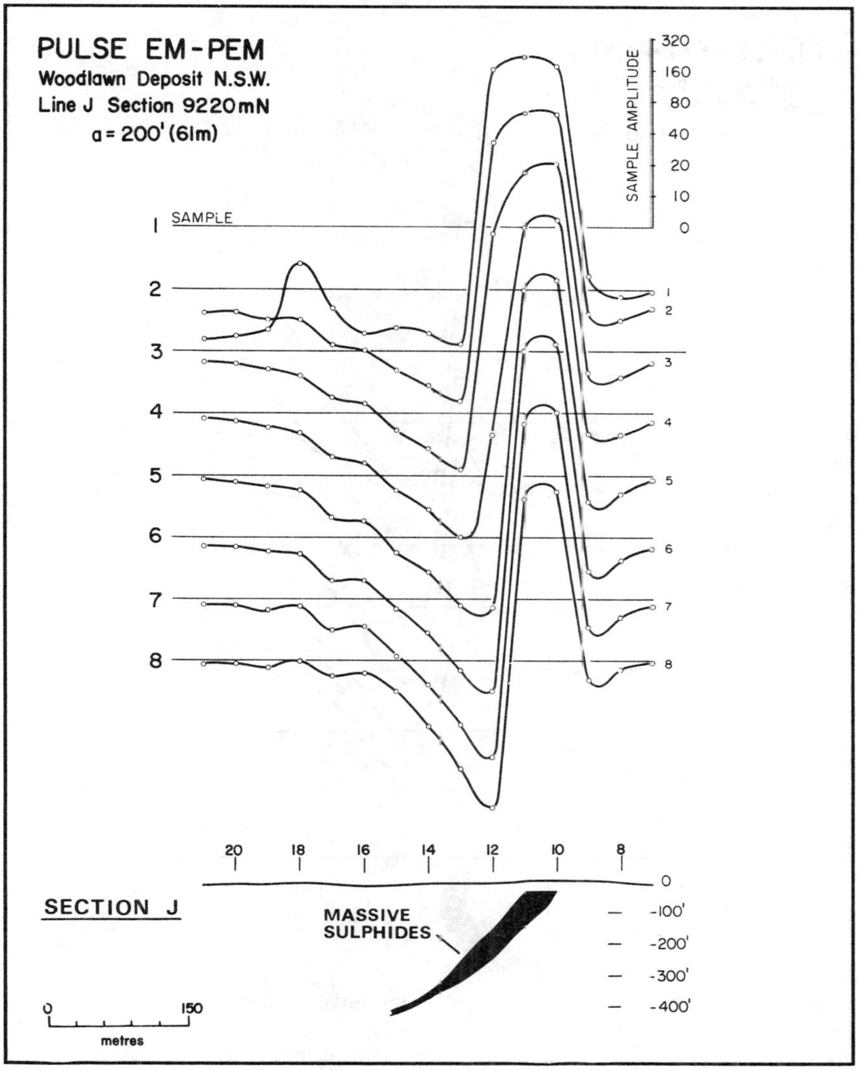

Figure 3 PEM Profile Line J (9220mN)

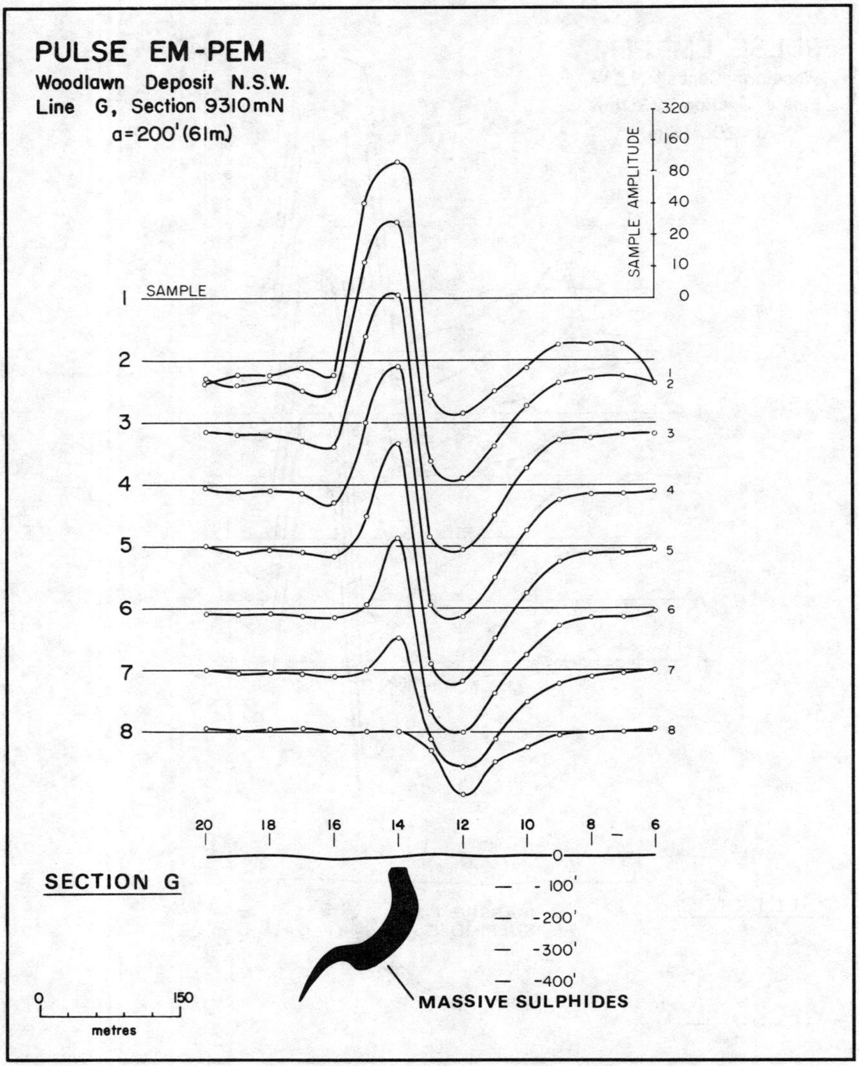

Figure 4 PEM Profile Line G (9310mN)

largely due to the orebody to the south of the traverse line rather than the shape of the orebody shown on the section. Also, the sample 8 response is completely negative indicating that the low frequency induced current paths do not cross the traverse line which would imply that the orebody terminates within 30m to the north of this traverse.

Conclusions

The Crone PEM method measured very large anomalies over the Woodlawn orebody indicating its very high conductivity. A conductivity-thickness product for the orebody, based on later time samples, was calculated at about 90 siemens. Samples at earlier times indicate a localized region of higher conductivity near the footwall of the orebody. Dips calculated from the responses were inaccurate when the profile was not orthogonal to the strike of the orebody or was near an abrupt changed strike direction.

Reference

Ketola, M. and Puranen, M. 1967. Type Curves for the Interpretation of Slingram (Horizontal Loop) Anomalies over Tabular Bodies. Geol. Survey of Finland

7.10. Electromagnetic Pulse Survey of the Woodlawn Orebody

G. O. DICKSON

Contents

Abstract

An electromagnetic pulse system which measures the secondary field decay in three mutually perpendicular directions is described, along with the test results obtained over the Woodlawn orebody. The results demonstrate the interpretational advantages of measuring three components of the secondary field.

Introduction

The electromagnetic pulse (EMP) concept was first suggested to Newmont Exploration Limited some twenty five years ago. An initial field system was constructed and used for one survey in Cyprus by Dolan (1969). Although results obtained with this equipment looked promising, it proved to be too cumbersome for general use. Several different designs of the original equipment were produced some ten years ago for use in drill holes. These were tested in Canada and Africa. However, it has only been in the last few years that advances in electronic technology have made it possible to construct a practical field system for exploration in the electrically conductive surface conditions commonly found in Australia and other parts of the world.

In the EMP method, (Dickson and Boyd, 1980) a rapid series of current pulses is transmitted through a large loop of wire laid out on the ground surface in a Turam type arrangement. Abrupt interruption of the primary field at the end of each current pulse tends to set up eddy currents in the ground. These in turn give rise to secondary magnetic fields, which can be detected as a voltage induced in a sensitive receiver coil. Ohmic losses in the ground cause the eddy currents to decay with time, and hence a transient response is observed at the receiver. If the receiving coil is sufficiently close to a conductor, the variation in the shape and amplitude of the decaying secondary fields caused by the conductor can also be detected.

In general, eddy currents in a conducting body will decay more slowly than eddy currents in a more resistive body.

Bulgakov (1971) and Lee and Lewis (1974) have derived expressions for the electromotive force (emf) induced in a coil over a half space. The response in the early and late times can be approximated by simple power law expressions. Although the results of Bulgakov and Lee and Lewis concern the single loop transient EM system, they should be identical to those for a dipole receiver in the case of a half space (Wait, 1972). Spies (1976) has used these and other early time expressions to satisfactorily interpret transient EM data from various parts of Australia.

Wait (1951), Nabighian (1970) and Bulgakov (1971) have demonstrated that transient response at later times can be represented as an exponential function of the form

$$\frac{e(t)}{I} = Ae^{-(t/\tau)}$$

where τ is a decay constant.

The amplitude of the transient response depends on the primary field strength and on the mutual coupling between the transmitter loop and the receiver coil with respect to the conductive body. The rate of transient decay depends on the size, conductivity and shape of the conductive body.

Exploration interest lies more with the exponential behaviour of the decay curves. Highly conductive overburdens are common in many parts of Australia and hence, to define deeper economic targets, measurements need to be taken over a sufficiently long time. The early time response, which is usually strong, contains information on surficial conditions and the late time response information from deeper conductors.

The relatively slow development of the EMP method has been caused by the need to have a field system which can adequately handle input data with a large dynamic range and maintain sensitivity at the late times. As conductivity and size are inseparable parameters in the decay constant, it is also desirable that the system provides as much information about the geometry of the conductive body as possible.

Newmont introduced the EMP system into Australia in August, 1976, when it underwent field tests over a number of sulphide deposits. The test traverses were completed over the Woodlawn deposit during September 1976.

Equipment, Field Procedures and Data Processing

The EMP system consists of three fundamental and integral parts, field equipment, transcriber and computer software.

The field equipment collects and digitally records the EMP data, the transcriber unit provides the means of interfacing these data to a computer, and the computer software processes and presents the EMP data for interpretation.

Within the field equipment the EMP transmitter is simply a switch which repetitively interrupts a voltage supply across the transmitter loop thereby passing a rapid series of current pulses through the loop. The on time pulse width may be varied between 20 and 160 ms. The off time between consecutive current pulses is 750 ms. This choice of timing cycle is designed to minimize battery drain while keeping reading time at each station reasonably short. The voltage power supply to this loop is a bank of 12 volt vehicle batteries. The amplitude of the current transmitted depends on the number of batteries and on the resistance of the loop.

The transmitter loop consists of a single uninsulated turn of wire laid out on the ground in the form of a square or rectangle. The size of the loop may be varied, up to a maximum size of 800m by 400m.

The EMP receiver records the secondary magnetic field as a decaying voltage induced in a receiver coil. The actual sensor is an air cored coil, designed to be very sensitive to small signals and yet practical for one man to carry. This coil allows measurements to be made in both horizontal and vertical directions.

The amplitude of the transient decay curve is measured at a number of discreet sample points after the cut off of the primary field. The range of these sample points is selectable, but can be extended to approximately 80 ms.

Readings at each sample point are recorded digitally on both a printer for visual inspection in the field and on magnetic tape for later computer processing. Each reading represents the stacked average of 14 actual decays. In addition to the measured data, certain key information, such as various instrumental settings, is also recorded on tape for use in processing.

The third essential piece of EMP equipment is the transcriber unit. This unit plays back field tapes and transcribes the data onto a computer compatible tape. The transcriber is, therefore, the vital link between data collection and processing.

EMP field equipment is normally operated with a two man field crew, consisting of a trained operator and one assistant.

The rectangular transmitter loop is usually oriented with its long side parallel to the expected strike of the target, and traverses are run with the receiver along lines perpendicular to strike. These traverses may be run inside or outside the loop, depending on the nature of the target. The spacing between successive traverses, and the receiver station interval, are varied according to the expected size and depth of the target.

In open terrain, wire for the transmitter loop is laid from the rear of a vehicle, but in rough terrain it must be laid manually.

If the EMP survey involves several closely spaced loop positions the wire is simply pulled from one loop position to the next.

All three components of the transient electromagnetic field may be measured at each station. Three component measurement is achieved by orienting the receiver coil in one vertical and two perpendicular horizontal directions.

Because of the large volume of data collected with the EMP system, it is usually impossible to say, in the field, whether an anomaly has been detected. Several thousand numbers are recorded on magnetic tape during a normal daily operation. These require computer processing before display and evaluation.

Processing involves several separate stages and includes editing, sorting, stacking, reducing to physical units, storing, retrieving, displaying and interpreting. Considerable effort has been expended in the development of efficient storage and retrieval software for EMP data.

Traverse Locations

Three lines were surveyed at Woodlawn, two over the actual orebody and a third to the north of the orebody. The lines surveyed relate to the Woodlawn Test Grid as follows:

9400mN (approx. Line D) from 9700mE to 10150mE
9300mN (approx. Line G) " 9700mE " 10150mE
9200mN (approx. Line K) " 9700mE " 10150mE

The position of the lines relative to the Woodlawn orebody are shown in Fig. 1.

Discussion of Results

There are many differences between the EMP system and the single loop system (Buselli, this volume, p 297 ; Spies et al., this volume, p 281). One of the major differences is the ability to measure all three components of the secondary field decay. The collection of three dimensional field data is of considerable advantage in the interpretation of the geometry of the conducting body.

At Woodlawn, sixteen samples of the secondary field decay were measured between 2.82 ms and 37.50 ms, for all three components of the field. For this survey, positive X was taken as grid north, positive Y as grid east and positive Z vertically downwards. The leading edge of the loop was oriented grid north

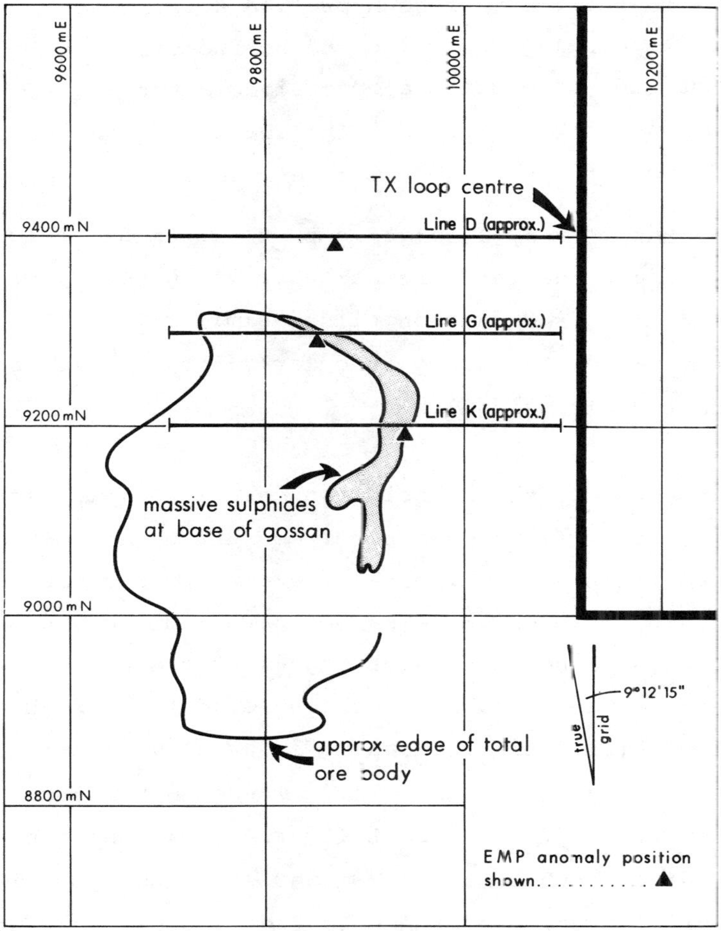

Figure 1 Location of EMP traverses over the Woodlawn orebody with the position of EMP anomalies shown.

along 10120mE, between 9000mN and 9800mN (Fig. 1). The size of the loop was nominally 800m by 400m.

EMP profiles for all three components on each traverse surveyed are shown in Fig. 2* with values plotted at the receiver position. For clarity, only every other channel has been presented between 3.44 ms and 37.50 ms, after primary field

*The folded Fig. 2 can be found in the envelope accompanying the book.

cut off. The sample time for each profile appears on the left hand side of the display, opposite the horizontal base line. On the right hand side, the scale compression factor and the microvolt level of the base line are shown. The scale is shown at the top of the display and the station values at the bottom.

In the EMP method, a large loop of Turam dimensions is generally used, and the interpretation of the geometry of the conducting body can proceed along lines similar to those outlined by Bosschart (1964). This is particularly the case if the body happens to be two dimensional.

On line 9200mN, the Z field response is asymmetric with the typical vertical field crossover occurring at about 9940mE. At the same position the horizontal field responses are noted to be symmetric, the X response being negative and the Y positive. On this particular line, excessive signal strength has caused amplifier overload and a subsequent loss of data near the anomaly centre (amplifier overloading is usually eliminated by reducing the primary field strength). These data may be interpreted in terms of a horizontal current element centred under 9940mE at a depth of about 20m and striking approximately 10° east of grid north. Reference to Fig. 1 shows that this interpretation agrees well with the strike of the orebody at this point.

A marked change in response is noted on line 9300mN. The vertical field crossover is centred at about 9850mE. The horizontal fields are symmetric, but in this case both X and Y components are positive. The tendency towards asymmetry in the H fields in the later times is due to the dip of the body. Again the data may be satisfactorily interpreted in terms of a horizontal current element centred under the crossover at a depth of approximately 25m and striking about 70° west of grid north. This interpretation also agrees well with the known geometry of the orebody. With the exception of the X field, the secondary field amplitudes are generally lower than those on line 9200mN.

The results from line 9400mN are probably the most interesting. This line is located about 80m north of any known mineralization. The secondary field strength of this line is much lower than obtained on the lines to the south. This is particularly true of the vertical field. The shape of the profiles is also decidedly different. In particular, both the vertical field and the Y field profiles are symmetric with an asymmetric X field profile. The anomalous secondary field observed on line 9400mN could be due to a current line plunging fairly steeply towards grid west, or grid northwest. The current line is centred approximately 80m grid south of 9870mE on line 9400mN. Again, this interpretation would fit the known orebody geometry fairly well.

For comparison with other electromagnetic methods, the secondary field decay has been plotted for station 9875mE on line 9300mN on a semilogarithmic graph. Clearly the decay is linear between 7 ms and 19 ms. Below 7 ms the response tends to be linear on a double logarithmic representation and may, therefore, be more appropriately represented as a power law decay. It is possible to fit a second linear section on a semilog representation beyond 19 ms to about 27.5 ms. The decay constant from 7 ms to 19 ms is 4.1 ms, and from 7 ms to 32.5 ms it is 5.3 ms. Both these values are in general agreement with the results given by Hone and Young (this volume, p 519).

Since the Woodlawn survey, considerable theoretical modelling has been attempted with the aim of being able to estimate the quality of the conductor from the EMP data. Although more work has yet to be done, the modelling indicates that the conductivity-thickness product of the Woodlawn orebody lies between 170 and 200 Siemens. An assumption of an average thickness

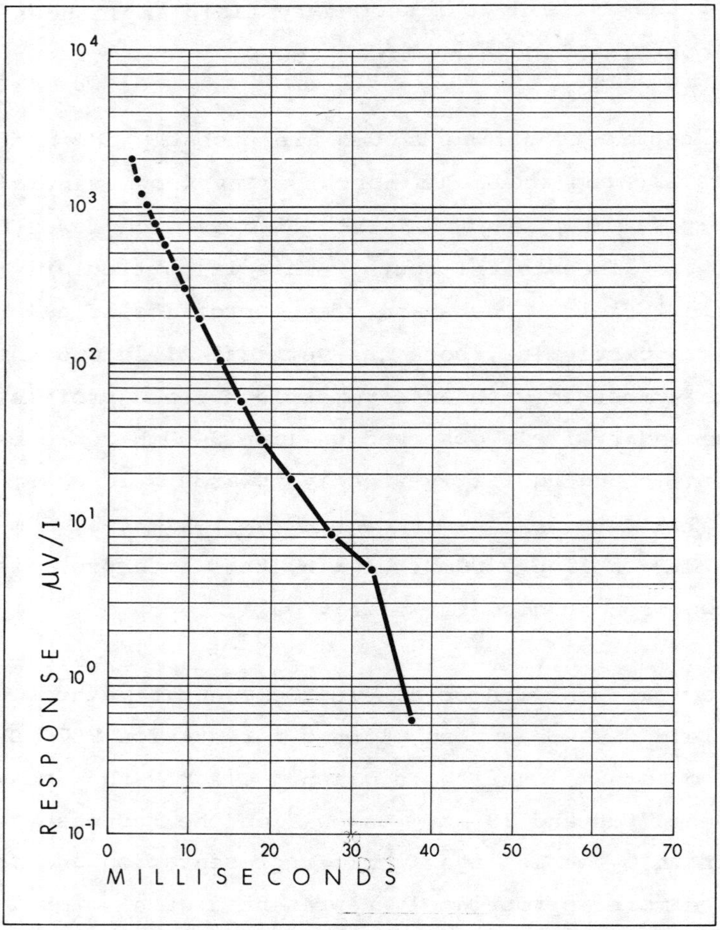

Figure 3 Time decay plot of the response for Station 9300mN, 9875mE.

of 10m for the massive conductive orebody gives an estimate for the conductivity of the orebody between 17 and 20 Siemens/m. Again this would appear to be in reasonable agreement with the results given by Hone and Young (this volume, p 519).

Summary and Conclusions

The results of the EMP survey at Woodlawn show the advantages of measuring all three components of the secondary field decay. The additional data obtained assist with interpretation of the geometry of the orebody, and hence lead to a clearer definition of drilling targets.

The EMP system is sufficiently sensitive to detect and define the attitude of the ore zone at Woodlawn, some 80m laterally removed from it. Signal levels at this point are such that detection would certainly be possible beyond 100m.

Digital recording in the field, hardware interfacing to a computer, along with all the necessary software to take the data through editing to the display and interpretation stages, provide all the sophistication necessary to handle the large data sets created with the EMP system and yet keep the field procedures simple.

The ability to estimate the conductivity of the orebody from theoretical modelling of EMP data is an important advantage since the rate of transient decay is not independent of the size of the conductor. Such interpretations allow targets to be graded on their conductivity independent of their size.

References

Bosschart, R.A., 1964, Analytical interpretation of fixed source electromagnetic prospecting data: Ph.D. Thesis, University of Delft, Holland.

Bulgakov, Yu.I., 1971, Interpreting surveyed results by the single loop transient process method. Paper 12 in papers on the transient method in the search for sulphide orebodies: 'NEDRA', Leningrad.

Dickson, G.O., and Boyd, G.W., 1980, Newent transient electromagnetic system, Bull. Aust. Soc. Expl. Geop. V.11, No. 1/2, p. 47-51.

Dolan, W.M., 1969, Mining and groundwater geophysics: Geol. Survey of Canada, Ottawa, p.336.

Lee, T., and Lewis, E., 1974, Transient EM response of a large loop on a layered ground: Geophysical Prospecting, v. 22 no.3, p.430-444.

Nabighian, M.H., 1970, Quasi-static transient response of a conducting permeable sphere in a dipolar field: Geophysics, v.32 no.2, p.303-309.

Spies, B.R., 1976, The transient electromagnetic method in Australia: B.M.R. Journal of Australian Geology & Geophysics, v.1, p.23-32.

Wait, J.R., 1951, A conducting sphere in a time varying magnetic field: Geophysics, v.16 no.4, p.666.

Wait, J.R., 1972, On the theory of transient electromagnetic sounding over a stratified earth: Canadian Journal of Physics, v.50 p.1055-1061.

7.11. *Transient Electromagnetic Test Surveys and Scale Model Studies of the Woodlawn Orebody with MPPO 1 Equipment*

B. R. SPIES, I. G. HONE AND J. W. WILLIAMS

CONTENTS

Abstract

Extensive transient electromagnetic (TEM) surveys were completed at Woodlawn with Russian built, single loop MPPO 1 equipment. Prior to the survey, major modifications to the equipment were required to improve stability.

Although the accuracy of measurements at late sample times was limited by electrical noise in the area, a regional survey with 100m and 200m square loops delineated a large TEM anomaly over the Woodlawn orebody at early sample times to 2.3ms. Detailed investigations of this anomaly using loop sizes from 25m to 200m indicated the westerly dip of the orebody and its closeness to the surface.

Apparent resistivities calculated from the TEM data gave values of less than 10 ohm m near the centre of the orebody and a background of 100 to 500 ohm m.

Analogue modelling of the TEM response of the orebody showed that the effective exploration depth of MPPO 1 equipment at Woodlawn is about 60m.

TEM is an efficient exploration tool in the search for buried massive sulphide deposits in a geological environment similar to that at Woodlawn.

Introduction

During February and April, 1973 the Bureau of Mineral Resources (BMR) conducted a transient electromagnetic (TEM) test survey over and to the south of the Woodlawn orebody using the Russian built, single loop, MPPO1 equipment. Young (1976) has described the survey, which was carried out to investigate the application of the TEM method to search for volcanogenic type massive sulphide deposits such as Woodlawn.

Physical Properties of Orebody & Host Rocks

The physical properties of the Woodlawn orebody are typical of volcanogenic massive sulphide deposits. Drill hole and laboratory measurements show strong contrasts in conductivity, density and chargeability, between the orebody and the unmineralized surrounding rocks (Hone and Young, this volume, p 519).

Unmineralized rocks have resistivities from 10^3 to 10^4 ohm m and the massive sulphide zone has a resistivity of about 0.1 ohm m.

The TEM Method

The TEM method measures the rate of decay of eddy currents induced into subsurface conductors. These eddy currents are produced by the collapse of a magnetic field created by a rectangular current pulse in an insulated loop of wire on the ground.

Equipment

MPPO1 equipment uses the same loop of wire for creating the magnetic field and for measuring the decay of eddy currents in the interval between successive current pulses. The decaying eddy currents produce a time dependent emf (e(t)) in the loop. This is measured by the MPPO1 receiver at various sample times (decay times) between 0.5ms and 16ms. To improve accuracy, measurements are averaged over several hundred cycles. The MPPO1 equipment is described in Appendix 1.

Field operation

Various loops sizes may be used in field operations but square loops with side lengths from 50 to 200m are commonly used. Increasing the loop size increases the depth of investigation but reduces resolution and discriminates against small sources. Loop overlaps are commonly used to improve the definition of anomalies. It is desirable that current in the loop should be as large as possible to increase the signal to noise ratio.

Noise

The MPPO1 receiver responds to a broad spectrum of electromagnetic radiation. Noise created by lightning, power lines, and other sources will be accepted and added to the signal produced by the induced eddy currents. If the transient signal is small, such as at late sample times or when conductors are absent, noise may cause erroneous readings. When noise is a problem its influence can be diminished by using larger currents or smaller loops.

At Woodlawn, the signal to noise ratio was acceptable at stations over the orebody at all the latest sample times employed. However, at stations away from the orebody, the signal to noise ratio was acceptable only at early sample times.

Presentation of results

The results of MPP01 surveys can be presented as decay curves or as profiles or contours of the response at different sample times. In practice, results are normalized by plotting e(t)/I, where I is the current in the loop.

TEM data can be converted to apparent resistivity values (or apparent conductivity values) by applying a homogeneous half space formulation given by Lee and Lewis (1974).

Interpretation of results

In general, eddy currents in good conductors decay more slowly than those in poor conductors, and a comparison of e(t) at different sample times can be used as a guide to the conductivity and dimensions of the sources of a TEM anomaly.

A more analytical method of interpretation is provided by studying the time constants, τ , of the decay curve. The transient decay curve is generally not of simple exponential form, but is a sum of exponential decays with a time constant $\tau(t)$ which varies with time. At late sample times the time constant of the transient decay often approaches a limit, τ_0 which can be shown to be a function of the conductivity and cross section of the buried conductor. τ_0 may be regarded as similar to the conductivity thickness product used as a parameter in the interpretation of many EM surveys. Veliken and Bulgakov (1967) suggest that economic massive sulphide bodies frequently produce a τ_0 of 10ms or more.

Wait (1956), Veliken and Bulgakov (1967), Negi and Verma (1972), and Morrison, Phillips and O'Brien (1969) have shown that the early part of the transient signal are equivalent to frequency domain measurements at high frequencies and small skin depths. The late part of the transient signal are equivalent to lower frequency measurements and greater skin depths.

Results

A regional TEM survey was carried out at Woodlawn over an area bounded by 9700mN, 7900mN, 9000mE and 10300mE. Square loops with 100m and 200m sides were used. In areas where significant anomalies were recorded, 50% and 25% loop overlaps were used for adjacent readings made with the 100m and 200m loops respectively. The survey covered an area of $2.5km^2$ at a cost which compared favourably with the cost of conventional EM or induced polarization (IP) surveys. The survey was executed most efficiently with a crew of four. Figs. 1, 2 and 3 show contoured TEM responses using 100m loops at sample times of 1.1ms and 2.3ms, and 200m loops at a sample time of 2.3ms.

The anomaly produced by the orebody was investigated in detail by surveys along traverse 9200mN. (Line K, approx). Data were recorded using loop sizes of 200, 100, 50, 25 and 10m. These results, normalized for loop size and current, are shown as profiles in Fig.6.

Regional Survey

The prominent features of the regional TEM survey are the very large TEM response located near the centre of the massive orebody, and the very small responses recorded south of 8900mN. A weak response was recorded over the volcanics and shales surrounding the orebody, and a small localized response is evident near 9250mN/10100mE over black shales. The weak localized response north and east of 9150mN/9400mE is associated with abnormally noisy TEM readings but is supported by the presence of a gossan and small SP and IP anomalies in this area. (Cifali and Whiteley, this volume, volume, p 321 ; Quick and Cifali, this volume, p 333).

Fig. 4 shows decay curves and time constants from stations in the markedly anomalous zone over the orebody (curves A to D) and from station 9550mN/10200mE in the weakly anomalous zone 350m northeast of the orebody (curve E). Curves A to D show larger decay constants than curve E, indicating that the transient signal lasts longer at the four stations over the orebody than at 9550mN/10200mE. Values of approximately 10ms were obtained from decay curves at stations close to the orebody.

Comparison with gradient array resistivity results

A comparison of the apparent resistivity data derived from the TEM survey with those obtained by employing a conventional gradient array survey (Quick and Cifali, this volume p 333) using a current dipole of 1830m and a potential dipole of 30m is shown in Fig. 5. The magnitude of the background resistivity is similar in both surveys. However, the TEM results show an intense low of less than 10 ohm m located over the massive ore in the northeast sector of the

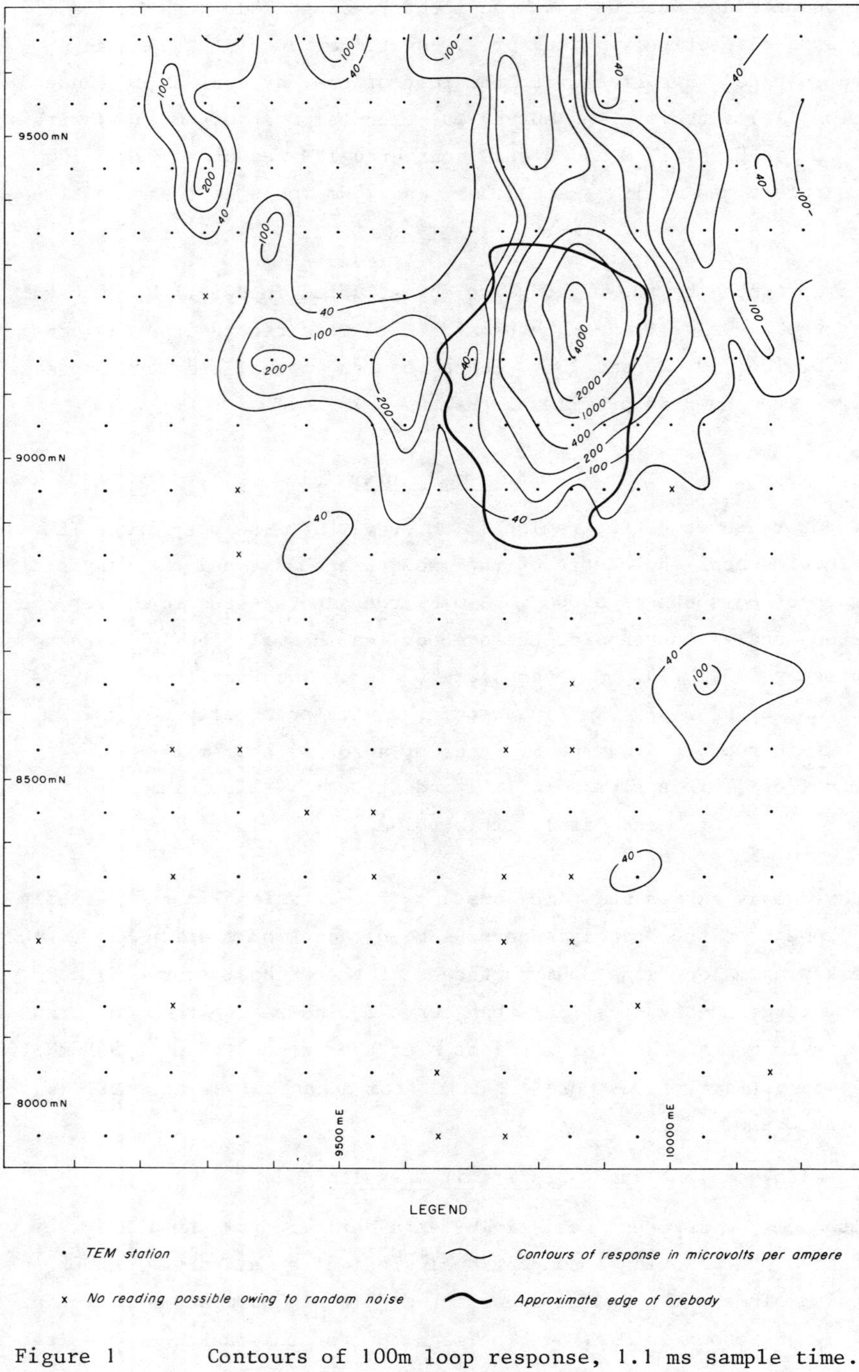

Figure 1 Contours of 100m loop response, 1.1 ms sample time.

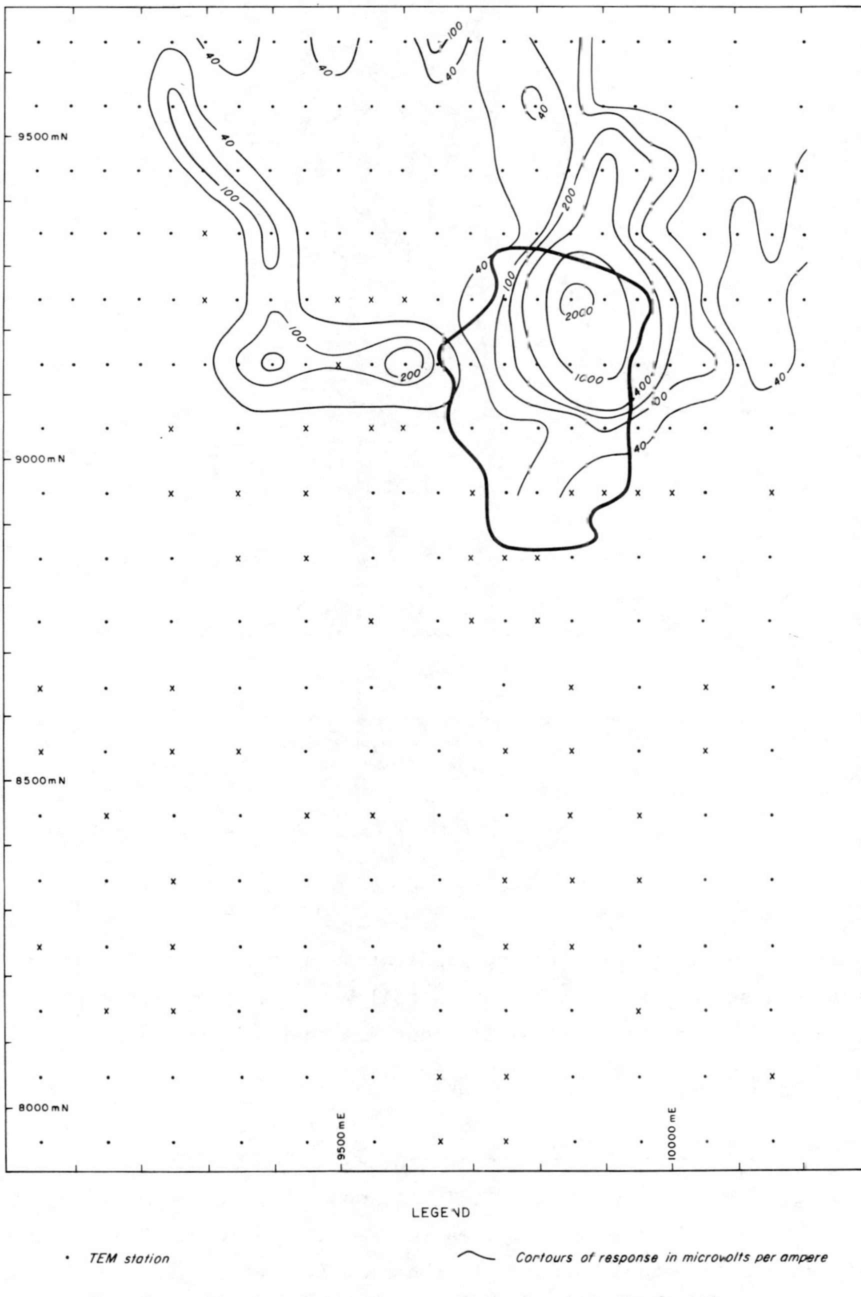

Figure 2 Contours of 100m loop response, 2.3 ms sample time.

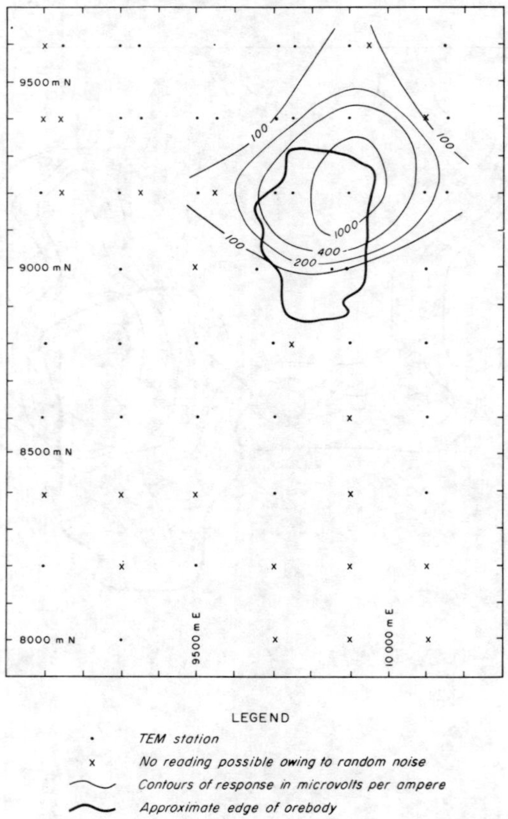

Figure 3 Contours of 200m loop response, 2.3 ms sample time.

orebody. This is in contrast with the gradient array survey results which show a less intense low of 100 ohm m which follows the surface expression of the orebody. This comparison of results confirms that EM methods are more sensitive to good conductors than are resistivity methods.

Detailed Survey, Line 9220mN (Line J)

The results for this line presented in Fig. 6, show that the anomaly centre migrates westwards as loop size is increased from 25m to 100m. Insufficient data were acquired to establish reliably the anomaly centre with 200m loops. The migration of the anomaly to the west with increasing loop size probably indicates the dip of the orebody because deeper parts of the orebody contribute more to the received signal as the loop size increases. However, the migration is in part due to the arcuate planar form of the orebody. As loop size is

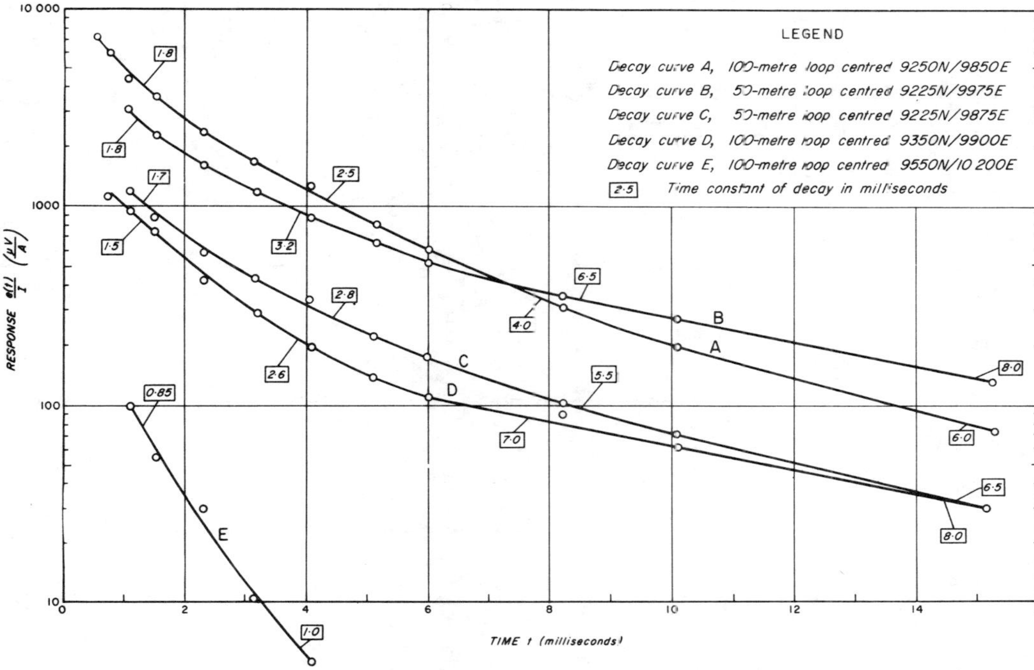

Figure 4 Selected transient decay curves.

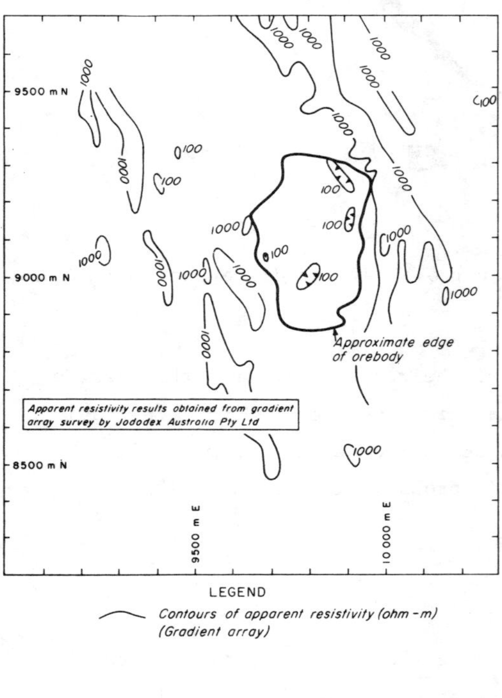

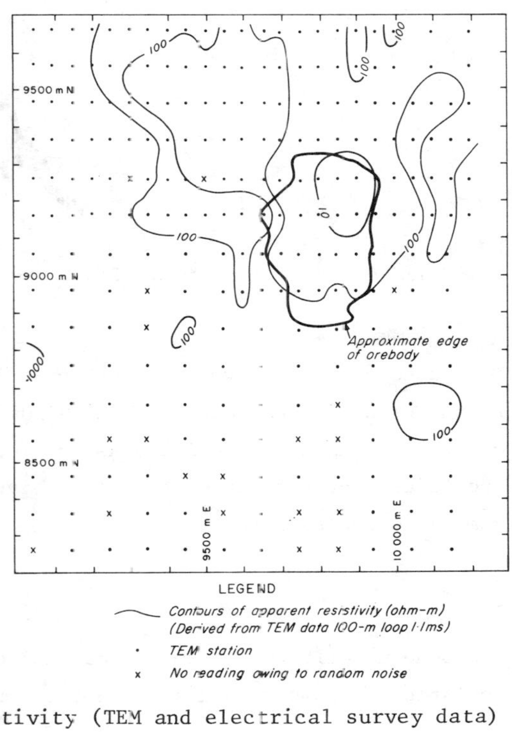

Figure 5 Contours of apparent resistivity (TEM and electrical survey data)

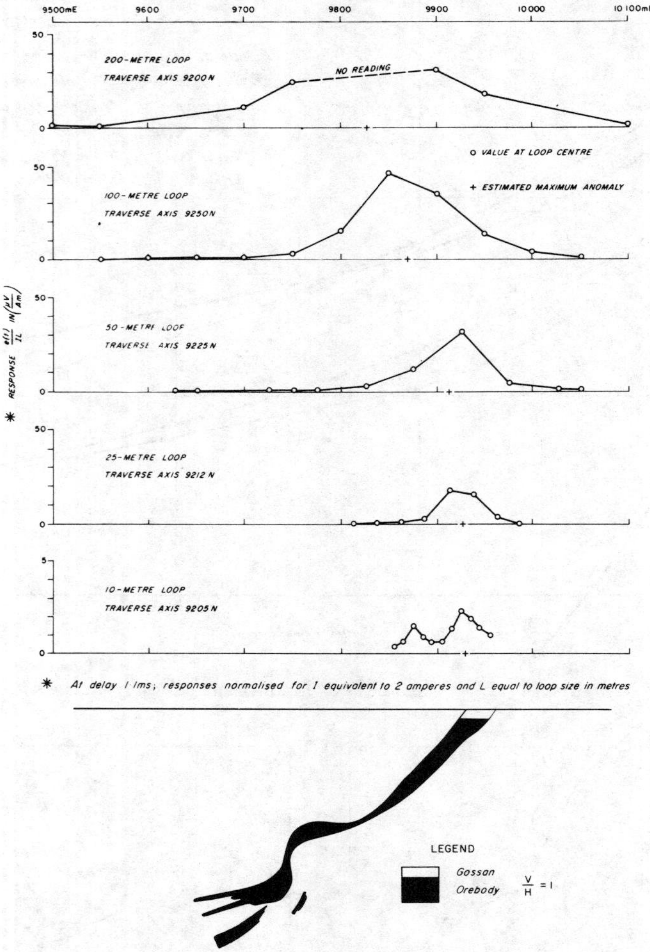

Figure 6 Profiles of TEM response over orebody for different loop sizes.

increased a progressively larger component of the response is from the massive sulphides north and south of Line 9220mN and mainly west of 9900mE. The rapid decrease in anomaly amplitude as the loop size is reduced from 25m to 10m indicates that the top of the conductive body probably lies between 10m and 25m below the surface.

Scale Model Study

Analogue modelling program was carried out using the MPP01 apparatus as the measuring equipment to assist in the interpretation of the field results.

For scale modelling, the decrease in size is compensated for by an increase in the conductivity or frequency, or both. For a time domain system similitude between the full scale and model situation occurs if

$$\sigma_m \mu_m L^2_m / T_m = \sigma_f \mu_f L^2_f / T_f$$

where the subscript f indicates a quantity measured in the field situations, the subscript f indicates a quantity measured in the model, and σ, μ, T and L refer to the conductivity, magnetic permeability, time and linear dimensions, respectively.

The expression relating measured response levels is

$$e(t)_m / e(t)_f = L_m T_f / L_f T_m$$

The theoretical basis of TEM modelling has been described by Spies (1977).

Description of Model

The model used was a tabular sheet of aluminium in air. A dimensional scaling factor of 1300 and a time scaling factor of unity were used to produce a scale model which represented an orebody having a thickness of 16.5m, a dip of 45°, a depth of burial of 12m, and a conductivity of 20 S/m. The conductivity value was chosen on the basis of downhole resistivity measurements at Woodlawn (Hone and Young, this volume, p 519).

Comparison of model and field results

Results of modelling for traverse 9220mN are presented in Fig.7 as profiles for 200, 50 and 25m loops at sample times t = 1.1, 2.3, and 5.2ms. Field results are shown as dots, open squares and triangles. The model profiles have been reduced in amplitude by 25 % to allow a better curve fit. The reduction in amplitude required implies that the chosen conductivity of 20 S/m was somewhat too high, or the model was too thick or too shallow.

The best curve fit is for the 100m loop, in which 50 % loop overlap was used. The results for the 200m loop show that the loop overlap was not sufficient to define the shape of the anomaly, the centre of which was missed entirely.

Results for the 50m and 25m loops show considerable scatter of field data from the model profiles. The tendency for field values to exceed model values can be explained by the model being too deeply buried. If some conducting

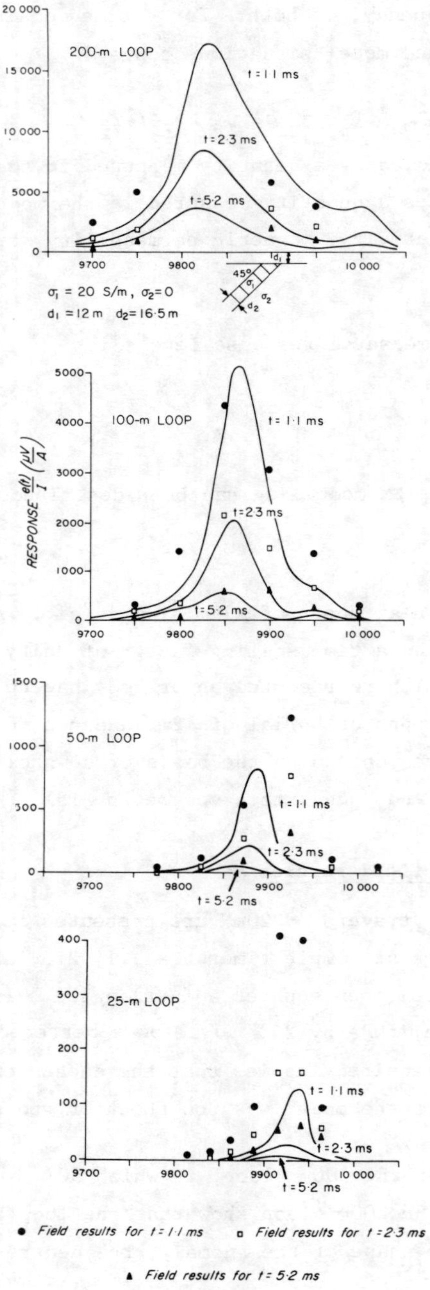

Figure 7 Comparison of field results and model study.

material were present in the zone of oxidation the response from the smaller loops would be increased without significantly changing the responses obtained from the larger loops. In addition to amplitude differences, some loop sizes produced lateral displacement of the anomaly.

The deficiencies in curve fits are caused by the simple model used. The model has a constant strike dip and conductivity whereas the orebody varies in strike, dip and conductivity.

Maximum exploration depth

To investigate the ability of the MPP01 to detect buried sulphide conductors, the response of the Woodlawn orebody at different depths of burial was modelled. The results of this study are presented in Fig. 8 for a sample time of 0.1ms, and different loop sizes. It is apparent that the larger loop sizes result in a greater response and hence potentially greater depth of exploration.

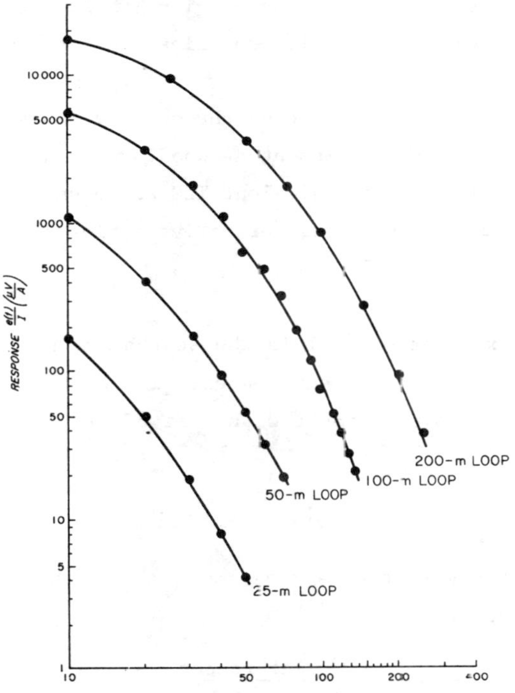

Figure 8 Model results for different depths of orebody.

Field results indicate that the response due to background and noise is about 200 μV/A at a sample time of 1.1ms. If the minimum anomaly amplitude resolvable is assumed to be twice the background level, the maximum search depth for a 100m loop is approximately 60m at 1.1ms.

At later sample times, background response will have reduced more rapidly than that due to conductive mineralization and a greater depth of penetration could be expected if the noise level is lower than that experienced during this survey. The search depth would also be greater for larger loops. If a 200m loop was used the orebody would produce a response of 70 μV/A when bureid at 200m depth.

Conclusions

A strong TEM response was recorded over the Woodlawn orebody but the accuracy of measurements at late sample times was considerably restricted by electrical noise. Values of apparent resistivity derived from the TEM data show a resistivity low of less than 10 ohm m near the centre of the orebody and a background resistivity of 100 to 500 ohm m. The TEM results indicate that the orebody has a westerly dip and that its top lies close to the surface.

Analogue modelling of the TEM response of the orebody shows that the effective exploration depth of the MPP01 equipment at Woodlawn is approximately 60m. It is considered that the use of equivalent TEM equipment with suitable noise rejection characteristics would permit effective exploration for a Woodlawn type orebody buried to depths of 200m.

TEM is an efficient exploration tool in the search for buried massive sulphide deposits in a geological environment similar to that of Woodlawn. The cost of TEM surveys is comparable with that of IP or conventional EM surveys.

Acknowledgement

This paper is published with the permission of the Director, Bureau of Mineral Resources, Canberra.

References

LEE, T. & LEWIS, R., 1974 - Transient EM response of a large loop on a layered ground. Geophysical Prospecting. 22(3), 430-444

MORRISON, H.F., PHILLIPS, R.J., & O'BRIEN, D.P., 1969 - Quantitative interpretation of transient electromagnetic fields over a layered half space. Geophysical Prospecting, 17, 82-101.

NEGI, J.G., VERMA, S.K., 1972 - Time domain electromagnetic response of a shielded conductor. Geophysical Prospecting, 20(4), 901-9.

SPIES, B.R., 1977 - Absolute electromagnetic scale modelling and its use in interpretation of TEM response. BMR Journal of Australian Geology and Geophysics, 2(2), 89-95.

VELIKIN, A.B., & BULGAKOV, Yu.I., 1967 - Transient method of electrical prospecting, (one loop version). In UNO International Seminar on geophysical methods of prospecting for ore minerals, July 1967, Ministry of Geology of the USSR, Moscow.

WAIT, J.R., 1956 - Shielding of a transient electromagnetic dipole field by a conducting sheet. Canadian Journal of Physics, 34, 890-893

YOUNG, G.A., 1976 - Drill-hole logging and transient electromagnetic test surveys, Woodlawn deposit, New South Wales, 1973. Bureau of Mineral Resources, Australia, Record 1976/52 (Unpublished).

Appendix 1

Fig. 9 shows a block diagram of the MPP01 equipment. A master oscillator generating pulses 20ms wide at 18Hz activates a current switch which applies the primary field pulses to the loop. A commutator switches in the receiver amplifier in the interval between the primary current pulses. The drive to this amplifier is delayed to ensure that large transient signals at early sample times do not saturate the receiver amplifier. A clamp timing monostable multivibrator produces a pulse 400 microseconds (μs) wide, 28ms after the current pulse to activate a clamper so that voltages e(t), read off the decay curve at different sample times, are determined with respect to an origin (zero value) at 28ms.

The sample timing monostable multivibrator has a variable pulse duration from 0.5ms to 15ms, which is set manually by a switch on the front panel of the instrument. At the fall of the pulse in this monostable multivibrator, the sampler switch is closed for 100 μs during which time capacitor C_1 transfers charge to the storage capacitor C_2. After several hundred cycles the full value of the voltage e(t) is developed across C_2 and may be read off the millivoltmeter. Measurements from 1 to 50,000 microvolts are made possible by an input attenuator. The voltmeter has a very large input impedance, in excess of 50M ohms, so that the voltage in C_2 does not vary between samples.

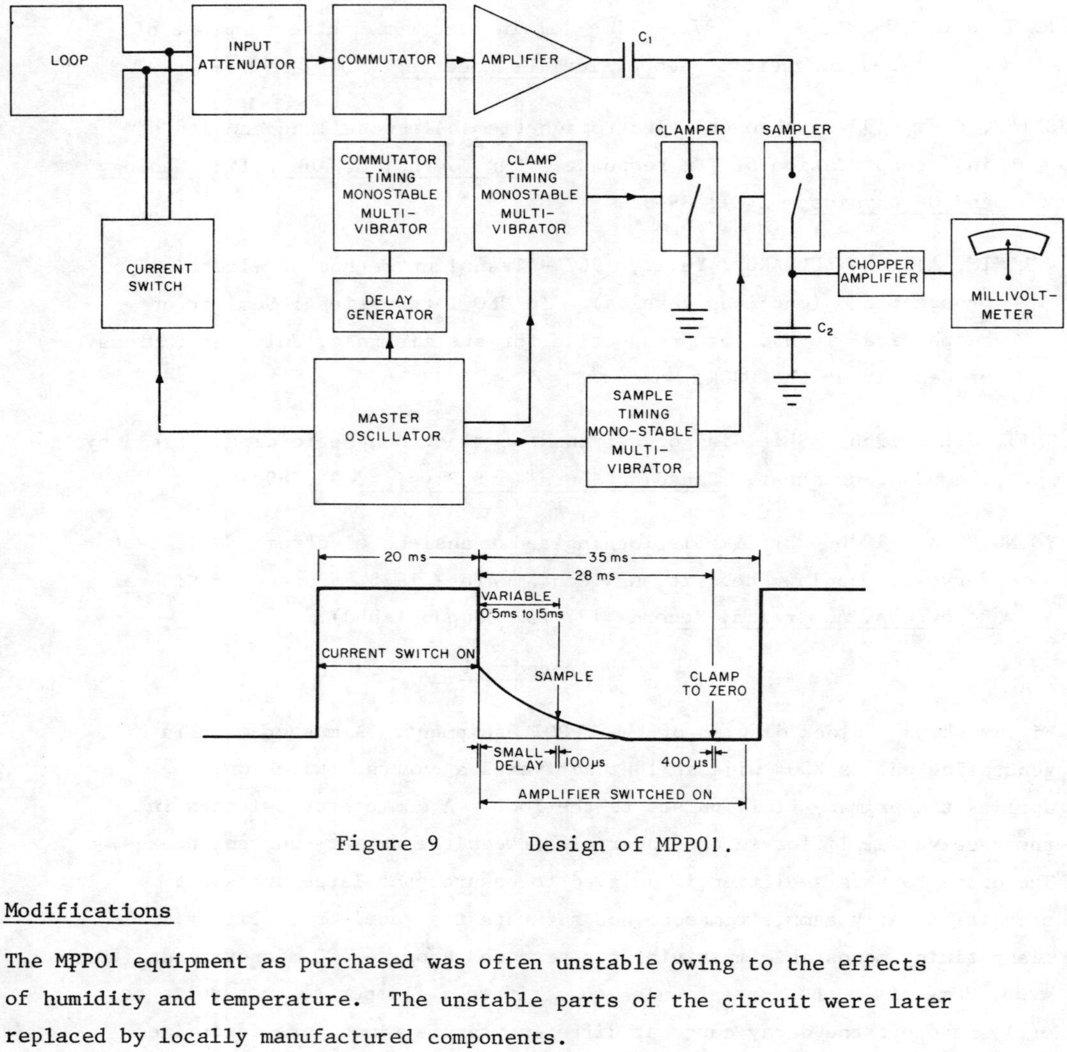

Figure 9 Design of MPP01.

Modifications

The MPP01 equipment as purchased was often unstable owing to the effects of humidity and temperature. The unstable parts of the circuit were later replaced by locally manufactured components.

A major modification was the redesign and replacement of all the original relays by field effect transistor (FET) switching and delay circuits.

The number of sample times available has been increased by a capacitor which can be connected in series with the sample timing circuit by a push button switch on the control panel.

7.12. *A Survey of the Woodlawn Orebody with an Experimental Transient Electromagnetic System*

G. BUSELLI

Contents

Abstract

The Woodlawn orebody was surveyed with the CSIRO transient electromagnetic (TEM) experimental system. Five traverses were completed across the orebody with loops of sides 45m and one traverse over the Black Shale about 2.5km south of the orebody. The results from this survey have been combined with those from other areas in Australia and used by CSIRO to design and construct an improved portable multichannel instrument (called SIROTEM) for TEM measurements.

Measurements of noise levels made during the survey show that the main sources of noise in the Woodlawn area were at frequencies of 0.05, 22, 44 and 800 kHz. The CSIRO system had adequate capabilities to reject all these noise signals. It is deduced that the most likely source of interference with the Russian made MPPOl unit used on a previous TEM survey of the orebody was a high level 800 kHz RF broadcast signal.

Results of the TEM surveys show that improved noise rejection capabilities of the CSIRO system enable measurements to be made with high signal to noise ratios. The responses differ on the various traverses from the north to the south of the orebody. The early and late time responses on each traverse also differ markedly. The late time responses more accurately delineate the position of the orebody. The decay curves exhibit the presence of at least three time constants. When the size of loops placed close to the orebody is increased beyond 45m, the coupling with the orebody appears to be decreased, and the time constant at intermediate sample times (between 3 and 20ms) is also decreased.

A comparison of TEM signal strengths from the Black Shale and the Woodlawn orebody shows that the shale response is very much less than that directly over the orebody. The Black Shale response is of the order of the response obtained at loop positions off the orebody. From the TEM response, the Black Shale therefore does not appear to be highly conductive.

Introduction

The Woodlawn orebody was surveyed as part of the test procedure for SIROTEM, a new instrument for transient electromagnetic (TEM) measurements being developed by the Commonwealth Scientific and Industrial Research Organisation (CSIRO). The technique had previously been investigated by CSIRO in Western Australia and Queensland, and the main objectives of the investigations at Woodlawn were:

1. To develop capabilities to extend the measurement of TEM signals to as late a delay time as possible, where mineralization response predominates over overburden response.
2. To work in a geologically well known area, to enable modelling of responses in the laboratory.
3. To investigate sources of noise interfering with TEM measurements.

Two field surveys were completed in the Woodlawn area. The first was carried out in September 1974, and the second in September 1975. Results of the first survey have been presented by Buselli (1977). This survey consisted of a single traverse across the orebody with loops of sides 45m, and an investigation of the effect of changing loop size at a fixed position over the orebody. In the second survey five traverses were completed with loops of sides 45m, four across the orebody and one across the Black Shale about 2.5km to the south of the orebody.

Earlier Spies et al. (this volume, p 281) made an extensive TEM survey of the Woodlawn orebody and reported considerable noise interference with measurements made with an MPP01 unit. Consequently, before the CSIRO TEM traverses were carried out, measurements were made of the most likely sources of this interference, and the noise rejection capabilities of the experimental TEM system were investigated.

The Experimental TEM System

Fig. 1 shows a block diagram of the experimental system used for TEM measurements at Woodlawn. A similar system was used at Cloncurry, Queensland and has been described in detail

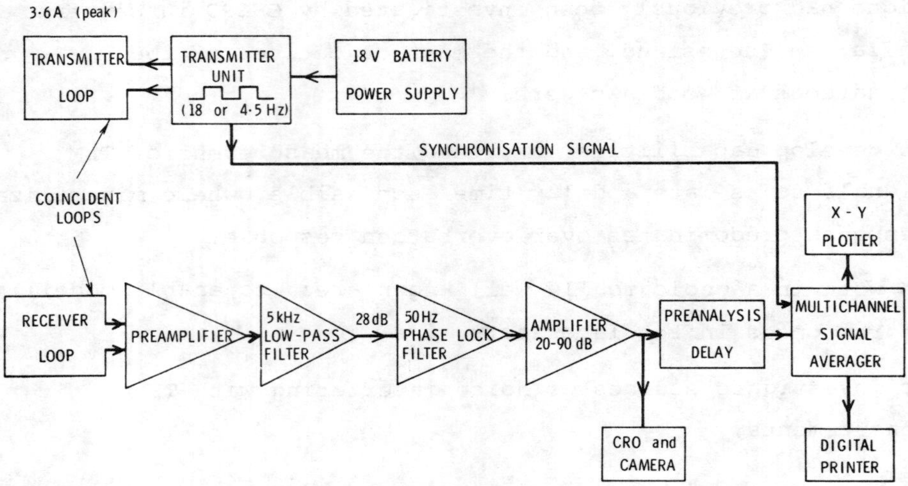

Fig. 1 Schematic representation of the experimental TEM system used at Woodlawn.

by Buselli (1974). The main modifications for use at Woodlawn, as described by Buselli (1977) were as follows:

1. A 50Hz phase lock filter was added to reject 50Hz noise at the input of the amplifier. This provided an attenuation factor of up to 30 in the 50Hz noise.

2. A printer was added to provide a digital readout of the decay curve, thus avoiding errors introduced by reading an analog plot of the decay curve.

3. The 5kHz low pass filter was improved to enable more efficient rejection of 16, 22 and 44 kHz frequencies of the main Very Low Frequency (VLF) sources detectable in different areas in Australia.

4. A square current pulse of 3A was driven into the transmitter loop at a pulse repetition frequency of 18hZ for measurements of responses out to about 25ms, and at 4.5Hz for responses out to late times from 20 to 100ms.

The transient decay curve was sampled, stored and averaged over 2^{12} repetitions by a multichannel signal averager. Because of the limited dynamic range of this averager, the decay curve was measured in four separate sections, using higher amplifier gain for the late sample times. Each decay curve was recorded both on an analog plotter and a digital printer, the numbers later being transferred to magnetic tape for computer analysis.

Separate but coincident transmitter and receiver loops were used for all the traverses.

The New CSIRO Instrument

SIROTEM, the instrument which CSIRO has developed, incorporates all the improvements suggested by experience with the experimental system. These include:

1. The use of multichannel averaging and the incorporation of a small printer with a capability of printing up to 32 channels.

2. An increased sample window width at late sample times to further reduce noise. For example, window widths of 26ms are used for sample times beyond 126ms.

3. A 50 Hz phase lock filter to reject 50 Hz noise and harmonics.

Further details of this new instrument are given by Buselli (1977) and Buselli and O'Neill (1977).

Electromagnetic Noise Interference at Woodlawn

Investigations of the main sources of electromagnetic noise at Woodlawn have been described in detail by Buselli (1977). Briefly, it was found that there were four sources of noise

GCSWO - L

interference: power lines at 50 Hz, two VLF stations at 22 and 44kHz, and radio frequency (RF) transmissions at about 800kHz. The levels of these signals were 0.05 to 0.07, 0.90, 5 to 20 and 20 to 40mV peak to peak loop input for the 45m loop. The 44kHz VLF and the 800kHz RF signals were therefore the most likely sources of interference. Since the 44kHz signal was transmitted only infrequently, the 800kHz signal was probably the only source to interfere significantly with measurements made using the MPP01 instrument (Spies et al., this volume p 281). Attenuation of this signal by the 5kHz cut off filter of the MPP01 was probably not sufficient for this high level signal. Any non-linearities in the circuitry would cause demodulation of the 800kHz signal, and the demodulated signal would saturate the amplifiers of the MPP01.

The 50Hz phase lock filter and 5kHz active filter of the CSIRO system sufficiently attenuate all electromagnetic noise signals to prevent any interference with the measurement of the transient response to late sample times.

Location of Traverses

Five traverse lines (I to V) of CSIRO TEM were completed across the Woodlawn orebody with 45m square loops and a station spacing of 45m. Each traverse as shown on Fig. 2 is represented by the line of loop centres for that traverse. Table 1 lists the position of the northern and southern sides of the loops for each traverse.

Survey Results

Because of the limited dynamic range of the multichannel averager, the transient response at each station has to be recorded in four separate sections, using higher amplifier gains for

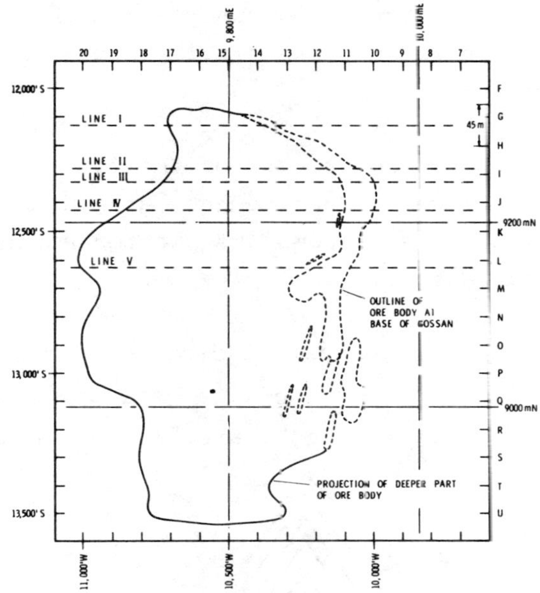

Fig. 2 Five traverse lines of CSIRO Woodlawn TEM survey using 45m loops. Each traverse line shows the location of the centre of the loops for that traverse.

Table 1 TEM Traverse numbers and loop positions on Woodlawn Test Grid

TEM Traverse No. (45m Loop)	Loop Position (Imperial Grid)
I	Northern side midway between Line F and G Southern side along Line H
II	Northern side along Line H Southern side midway between I and J
III	Northern side midway between Line H and I Southern side along Line J
IV	Northern side midway between Line I and J Southern side along Line K
V	Northern side between Line K and L Southern side along Line M

samples at later delay times. The manner in which these four sections are reconstituted into one single decay curve is described by Buselli (1977).

Figs. 3 to 7 show the results of the five traverses across the Woodlawn orebody. Measurements at a number of delay times in the interval 0.9ms to 80ms are plotted at the position of the relevant loop centre.

The shape of the orebody as indicated by drilling, on east to west section closest to the line of loop centres, and the Woodlawn Test Grid stations, are also shown on Figs. 3 to 7.

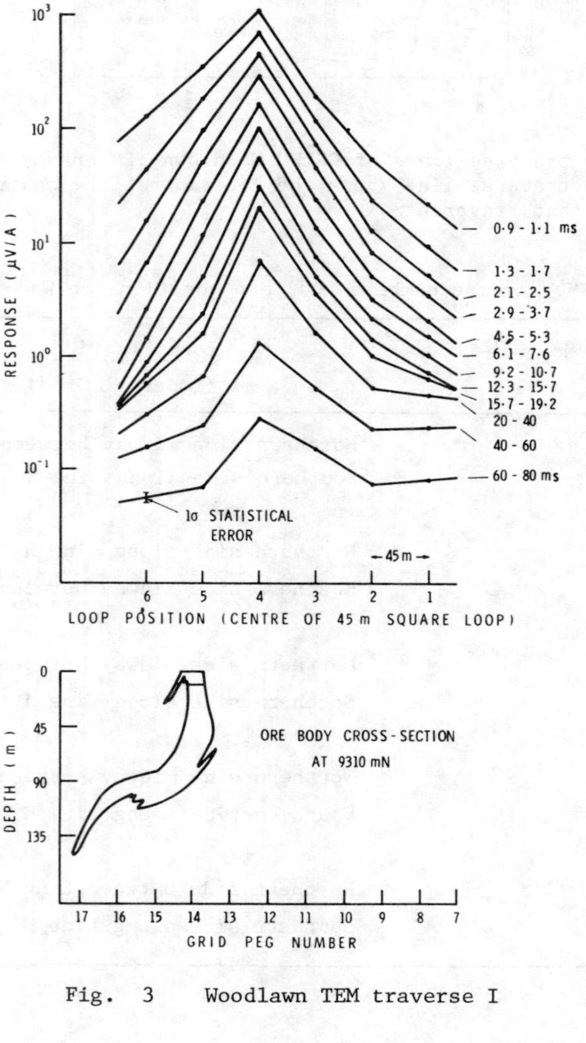

Fig. 3 Woodlawn TEM traverse I

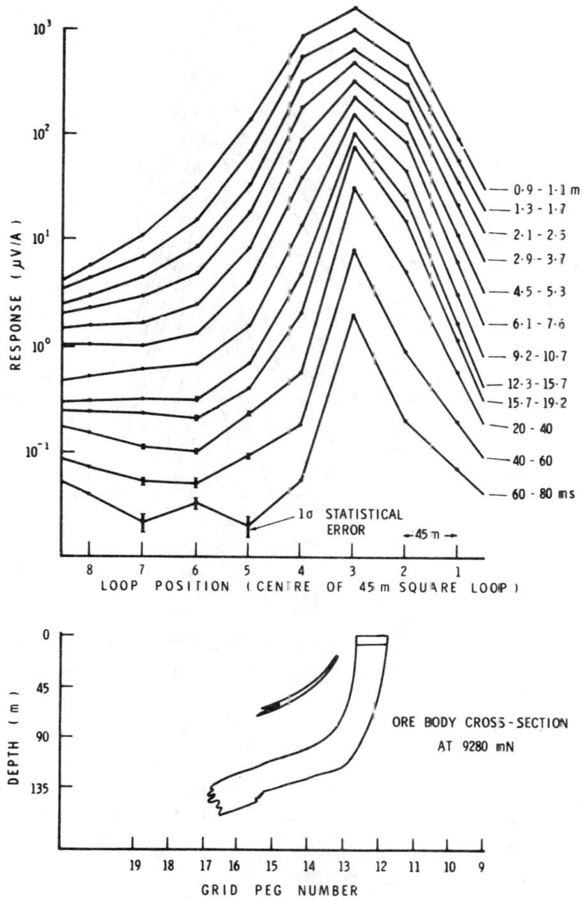

Fig. 4 Woodlawn TEM traverse II

The statistical measurement error (σ) shown on the traverse plots is of the order of 0.5×10^{-2} μV/A. This very low noise level (see Buselli, 1974) means that the experimental system has a high signal to noise (S/N) ratio even at late delay times. For example, on traverse II shown in Fig. 4, the S/N ratio for the 60 to 80ms channel is 280 directly over the orebody and varies from 4 to 12 at loop positions off the orebody. Signals less than 0.1 μ V/A can therefore be reliably measured. This compares to the detection limit of about 1 μ V/A for the MPP01 instrument (Spies et al., this volume p 281).

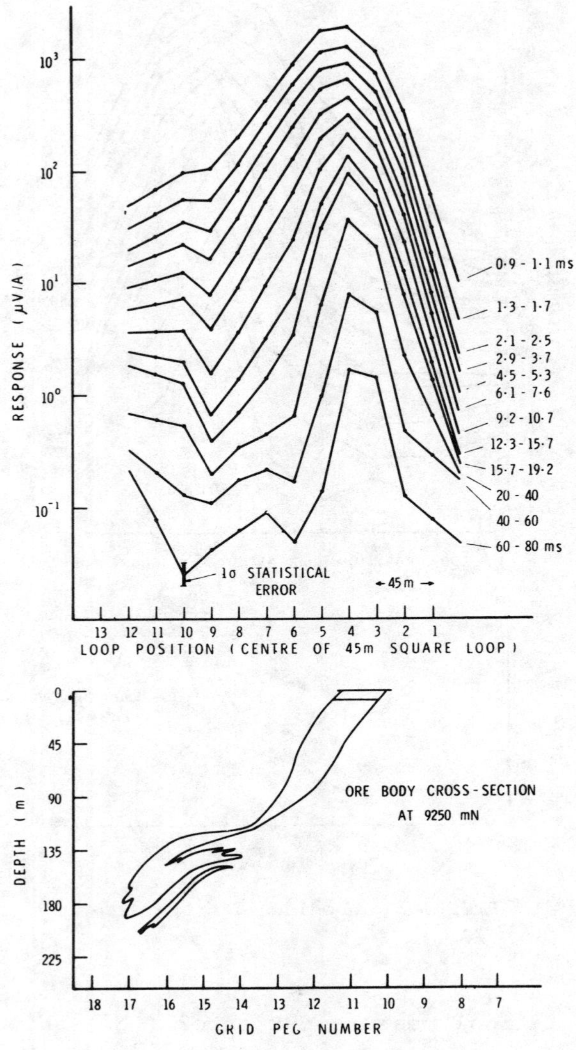

Fig. 5 Woodlawn TEM traverse III

The peak response on the northernmost traverse (I), shown in Fig. 3, is shifted to the east of the orebody. This is partly because the cross-section shown is about 6m north of the line of actual loop centres, resulting in the drawn position of the orebody being about 10m west of its actual position on

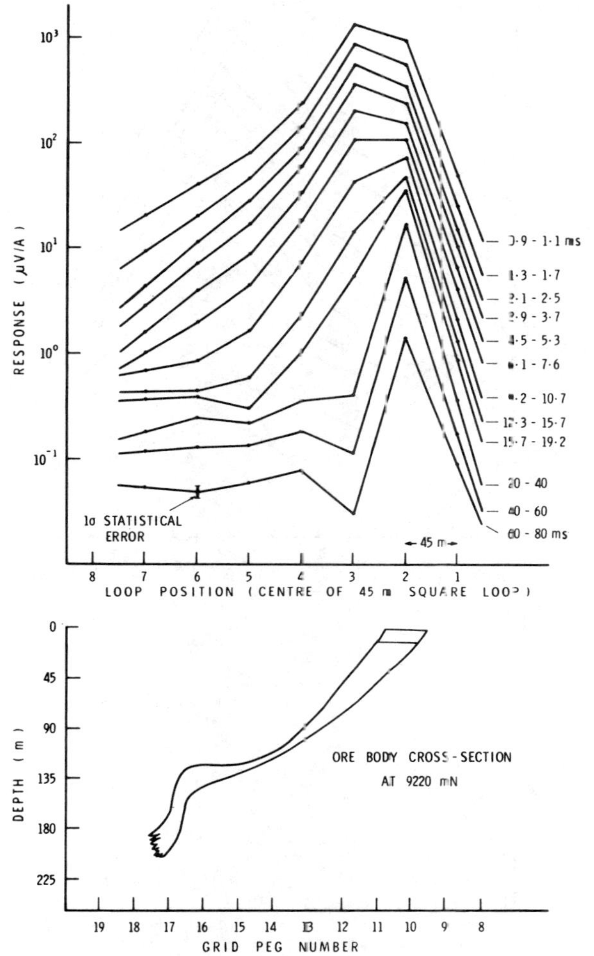

Fig. 6 Woodlawn TEM traverse IV

the traverse line. Also, the orebody here strikes about 300°M, whereas the traverse line is approximately east to west. This results in most of the orebody energized by the loop being to the south of the traverse, and the peak response being displaced to the east. The use of a large transmitter loop to retain a reasonable depth penetration and a small multiturn receiver loop with closer station spacing, would increase the resolution and may reduce this displacement of the peak response when using a large receiver loop.

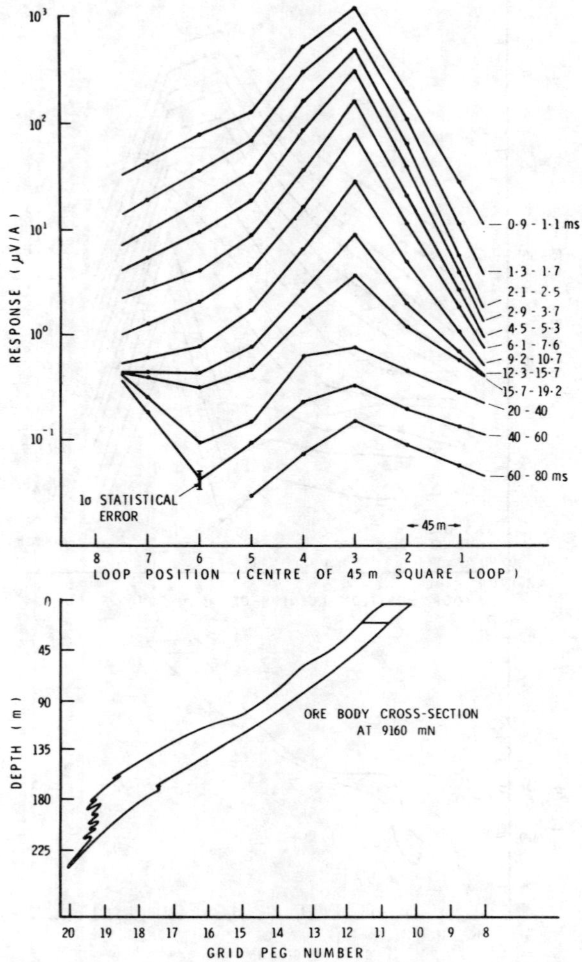

Fig. 7 Woodlawn TEM traverse V

A number of the traverses show marked differences between the early and late time responses. For example, in Fig. 4, the signal at late times (greater than 10ms) decreases by nearly two orders of magnitude two loop positions (90m) away from the position directly over the orebody, whereas that at early times decreases by one order of magnitude. The latter eventually decreases by two orders of magnitude at a position nearly four loop positions (180m) away from that directly over the orebody. Theoretical considerations (Lee, 1974) show that early time responses are influenced more by conductive material near the surface, whereas responses at later times are affected more by the deeper parts of the orebody. The orebody is therefore more accurately delineated by the response at delay times later than about 10ms. Note that theoretically, even in the optimum case of low overburden conductivity and large orebody size, the depth of investigation is limited by the loop size. For this case, this would not be more than about 45m, i.e the length of one loop side.

TEM traverse IV shown in Fig. 6 indicates a very distinct difference between the rate at which the amplitudes of the early and late time samples decrease as the loops are moved away from the orebody.

This traverse also shows a migration of the peak response from loop position 2 (directly over the orebody) for late delay times (those greater than 10ms) to position 3 for the early delay times.

The late time anomaly peak is broader for the southern traverses reflecting both the width of the orebody and the shallower westerly dip. This is shown on the most southerly traverse V in Fig. 7, where the anomaly peak for delay times beyond 20ms extends over about 200 to 250m compared to about 140 to 180m for the most northerly traverse I shown in Fig. 3. The late time response for Traverse V is also weaker than those from the northern traverses, because the depth to the top of the massive sulphide deposit is greater in the south than the north. This variation of response from north to south along the orebody is discussed in greater detail below.

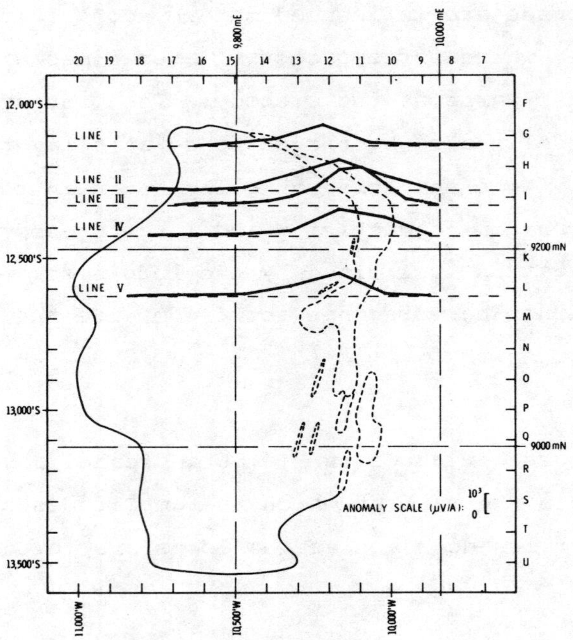

Fig. 8 Results of five TEM traverses with 45m loops across the Woodlawn orebody for the 0.9 to 1.1 ms channel

Variation of the TEM Response from North to South along the Orebody

Figs. 8 to 10 summarize the results of the five traverses for the time channels 0.9 to 1.1, 9.2 to 10.7 and 60 to 80ms, respectively. On each, a scale for the amplitude of the response in μ V/A is shown. Each of these figures shows that the peak response of the most northerly traverse (I) appears to be shifted to the east of the orebody. The responses are plotted at the centres of the transmitter receiver loops, but for this traverse most of the orebody coupled with the loops lies to the southeast of the intersection of the loop centre line and the orebody. As discussed earlier, the peak response consequently appears to be shifted along the centre line towards the east. (It is interesting to compare this distribution with the gradient array PFE data in this region, Quick and Cifali, this volume, p 333 - Ed).

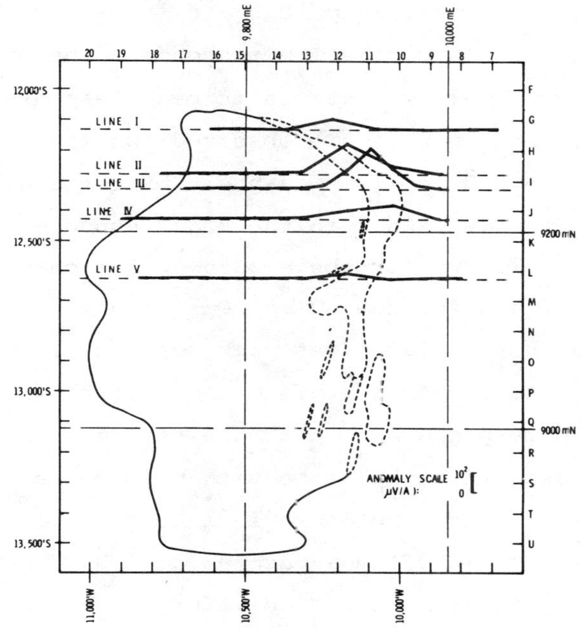

Fig. 9 Results of five TEM traverses with 45m loops across the Woodlawn orebody for the 9.2 to 10.7 ms channel

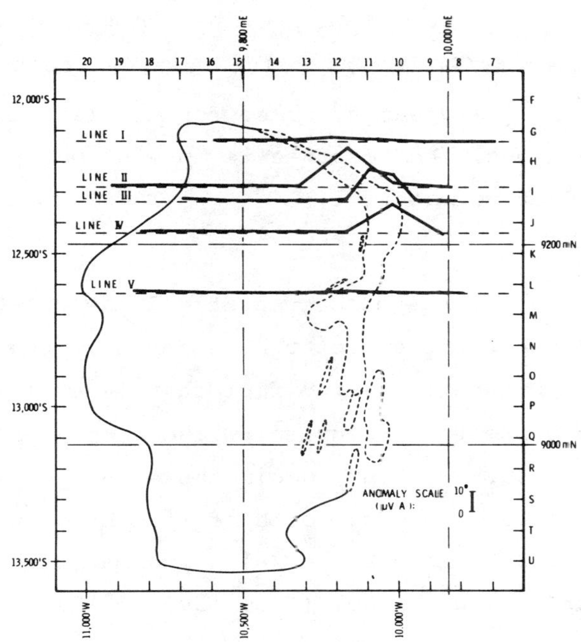

Fig. 10 Results of five TEM traverses with 45m loops across the Woodlawn orebody for the 60 to 80 ms channel

Comparison of Figs. 8 to 10 again shows a marked difference between the early and late time TEM response. The early delay times (Fig. 8) show small variations in peak response from north to south while marked changes are observed for late delay times (Fig. 10). For the 0.9 to 1.1ms channel, there is a variation by a factor of not more than 2 in the peak response of the five traverses. However, for the 9.2 to 10.7ms channel, this factor is about 7, and for the 60 to 80ms channel, it has increased to 14. For both these delay times peak response is greater on the central traverses. Hence, the north to south variation in the late time TEM response reflects more strongly the position of the loops with respect to the orebody than does the early time response. In particular, the decrease in the 60 to 80ms response south of line H (Fig. 10) from a maximum of 2.02 μV/A to a minimum of 0.15 μV/A is most probably due to the massive sulphides being deeper in the south. This depth is about 16.5m compared to about 10m for the depth of the massive sulphides near Line H. For the early time channel of 0.9 to 1.1ms, the peak response south of Line H only varies from a maximum of 1910 μV/A to a minimum of 1220 μV/A. The relative changes in signal level are not as great as those for the late delay times because at early times, the response is more strongly influenced by shallower conductive material.

The decrease in response seen at all the delay times north of Line H is not the effect of the depth of the massive sulphide deposit, but reflects the fact that less of the massive sulphide deposits is coupled to the loops for the northernmost traverse. However, again there is a greater relative change in the response north of Line H for the 60 to 80ms channel (from 2.02 to 0.28 μV/A) compared to the relative change in the early time channel (from 1590 to 1050 μV/A).

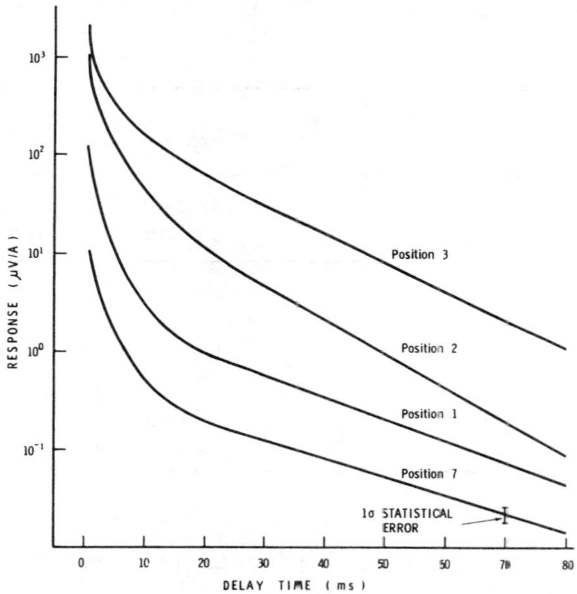

Fig. 11 Transient decay curves at a number of positions along traverse II

Characteristics of the TEM Decay Curves

An example of the decay curves obtained from traverse II (Fig. 4) is shown in Fig. 11. These decay curves were obtained directly over the orebody (position 3), near the orebody (position 2) and away from the orebody (positions 1 and 7).

Clearly, more than one time constant is present in each of the decay curves. Table 2 shows the time constants in three time intervals (less than 3, 3 to 10, and greater than 20ms) read from each of the decay curves obtained directly over the orebody (position 3) and away from the orebody (position 7). It can be seen that at least three different time constants are present on each decay curve, and in both cases, the value of the time constants increases progressively from the early to the late delay times. Surprisingly, the time constant obtained for the late sample times off the orebody is 9ms larger than that directly over the orebody. This larger time constant is possibly

Table 2 Example of time constants in three time intervals for decay curves at two positions of traverse II

Time interval (ms)	Time constant (ms)	
	Loop position 3	Loop position 7
< 3	1.0	1.6
3-10	5.2	3.5
> 20	14.2	23.5

due to the response of the deeper part of the orebody whereas at position 3 the highly conductive supergene zone and the top of the orebody effectively shield the deeper parts of the orebody.

These results indicate that a number of aspects of the decay curve characteristics need further study. These include:

1. The effect of choice of the zero level channel on the amplitudes and time constants.

2. The late sample times which will enable the third time constant to be accurately determined.

3. The optimum times at which the decay curve should be sampled.

The Effect of Loop Size on the Decay Curves

The effect of loop size on TEM decay curve characteristics near the Woodlawn orebody has been discussed in detail by Buselli (1977) for square loops with sides of 11.3, 22.5, 45, 90 and 150m. Fig.12 shows the positions of these loops in relation to the orebody. Fig. 13 shows the decay curves obtained. The

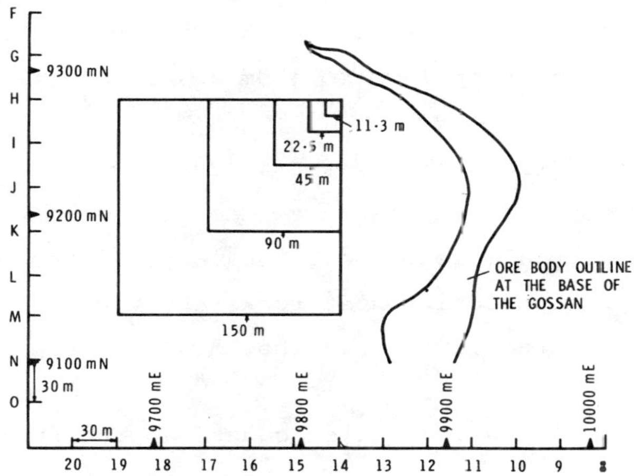

Fig. 12 Layout of loops of different sizes shown in relation to the Woodlawn orebody outline at the base of the gossan

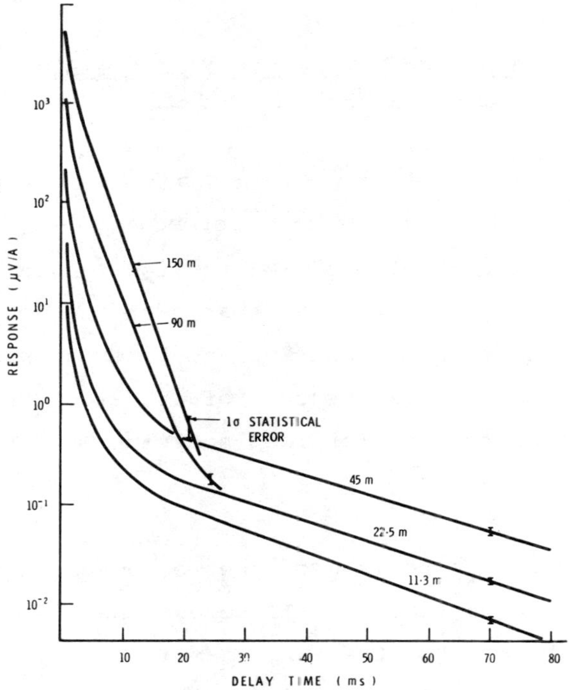

Fig. 13 Decay curves obtained using loops of different sizes over the deeper part of the Woodlawn orebody.

loops were located away from the shallower part of the orebody, over the deeper part to the west. In this region depths to the orebody were from about 110m to 150m.

It was found that an increase in the loop size beyond a side length of 45m reduced response from the orebody to a lower percentage of the overall response. The larger loops were better coupled to the less conductive host rocks and the responses obtained with the 90 and 150m loops showed a more rapid decay.

The increased area of the loops also caused an increase in the noise level which was not accompanied by a corresponding increase in the TEM signal response from the orebody. The consequent decrease in the S/N ratio by factors of up to 8 and 27 for the 90m and 150m loops, respectively, prevented measurement of TEM signals with these loops at sample times beyond about 20ms.

Comparison of the TEM Response of the Black Shale and the Woodlawn Orebody

The decay curves from the Black Shale are not presented because of instrumental problems at intermediate sample times. However, it was apparent that the TEM signal strength over this shale was very much weaker than that over the Woodlawn orebody. For samples earlier than 2ms and later than about 20ms the amplitude of the TEM response of the Black Shale appears to be similar to that for loop positions off the orebody.

Table 3 shows a comparison of the responses at two sample times obtained on traverse III across the orebody and across the Black Shale. For the early sample window (0.9 to 1.1ms), the response over the orebody is 95 times greater than that of the Black Shale, while the response off the orebody is only about 3 times greater. For the late sample time (60 to

Table 3 Comparison for two delay times of the Woodlawn orebody TEM response (for traverse III) and that of the Black Shale

Sample time (ms)	Over orebody response (μV/A)	Off orebody response (μ V/A)	Black Shale response (μ V/A)
0.9-1.1	1.9×10^3	6×10	2×10
60-80	1.7	8×10^{-2}	6×10^{-2}

80ms), the response of the orebody is 28 times greater, while the response off orebody is only 1.3 times greater.

There was no obvious TEM anomaly over the Black Shale except possibly the response at the earliest sample times from 0.6 to 0.9ms. Based on these results the Black Shale would not therefore be interpreted as a highly conductive body.

Conclusions

The Woodlawn orebody is an excellent target for single loop TEM methods because of its large conductivity contrast with the host rocks, its relative shallowness, and the absence of a highly conductive overburden. This is in spite of its relatively short strike length. In particular, the Woodlawn orebody gives a strong response to the single loop method when using loop side lengths of less than 100m.

The development of equipment with an improved capability to reject electromagnetic noise has enabled TEM measurements to be made at Woodlawn without significant interference from nearby noise sources which adversely affected earlier MPP01 results (Spies et al., this volume, p 281). Measurements with a high signal to noise ratio can be made to samples times as late as 80ms, 5 times later than with the MPP01 system. Comparison of the early and late time TEM responses show that late sample time measurements, which are less sensitive to the responses from lower conductivity host rocks and overburden material, delineate the orebody more accurately. The extension of the period of observation to 90ms reveals that at least three time constants are present in the decay curves.

The TEM response from the Black Shale south of the Woodlawn orebody, although not completely defined, is weak and is similar in magnitude to that obtained for loop positions off the Woodlawn orebody. The TEM results do not imply the presence of a highly conductive body in the area of the Black Shale and hence clearly distinguish it from the Woodlawn orebody.

References

Buselli, G., 1974, Multichannel transient electromagnetic measurements near Cloncurry: Bull. Aust. Soc. Explor, Geophys. v.5, no.2, p31-47.

Buselli, G., 1977, Transient electromagnetic measurements to late delay times over the Woodlawn orebody: Bull. Aust. Soc. Explor. Geophys., v.8, no.1, p.1 to 5.

Buselli, G., and O'Neill, B., 1977, SIROTEM: A new portable instrument for multichannel transient electromagnetic measurements, Bull. Aust. Soc. Explor. Geophys., v.8, no. 3, p. 82 to 87.

Lee, T., 1975, Transient electromagnetic response of a sphere in a layered medium: Geophys. Prosp. 23, 492 to 512.

CHAPTER 8

GROUND GEOPHYSICAL SURVEYS: ELECTRICAL METHODS

8.1. Self Potential Surveys of the Woodlawn Orebody

G. CIFALI AND R. J. WHITELEY

Contents

Abstract

A well defined self potential anomaly with a magnitude of about 250 millivolts occurs above the Woodlawn orebody. Steep potential gradients around this anomaly closely outline the shape of the orebody and reflect changes in dip and depth of burial.

Other prominent sources within the Woodlawn area are conductive zones within black shale units. One such zone selected for comparative studies produced a self potential anomaly of similar magnitude to that over the Woodlawn orebody. It does not appear possible from the self potential data alone to distinguish such anomalies from those produced by economic massive sulphide orebodies or to tell if the black shale is the sole source of the self potential anomaly.

Introduction

The existence of natural voltage gradients in the vicinity of certain ore deposits is well known. Measurement of these, so called, self potentials is widely used in exploration for shallow massive sulphide deposits (e.g. Parasnis, 1970).

At the present time the causes of self potential (SP) anomalies are not completely understood (Bolviken and Logn, 1975) although in the vicinity of sulphide orebodies, they are explained in terms of an electrochemical mechanism.

The most widely accepted mechanism has been proposed by Sato and Mooney (1960) who believe that SP anomalies associated with sulphide orebodies result from the voltage drop within resistive country rocks. Electric current is produced by separate but simultaneous reduction of oxidizing agents near the surface and the oxidation of reducing agents at depth. SP anomalies are characterised by a prominent negative voltage anomaly often hundreds of millivolts in magnitude, near the shallowest region of the orebody. The ore itself does not participate directly in either reaction, but serves as a conductor to transfer electrons from the reducing agents at depth to the oxidizing agents near the surface. This mechanism usually requires the presence of differing oxidation potentials of groundwaters at different depths. Recent research (Govett, 1973) suggests that such electrochemical mechanism might influence the composition of ore and country rock providing additional guides for exploration.

Within the eastern Lachlan Fold Belt volcanogenic sulphide deposits such as Woodlawn are known to produce prominent SP anomalies (Ellis et al, 1973; Davis, 1977). Unfortunately, one of the major limitations to the use of the SP method in the region is that graphitic rocks and conductive black shale units also produce prominent anomalies, and within the region such rock types are common. Malone (this volume, p71) believes that SP would have been the most cost effective technique at Woodlawn after soil geochemistry.

At Woodlawn SP surveys were applied extensively in early exploration work during 1970. More detailed surveys were completed in June 1974.

Equipment and Field Procedures

One of the major advantages of the SP method is the simple equipment requirement. This basically consists of a high impedance voltmeter, nonpolarizing electrodes and connecting wire. At Woodlawn a Scintrex VP7 Self Potential Meter was used together with nonpolarizing, copper in copper sulphaste solution porous pot electrodes. Voltage measurements were made on successive east to west lines 61m apart between a fixed remote electrode and a mobile electrode. A station spacing of 15.2m was used throughout.

TABLE 1

Self Potential Anomalies in the Woodlawn Area

Anomaly No.	Approx. Magnitude (millivolts)	Associated Rock-type
1	250	Woodlawn orebody
2	125	Black shale
3	125	Black shale
4	200 to 350	Black shale
5	100	? Black shale
6	25	Fine grained acid volcanics

Results and Discussion

The initial (1970) survey defined several notable anomalies. Fig. 1 shows the location, approximate magnitude and extent of these anomalies. They are also listed in Table 1 with associated rock types.

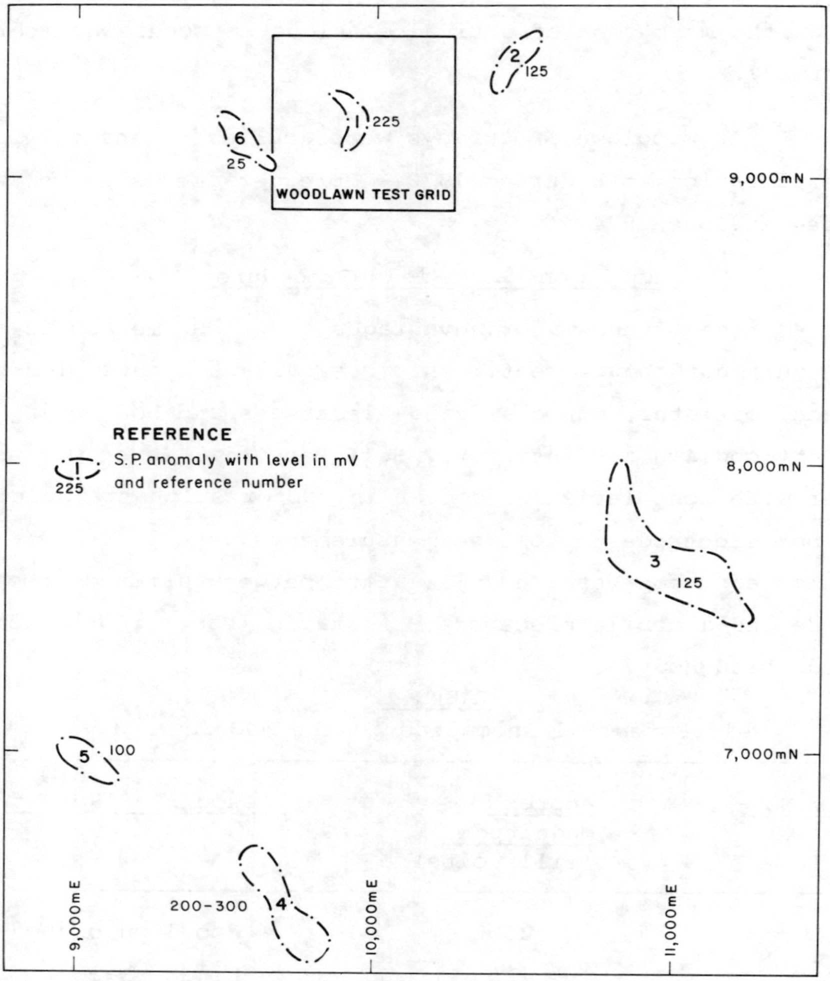

Fig. 1 SP anomaly location map.

The Woodlawn orebody produced one of the largest SP anomalies encountered (Anomaly 1), although several zones of black shale produced anomalies of comparable magnitude (Anomaly

2, 3, 4 and 5). In particular Anomaly 4 was of very similar magnitude and was drilled.

The smallest anomaly found in the region was Anomaly 6 having a magnitude of about 10 per cent of that over the Woodlawn orebody. Drilling on this anomaly encountered acid volcanics containing disseminated pyrite and subordinate base metal sulphides at depth.

The major problem in the interpretation of SP surveys in this geological environment is distinguishing between anomalies over massive sulphides (e.g., Anomaly 1) from those over uneconomic black shales (e.g., Anomaly 4). Also, is the black shale at Anomaly 4 the sole source of the SP anomaly or is it associated with massive sulphides?

This question is particularly important since the SP anomalies are associated with local zones within black shale units, having much shorter strike lengths than do the black shale units themselves. The SP anomalies appear to be confined to zones where the conductivity of the black shale has been enhanced, presumably mostly by pyrite.

As a result of this ambiguity, the black shale zone producing Anomaly 4 was selected for additional geophysical testing as reported elsewhere in this volume.

Woodlawn orebody (Anomaly 1)

The second SP survey in 1974 was confined to the Woodlawn Test Grid (Fig.1). The magnitude of the SP anomaly obtained was comparable to that measured in 1970 and is shown in Fig. 2. This figure shows that a well defined SP anomaly occurs above the Woodlawn orebody. This anomaly has an overall amplitude

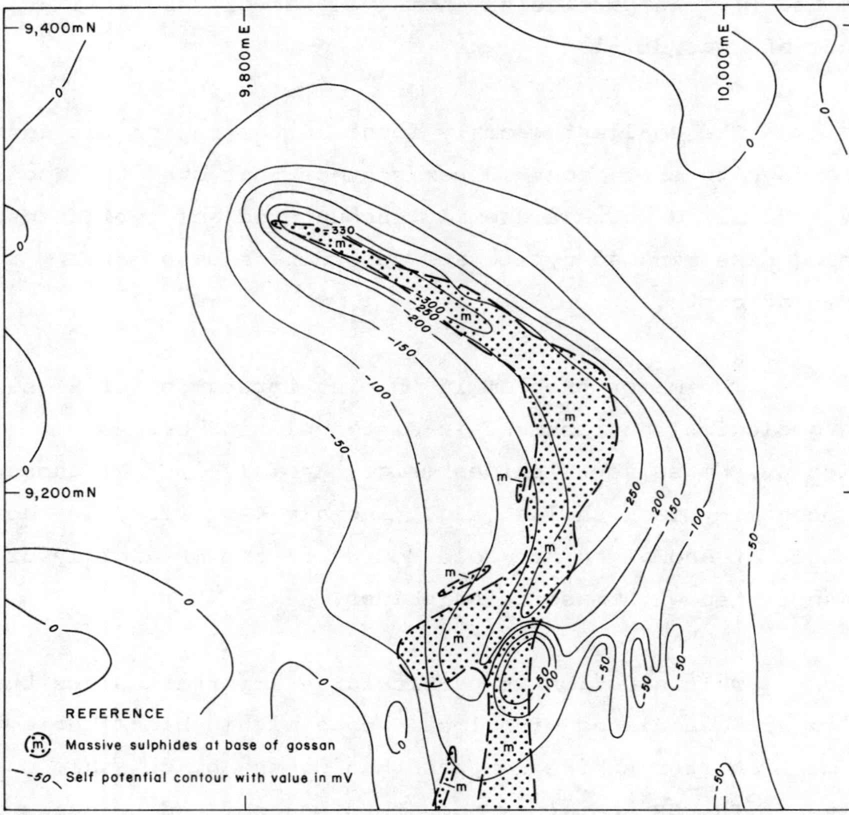

Fig. 2 SP anomaly over Woodlawn orebody.

of about 250 millivolts (mV) below the background and a maximum amplitude of -330mV near the northern end of the orebody.

The steep voltage gradients on the flanks of this anomaly closely reflect the complex shape of the massive sulphides at the base of the gossan and the change in dip and depth of the Woodlawn orebody revealed by drilling (Malone, this volume, p 49).

The lower voltage gradients on the western side of the orebody above the hanging wall reflect the general westerly dip of the bulk of the orebody. This asymmetry may, however, be enhanced by the differing electrical resistivities in the host rocks on either side of the orebody measured in subsequent electrical surveys (e.g. Quick and Cifali, this volume, p333).

The overall consistency of the SP contours reflects the continuity of the massive sulphides. Local disturbances near the southern end of the orebody (-50mV closures) may be due to attenuation of the SP anomaly by waste from an earlier prospecting shaft in this area.

Fig. 3 shows the SP profile along Line G, (9312mN) across the most intense part of the anomaly near the northwestern end of the orebody. In this area the gossan is thinnest (about 8m) but supergene alteration has penetrated to the greatest depth known in the deposit. As shown in the geological section the orebody is essentially vertical to a considerable depth. The isolated high on the eastern side of the profile is probably due to a near surface feature and is not related to the Woodlawn anomaly.

Black Shale (Anomaly 4)

The SP profile across a pegged line near the centre of the Black Shale at Anomaly 4 is shown in Fig. 4. Fig. 4 shows an asymmetric anomaly with a magnitude of about 250 mV, centred at about Station 10.5. The drilled section from W43 is also shown. Drilling encountered unmineralized tuff before entering the Black Shale which contained finely disseminated pyrite and traces of base metals sulphides. Several pods or stringers of pyrite were also encountered. These undoubtedly enhanced the conductivity of the Black Shale and contributed to the large SP anomaly in this region.

Conclusions

A well defined SP anomaly occurs over the Woodlawn orebody. Steep potential gradients around the flanks of this

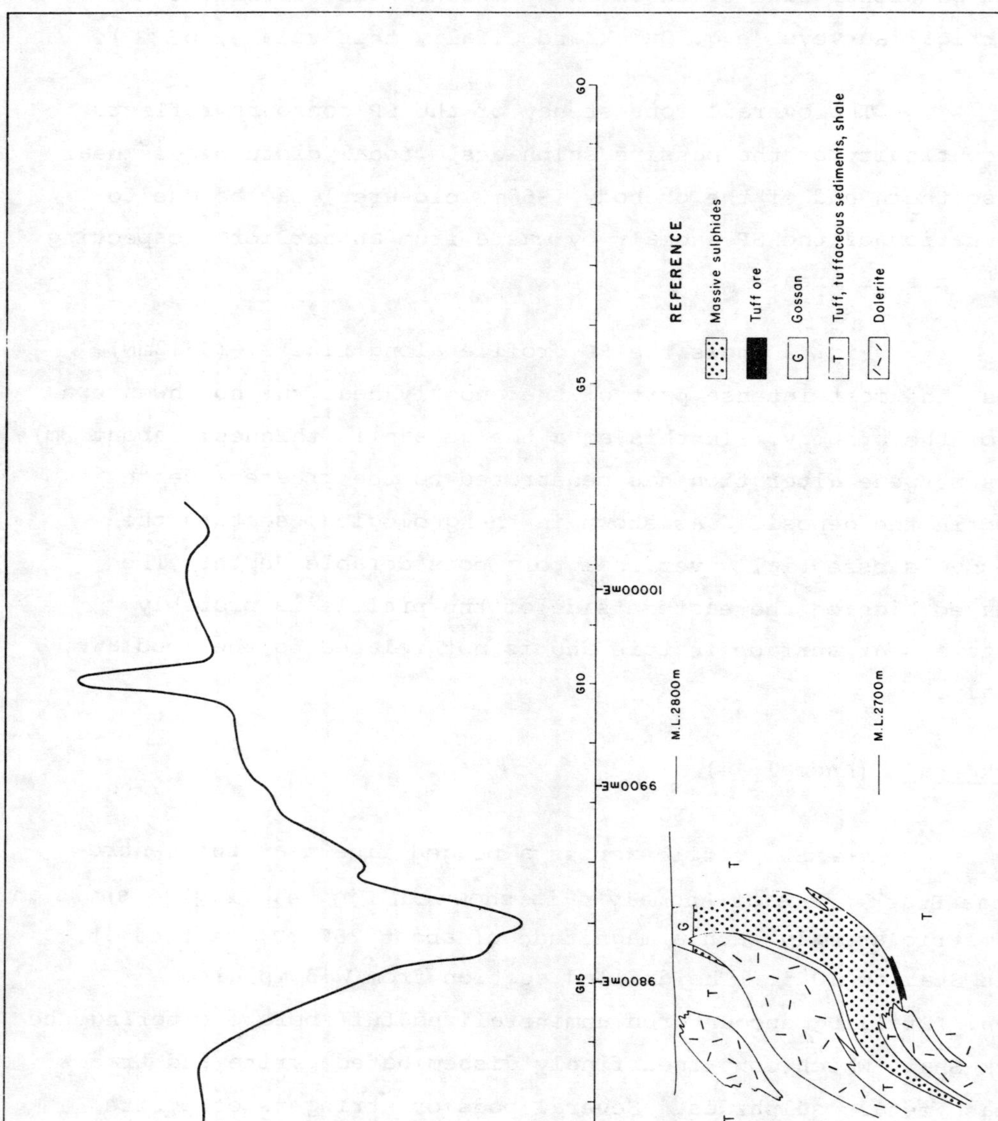

Fig. 3 SP profile, line G (9312mN)

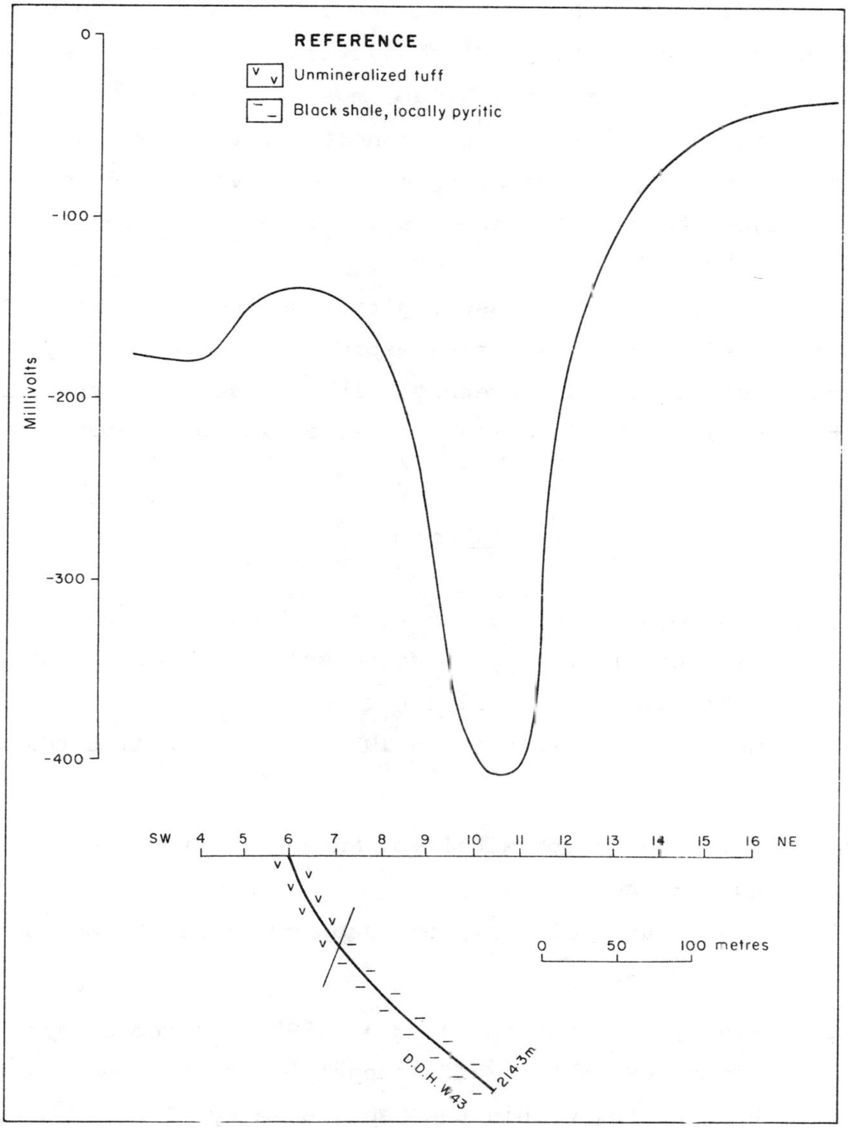

Fig. 4 SP anomaly over the Black Shale.

anomaly closely outline the shape of the orebody and reflect changes in dip and depth of burial. This SP anomaly helped to confirm the continuity of the orebody between the massive sulphide intersections in drill holes.

Other sources of prominent SP anomalies include black shale. These produced anomalies similar in magnitude and character to those over the Woodlawn orebody and appear to be confined to areas where the local conductivity of the shale has been enhanced by uneconomic sulphides, mostly pyrite. It does not appear possible on the SP data alone to distinguish such anomalies from those produced by economic massive orebodies or to tell if black shale is the sole source of SP anomalies.

Notwithstanding these limitations the SP method is felt to be a useful and inexpensive exploration method for possible shallow massive sulphide occurrences, although such occurrences could not necessarily be inferred from SP anomalies alone.

References

Bolviken, B. and Logn, O. 1975. An electrochemical model for element distribution around sulphide bodies. in Elliott, I.L. and Fletcher, W.K. (eds.) Geochemical Exploration 1974. Elsevier Amsterdam, p. 631-648.

Davis, L.W. 1977. The geophysical volcanogenic target in eastern Australia. Bull. Aust. Soc. Explor. Geophys. Vol. 8, No. 3, pp. 50-69.

Ellis, A.H., Gould, I.G. and Smith, M.J. 1973. Three Case Histories of the Self Potential Method of Base Metal Prospecting within the Lachlan Geosyncline, N.S.W. Proc. of Aus. Inst. Min. and Metall. No. 246. pp. 21-32.

Govett, G.J.S. 1973. Differential secondary dispersion in transported soils and post-mineralized rocks: an electrochemical interpretation in: M.J. Jones (Ed.) Geochemical Exploration 1972. Inst. of Min. and Metall. London. pp. 81-91.

Parasnis, D.S. 1970. Some recent geoelectrical measurements in the Swedish sulphide ore field illustrating scope and limitations of the methods concerned. Min. Groundwater Geophys. 1967 (Ottawa) pp. 290-301.

Sato, M. and Mooney, H.M. 1960. The electrochemical mechanism of sulphide self potentials. Geophysics. Vol 25, p 226-249.

8.2. *Resistivity and Induced Polarization Measurements with Dipole-Dipole and Gradient Arrays, Woodlawn*

K. P. QUICK AND G. CIFALI

Contents

Abstract

Extensive frequency domain resistivity and induced polarization (IP) measurements over the Woodlawn orebody were completed using the dipole-dipole and gradient arrays at various electrode spacings. Most of these surveys were completed after the orebody had been defined by drilling and examined, in detail, the response of the Woodlawn deposit in order to assist exploration of the surrounding regions.

The resistivity data reflect the shape and distribution of the massive sulphide body reasonably well. The IP data are more affected by host rock types in the vicinity of the orebody such as black shale, pyritic shale and volcanics, and volcanics containing disseminated mineralization, than by the massive sulphide body itself. The complexity of the resistivity and IP data reflects the geometry of the deposit, the nature and distribution of mineralization, and the contrasts and relationships between the orebody and host rocks.

Introduction

The frequency domain resistivity and induced polarization (IP) methods using the dipole-dipole array played an important role in initial exploration of the Woodlawn area. This method was once thought to be insensitive to truly massive sulphides, but has proven extremely useful in locating such bodies either directly or indirectly by detecting an associated halo of disseminated sulphides.

The initial IP survey at Woodlawn (Smith, this volume, p 79) was reinterpreted after drilling had confirmed the presence of the orebody. The grounded fences which had adversely affected the initial data were removed and more detailed dipole-dipole IP surveys were completed over the orebody in February, 1972 and in July, 1974.

The gradient array, although not applied during the early electrical surveys at Woodlawn, was later used extensively. The decision to use this array was taken after encouraging results were obtained over the main massive sulphide orebody in a test of the method in December, 1971. Complete coverage of the immediate surrounds of the orebody using this array proved useful in delineating trends and establishing continuity (or discontinuity) of many features of interest.

In April, 1974, the gradient array grid covering the ore zone was reconstructed and a series of more detailed electrical measurements were taken. This grid was later termed the Woodlawn Test Grid and was used extensively during the geophysical study.

Dipole-Dipole Survey

Locations of the dipole-dipole resistivity and IP lines surveyed are shown in Fig. 1. Line 2X was oriented in an east to west direction and Lines 2A and 2F were oriented orthogonal to the strike of the northern and southern parts of the orebody respectively to maximize response and resolution.

Equipment and field procedures

A Heinrichs Geoexploration, frequency domain, IP system was used for the survey; this comprised a Mark 4C receiver, Mark 4 transmitter and power source. Frequencies used were 0.1 and 3.0 Hz.
Surveys were completed with dipole lengths of 50 ft (15.2m), 100 ft (30.5m), 200 ft (61m), and 400 ft (122m).

Discussion of results

The resistivity and IP pseudo-sections obtained are shown in Figs. 2*to 7 together with the drilling cross-sections at the same scale. Apparent resistivities (ρa) were calculated in ohm ft/2 π (1 ohm ft/2$\pi \simeq$ 0.5 ohm m) and since the original

*The folded Figs. 2-14 and 16 can be found in the envelope accompanying the book

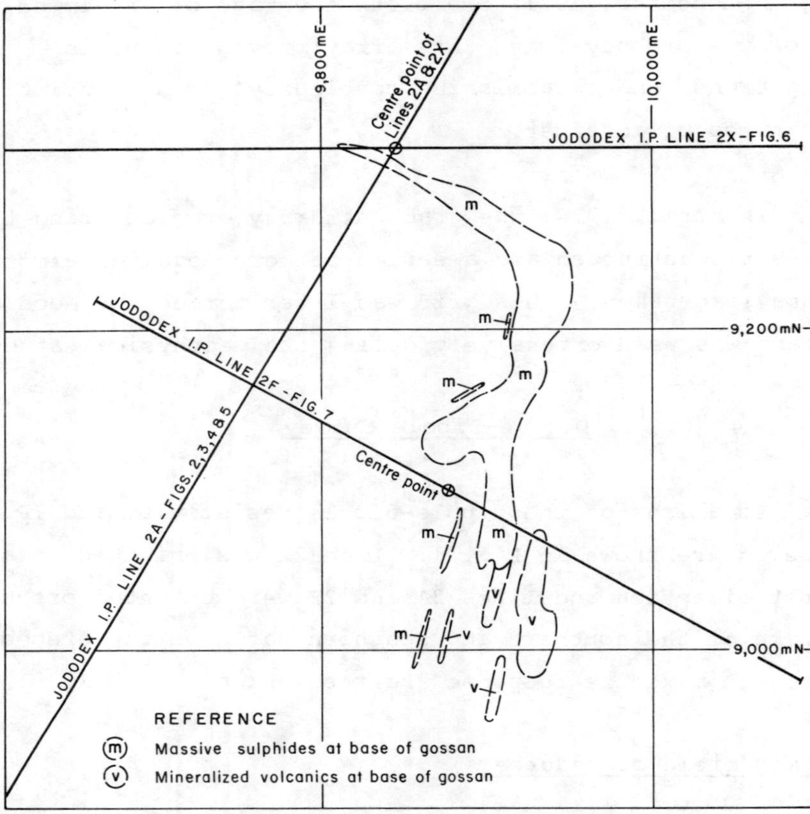

Figure 1 IP line location map

interpretation used these units they are retained. Metal conduction factors (MF) were calculated using percent frequency effects (PFE) as:

$$MF = \frac{1000\ PFE_{0.1}^{3}}{\rho a}$$

where ρa = apparent resistivity at 0.1 Hz

$PFE_{0.1}^{3}$ = percent frequency effect between 0.1 Hz and 3.0 Hz

Figs. 2 to 5 show the data from line 2A perpendicular to the strike of the northern part of the orebody and illustrate the changing character of the anomaly as the dipole length is increased from 15.2m through 30.5m and 61m to 122m. The depth to massive sulphides on this line is 7m and the shallow part of the orebody is approximately 8 to 9m wide but is wider at depth.

It is of interest to note two particular features of the electrical anomalies on line 2A (Fig. 2):

(i) the approximately 15m offset of the strongest responses of the resistivity and the PFE anomalies.

(ii) the better definition of the massive sulphide orebody by the resistivity anomaly.

These features are more obvious in the data obtained with the shorter dipole lengths (Figs. 2, 3, 4).

The resistivity pattern indicates a narrow, well defined conductor, very close to the surface. The PFE pattern is broader and less distinct. This diffuse PFE response is almost certainly due to a smaller frequency effect contrast than the resistivity contrast between massive sulphides and host sequence. (This was substantiated by later measurements on cores and drill hole logging: Tyne, Idnurm and Malone, this volume, p 99; Tyne, this volume, p 375 - Ed). In fact, a considerable amount of sulphides is present in the footwall sequence close to the massive sulphides, and there is sufficient geological evidence to indicate the presence of a bed of pyritic black shale, 50m to 100m or more stratigraphically below the orebody at this point (Malone, this volume, p67). The ultimate effect of these extraneous sources is to widen the PFE response beyond the limits of the resistivity anomaly. The contrast between the massive sulphide body and its host sequence is therefore better defined by the resistivity than the IP method.

For exploration of similar targets the optimum dipole length would appear to be 30.5m for the particular source conditions beneath line 2A although the anomaly is detected at all the dipole lengths.

An interesting feature of the 61m dipole resistivity pseudo-section (Fig. 4) is the zone of higher values cross-cutting the main low resistivity diagonal at n = 5 and 6, between station 4S and 8S. This is thought to be due to the dolerite sill which intrudes the ore zone.

The greater depth of investigation of the 122m dipole (Fig. 5) is reflected in the lower values of apparent resistivity at n = 3 and 4. These are probably due to the considerable thickness of flat-lying massive sulphides below the 2680m level on the southwestern side of this line. However, some contribution from other parts of the complexly folded body cannot be ignored.

The relative depths of investigations achieved with the various dipole lengths are better reflected in the resistivity data. The responses obtained at the shorter lengths (Figs. 2 and 3) reflect the near surface conditions, particularly the shallow geometry of the orebody. As depth of investigation increases with increasing electrode spacings (Fig. 4) the data exhibit almost equal gradients on both arms of the anomaly which, after neglecting the interference from the dolerite intrusions, reflect the nearly vertical attitude of the orebody. Finally, the data from the largest current dipole (122m) in Fig. 5 is effective in reflecting the flat-lying portion of the orebody beneath that line and possibly other portions of the orebody to the southeast of the line. In this case the northeastern side of the anomaly displays steeper gradients.

On the resistivity pattern of Figs. 2 and 3 it is interesting to note the steeper gradients on the southwestern side of the anomaly.

Figs. 3 and 6 afford an interesting comparison. The 30.5m dipole was used for both lines and they intersect the projection of the orebody at the same point but at different angles. Line 2X (Fig. 6) is oblique to strike. Anomalies on this line exhibit higher PFE values and lower resistivity values than on line 2A (Fig. 3) which is oriented orthogonal to strike.

The anomalies on line 2X are not only larger in amplitude than those on line 2A but they also show larger relief. However, resolution on line 2X is poorer than on line 2A. This can be seen by comparing the gradients at the anomaly boundaries on both lines. On line 2X these are smaller than on line 2A and the anomaly is broader. Thus, in this case, both resistivity and PFE responses are enhanced by orienting current lines at an oblique angle to the general strike of the target, sacrificing, however, some degree of resolution.

Line 2F (Fig. 7) shows the broader PFE response over the southern portion of the ore zone where the depth of oxidation has increased to 19 metres. The PFE anomaly is at least 120 metres wide and is open to the east. Obviously, sources other than the massive sulphides are contributing to the response. These are probably mineralized tuffs to the west and the chlorite schist to the east of the orebody. The steep gradients on the western side of the anomaly appear to coincide with the contact between a tuff dolerite. The n = 1 data reflect shallow material since they locate the anomaly centre between stations 1E and 2E with weaker fringes on either side. The larger spacing data (n = 2, 3, etc.) are influenced by the downdip extension of the orebody and the centre of the response is moved some 15 metres to the west to between stations 0.5E and 1.5E.

The apparent resistivity pattern shown on Fig. 7 also reflects the distinct westerly dip of the orebody by exhibiting much steeper gradients to the east. Again, the n = 1 data correctly indicate the near shallow location of the conductive orebody between stations 1E and 2E. The larger spacing data are again dominated by the downdip extension of the sulphides.

Gradient Array Survey

As for the dipole-dipole survey, Heinrichs Geoexploration equipment was again used for the frequency domain gradient array survey.

The current line joining the current electrodes was oriented in an east to west direction approximately orthogonal to the general strike of the orebody. The current electrode separation was 1829m (6,000 ft), extending along line J (9220mN) and centred above the orebody. A potential dipole length of 30.5m (100 ft) was used and data were recorded on lines 30.5m apart covering the Woodlawn Test Grid.

Measurements were made at frequencies of 0.1 and 1.0 Hz. These frequencies were chosen to minimize the electromagnetic coupling effects which can be troublesome with this array. As an additional precaution, the current cable was laid out in a large U-shape so as to ensure a minimum distance of approximately 300 metres to the nearest line surveyed.

As overburden in the Woodlawn area is relatively thin and usually between 2 and 10 metres and, in addition, is poorly conductive with resistivities of approximately 30 ohm m serious coupling problems were not anticipated. Nevertheless, on occasions during the survey, field measurements were made at a number of frequencies up to 10 Hz. These were closely referenced to theoretical gradient array electromagnetic coupling data ensuring that results were not being distorted by coupling effects. At no time during this survey did electromagnetic coupling appear to be a problem.

Discussion of Results

Contoured results obtained over the Woodlawn Test Grid are shown in Figs. 8, 9, and 10 and are discussed below.

Resistivity

The most significant aspect of the resistivity data shown in Fig. 8 is the accuracy with which the apparent resistivity anomaly reflects the shallow region of the massive sulphides. Considering the length of the potential dipole with respect to the depth and width of the source, the anomaly amplitude is quite substantial. Changes in dip and strike are faithfully reproduced, the only discrepancy being at the northern end of the orebody where the strike of the massive sulphides and the trend of the apparent resistivity contours are about 30° apart.

This divergence is difficult to explain. It may be due to an interaction between the geometry of the orebody and the current electrodes. These electrodes were positioned in an east to west direction so that the line joining them was approximately orthogonal to the body as a whole. This configuration would tend to produce north to south equipotential lines in a uniform ground in the central region of the array. The northern limb of the orebody strikes about 300°M, that is, at an angle of 60° to the theoretically expected north to south orientation of equipotential lines in the host rocks north of the orebody. This may have resulted in the anomaly being displaced away from the body.

This explanation is not entirely convincing because the anomaly follows the 300°M striking limb for most of its length and only diverges over the last 50m. The real explanation may become apparent as mining reveals the actual distribution of conductive rocks in the ground.

PFE results

In contrast with the resistivity results, the PFE data (Fig. 9) show a broad, diffuse anomaly which, while generally reflecting the shape of the orebody, appears to be offset somewhat to the east and to extend to the north and south beyond the limits of the known orebody. In fact, the central region of the massive sulphides, south of Line K (9220mN), appears to be associated with relatively low values. This particular aspect will be discussed later.

The broadness of this anomaly is thought to be due to a number of additional polarizing sources in the immediate vicinity of the massive sulphides. Both tuffs and schists of the hanging and foot walls contain sulphides in sufficient quantity to produce a significant response.

A continuous black shale unit stratigraphically beneath the main mineralized horizon also contributes to the anomaly, particularly in the northern part of the orebody. Geological information from drill holes and outcrop east of the ore zone is still incomplete but these pyritic and tuffaceous sediments appear to be the major contributing source of the eastern response near the central and northern regions adjacent to the orebody.

The isolated PFE low at 9100mN, 9900mE (station N11.5) was subsequently confirmed by detailed electrical measurements. An increase in apparent resistivities was found to be present in the same area in the detailed investigation. The PFE low and the higher apparent resistivity values may be caused by the presence of an old open exploration shaft at 9100mN, 9915mE (station N12).

South of Line N larger PFE values were measured. These are almost certainly caused by the less continuous and more disseminated pattern of mineralization in this area where the ore occurs primarily in unconnected lenses of mineralized tuff and chlorite schist. These have a lower sulphide content than the massive, complex ore. The peak PFE response coincides with the surface projection of this mineralized zone. However, the anomaly is considerably wider than the ore zone. This is probably due to the pyrite bearing sediments to the east and the 45° dip of the orebody to the west. The latter may further be exaggerated by topographic effects since, in this area, the ground level drops in the same direction as the subsurface dip. A black shale unit at about the same stratigraphic level as the main massive sulphide horizon may also be contributing to the response but this unit does not extend north of Line S (8950mN). No other obvious shallow sources have been identified in outcrop or in drillholes.

MF

The MF contour pattern (Fig. 10) generally reflects the strong resistivity contrast between the orebody and host rocks with a well defined anomaly.

Detailed Gradient Array Survey

In April, 1974, the gradient array covering the ore zone was reconstructed and a series of more detailed measurements were taken. Equipment and electrode arrangement were the same as used previously and measurements were again made at frequencies of 0.1 and 3 Hz with potential dipole lengths and corresponding station spacings of 15.2m (50 ft) and 7.6m (25 ft).

Figs. 11, 12 and 13 show the responses obtained with various dipole lengths. These figures include data from the earlier gradient array survey using a 30.5m potential dipole. As can be seen the most distinctive anomaly was obtained with the shorter potential dipole, particularly when judged in relation to the previously noted PFE low over a portion of the massive sulphide body obtained earlier with the 30.5m dipole.

Fig. 11 shows an accentuation of the resistivity response with decreasing dipole spacing. However, the optimum spacing would appear to be 15.2m since the 7.6m data exhibits unwanted noise. The 30.5m response lacks detail, which is not surprising considering that the orebody is 32m wide and is overlain by 11m of overburden.

The minimum of the apparent resistivity low occurs somewhat downdip from the subcrop of the massive sulphide body.

The change in the shape of the PFE response at different dipole lengths (Fig. 12) is more critical. In addition to an increase in amplitude with decreasing length, the data reveal a distinct splitting of the anomaly which clearly reflects two discrete sources. The larger response stems from pyritic tuffs and sediments to the east and the lower amplitude anomaly stems from the orebody. Again the peak of the orebody response is somewhat offset down dip from the top of the orebody.

Detailed information over the northern part of the orebody shown in Figs. 14, 15 and 16 is equally revealing. The splitting of the massive sulphides below the gossan is reflected in the resistivity response (Fig. 14) which shows only a small rise in the major low. The near vertical orebody is clearly reflected in the symmetry of the anomaly pattern. This may be contrasted to Fig. 11 which shows an asymmetric anomaly caused by the 45° dip to the west.

The two distinct lenses of massive sulphides shown on the geological cross-section produce two separate PFE anomalies (Fig. 15). An additional prominent pronounced anomaly to the east of these anomalies reflects the presence of pyritic sediments. The responses appear broadened due to the obliqueness of the line to the strike of the orebody.

Of the three parameters presented, MF (Fig. 16) appears to be the best discriminator of the massive sulphide orebody, giving a clear, uncomplicated anomaly. The apparent resistivity parameter in general is also very sensitive to massive mineralization. In practice it has been found that all three parameters need to be used in interpretation of gradient array data.

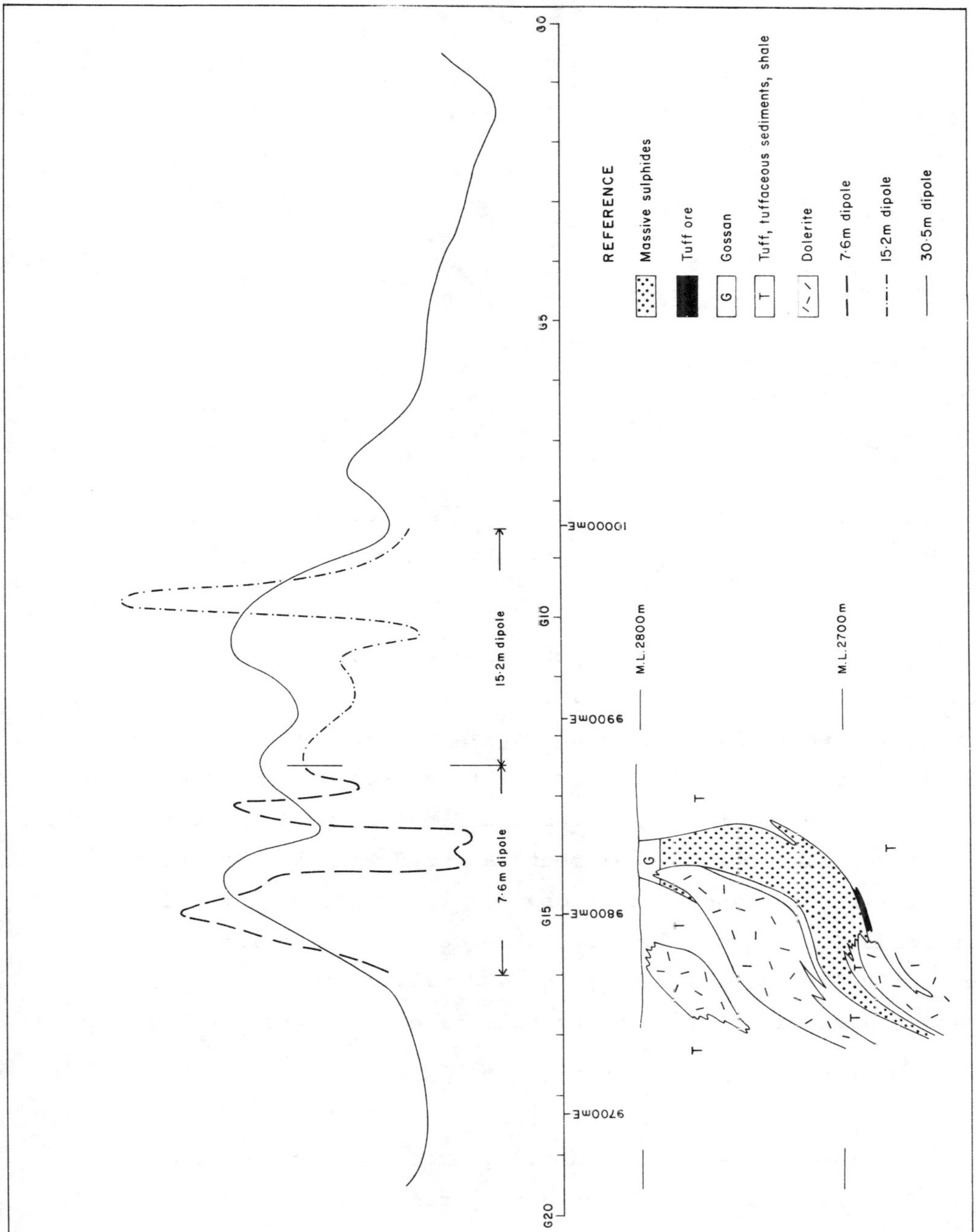

Figure 15 Line 9312mN (Line G, approx.), Gradient Array PFE Profiles

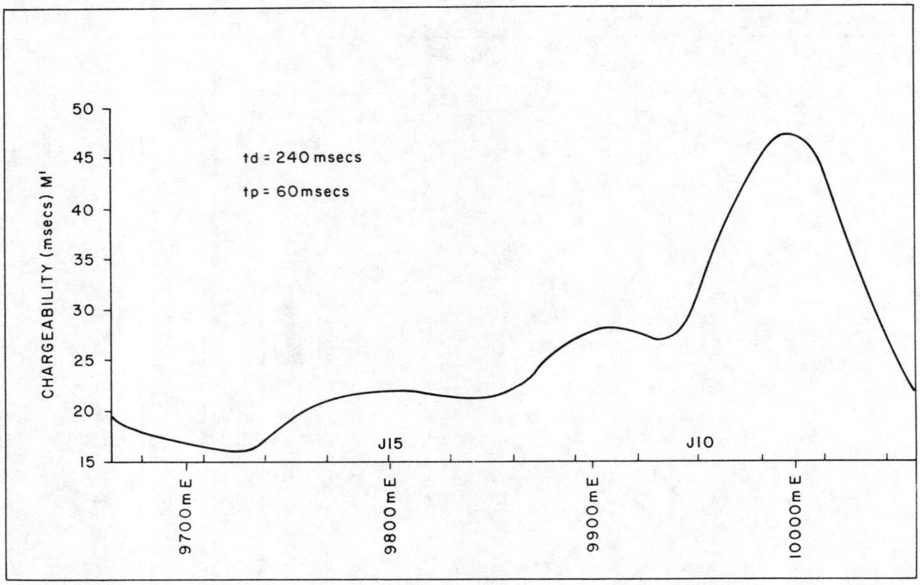

Figure 17 Line 9220mN (Line J), Apparent Chargeability Profile (milliseconds)

For comparison with the frequency domain IP data a profile of apparent chargeability is presented in Fig. 17. These data were collected by the Bureau of Mineral Resources (BMR) in April, 1974, using a gradient array with a current electrode separation of 701 metres and potential dipole length of 30.5 metres. Huntec Mark 3 Time Domain equipment was used with a delay time (t_d) of 240 milliseconds and a first channel (M1) sample width of 60 milliseconds (tp). The width of subsequent channels was consequently 120 (M_2), 240 (M_3) and 480 milliseconds (M_4) with apparent chargeability (M') being calculated from the relationship.

$$M' = t_p(M_1+2M_2+4M_3+8M_4) \times 0.01 \text{ msec.}$$

The apparent chargeability profile obtained closely follows the PFE response for the same potential dipole length (Fig. 12). This demonstrates the practical equivalence of the two techniques. However, the anomaly attributed to the Woodlawn orebody does not appear to be as distinct on the chargeability curve.

Conclusions

Field measurements have shown the resistivity and IP methods have obvious potential for the discovery of orebodies of the same type and general setting as the Woodlawn deposit. The actual response, however, can rarely be expected to be uniform or uncomplicated and must be interpreted by considering a range of geological possibilities.

A comparison between the electrical anomalies delineated and the geology from drilling shows that some of these anomalies are offset with respect to the orebody. This is thought to be due to the presence of sources distinct from the massive sulphide body. These extraneous sources lie in the immediate surrounds of the orebody, particularly in the footwall sequence, and are close to the orebody since their IP response can be resolved only by using very short dipole lengths and station spacings.

Tuffs, tuffaceous shale and black shale, all locally mineralized with pyrite, with or without base metal sulphides, appear to be the most likely sources of extraneous IP response. However, drillhole information from the stratigraphic interval just below the ore zone is insufficient to resolve this problem. It is anticipated that further geological knowledge, gained during mining, will determine the various sources contributing to the resistivity and IP response.

The complexities inherent in the anomalies are, amongst other things, the result of three important variables. These are the geometry of the deposit; the nature, grade, and distribution of the mineralization; and the contrast and relationship between orebody and host rocks.

Although the resistivity and IP methods are capable of detecting targets such as the Woodlawn orebody, care must be taken when moving to the more refined levels of interpretation.

The responses obtained with both the dipole-dipole and the gradient array are complex, owing mainly to the small width of geological formations, the presence of extraneous IP sources in close proximity and the fact that a single geological formation may produce different resistivity and frequency effects along its length.

8.3. Electrical Profiling with a Number of Arrays at Woodlawn

E. D. TYNE AND R. J. WHITELEY

Contents

Abstract

A large number of electrode arrays were tested over the Woodlawn orebody. All arrays produced simple resistivity responses which were consistent with the shallow, highly conductive orebody. The time domain induced polarization (IP) responses were, however, complex due to the proximity of a number of polarizable sources.

The major IP sources are the orebody and a polarizable region with the footwall rocks to the east of the orebody.

Frequency and time domain IP data were compared for the gradient array and were shown to be essentially similar. The response of this array was also computer modelled substantiating the multiple IP sources in close proximity.

The three and dipole-dipole arrays produced the most discernable resistivity anomaly over the orebody.

The maximum IP response was similar for the gradient, three and dipole-dipole arrays with the most discernable IP anomaly produced by the dipole-dipole array. The Wenner array produced a maximum IP response which was twice that of the gradient array. The equispaced Unipole array appeared to define, with large anomalies, the edges of the broad polarizable region associated with the orebody and footwall rocks.

A downhole to surface array using a buried current electrode within the massive sulphide orebody gave IP profiles similar in shape to the three array but with an enhanced IP anomaly amplitude over the orebody.

Introduction

Resistivity and induced polarization (IP) surveys may be carried out in either the time or frequency domain using a wide variety of electrode arrays. Many of these have been described by Whiteley (1973) and may be classified either as conventional or compensation arrays.

Six different electrode arrays were tested over the Woodlawn orebody and their responses compared. The electrode arrays tested were the conventional gradient array, dipole-dipole array, equispaced three array (Sumner, 1976, p 220), tripotential array (Carpenter and Habberjam, 1956) and the compensation, unipole array (Gupta and Bhattacharya, 1963). In addition, a surface to downhole array using a buried current electrode was also tested.

Electrical profiling was carried out on line G (9312mN) and line J (9220mN) of the Woodlawn Test Grid as shown on Fig. 1. The downhole current electrode for the downhole to surface array was

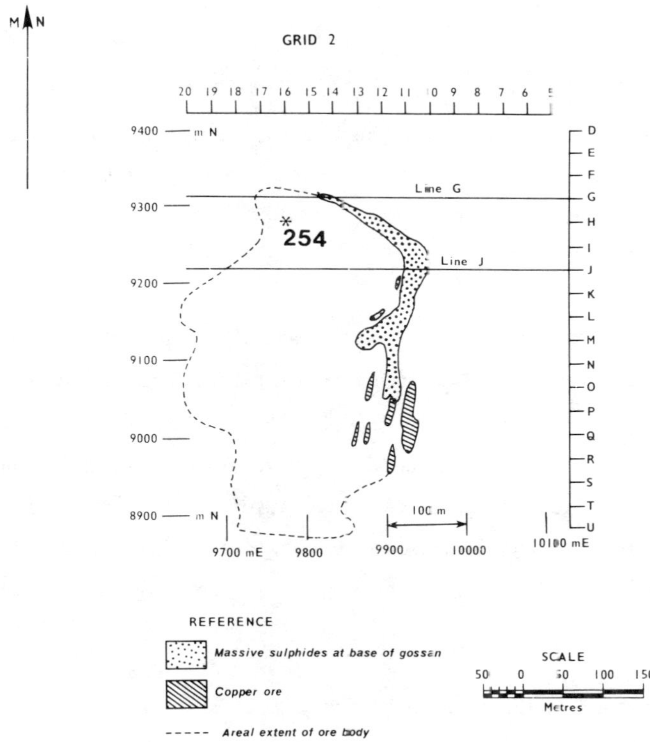

Fig. 1 Location of electrical profiles and downhole current electrode.

placed within the massive sulphides at a depth of 125m in the vertical drill hole W254 (Fig. 1).

Time and frequency domain IP measurements were compared for the gradient array whose response was also modelled.

An Austral T842 current transmitter and a Huntec Mark 3 IP receiver were used to measure both apparent resistivity and time domain IP parameters. Apparent chargeabilities M_{1530}^{480} were obtained using an 8 second time domain current waveform on a 50% duty cycle by integrating the IP transient decay curve over the period from 480ms to 1530ms.

Gradient array, Line J

Gradient array resistivity and IP data in both the time and frequency domain, obtained along line J, are shown in Fig. 2. The Woodlawn orebody is almost orthogonal to line J and has a westerly dip of about 45°.

The frequency domain IP data were gathered by Jododex Australia Proprietary Limited (Quick and Cifali, this volume, p333) using frequencies of 0.1 and 1.0 Hz. For both surveys the same current electrodes were used with a spacing of 1829m. A potential dipole spacing of 15.2m was used. The centre of the current electrode pair was approximately over the orebody and the line joining the electrodes was along line K (9190mE).

As expected, the apparent resistivity profiles obtained with both systems are almost identical with a distinct low over the orebody. This low anomaly exhibits some asymmetry, characteristic of the westerly dipping, highly conductive orebody. This asymmetry is, however, accentuated by the large resistivity contrast between the very resistive footwall rocks with resistivities of about 2000 ohm m and the less resistive hanging wall rocks with resistivities of about 500 ohm m.

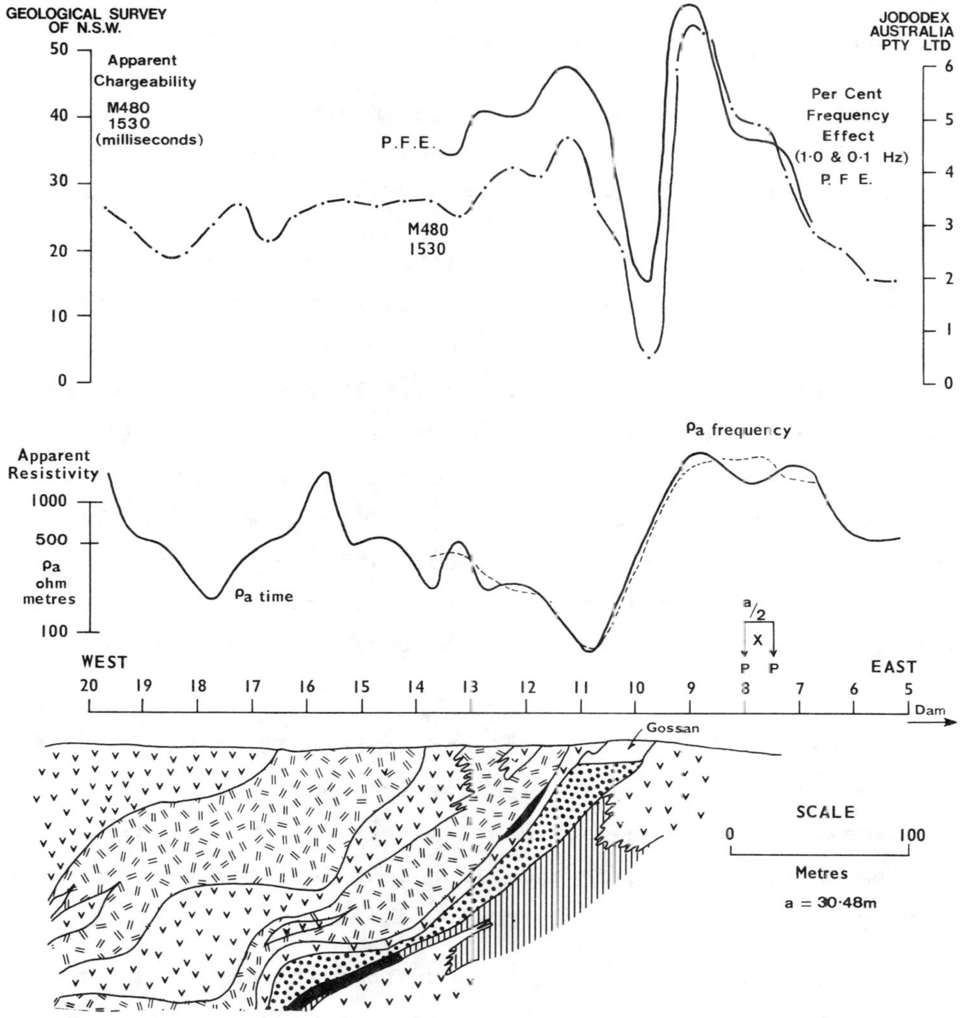

Fig. 2 Frequency and time domain gradient array IP profiles, Line J.

The IP profiles are also shown on Fig. 2. There is excellent correlation between the profiles although the amplitudes of the apparent chargeability and percent frequency effect (PFE) anomalies cannot be directly compared due to inherent differences in the nature of these parameters. This correlation indicates that, in the Woodlawn environment, the IP effect may be efficiently measured in either the time or frequency domain, provided integration periods or frequencies are suitably chosen to reject a major portion of extraneous influences such as electromagnetic (EM) coupling.

The excellent resolution of the features on the gradient array profiles is due to the small spacing of the potential dipole relative to the width of the anomalous sources present.

The IP data show two peaks attributed to two polarizable sources in close proximity, separated by an abrupt low above the orebody at J9.75. The eastern IP anomaly has a larger amplitude and correlates with the resistive footwall rocks which are tuffs and pyritic black shales. The lower amplitude, western IP anomaly correlates with the highly conductive Woodlawn orebody. The asymmetry of this western anomaly is consistent with the westerly dip of the massive sulphides, although the abrupt low at J9.75 causes some distortion to the eastern flank of the anomaly, which makes it difficult to assess the true dip of the source.

Although the amplitude of the western IP anomaly associated with the massive sulphides is smaller than that over the footwall rocks these data confirm that massive sulphides give rise to significant IP anomalies despite a common misconception that they are not particularly polarizable.

It is also of interest to note that the amplitude of the IP anomaly over the footwall rocks is similar on both the apparent chargeability and PFE profiles whereas there is a noticeable difference in anomaly amplitude over the massive orebody. This may be partly due to different ground conditions at the times of the two surveys, however, it may also be related to the different maximum current charging times which were only 2 seconds for the time domain survey compared with 5 seconds (0.1 Hz) for the frequency domain survey. It is likely that the highly conductive massive sulphide orebody requires longer charging times to reach polarization saturation compared with the resistive, polarizable footwall rocks.

The abrupt low at J9.75 between the two IP anomalies is of particular interest and is discussed further by Tyne (this volume, p 414) and in conjunction with the computer modelling results in the next section. This low above the orebody appears to result from a geometric effect arising from the proximity of the two polarizable sources which cause subtractive interaction of the respective IP voltage discharges (or frequency effects).

Computer model of gradient array response on Line J

Simple computer modelling of the gradient array electrical data along line J was undertaken. The dipping Woodlawn orebody was simulated by three discrete prisms each with a resistivity of 1 ohm m offset to the west with increasing depth. The resistive footwall zone was modelled as an additional prism with a resistivity of 9000 ohm m.

Apparent resistivity and chargeability anomalies were calculated for each prism using the surface integral method of Pratt (1972) with a computer program written by Templeton (1975). Electrical interaction between the prisms and the ground surface was taken into account in the iterative solution for each prism, however, electrical interaction between the prisms was not considered. Although mutual interaction between the conductive prisms simulating the Woodlawn orebody would not greatly alter the shape of the resultant IP anomalies, they were separated to reduce any such effects.

The individual anomalies of the four prisms simulating the Woodlawn situation were combined using the rules of composition (Habberjam and Jackson, 1974) to give profiles for comparison with field data.

Fig. 3 shows the computed apparent resistivity and IP profiles together with the field profiles and model for the

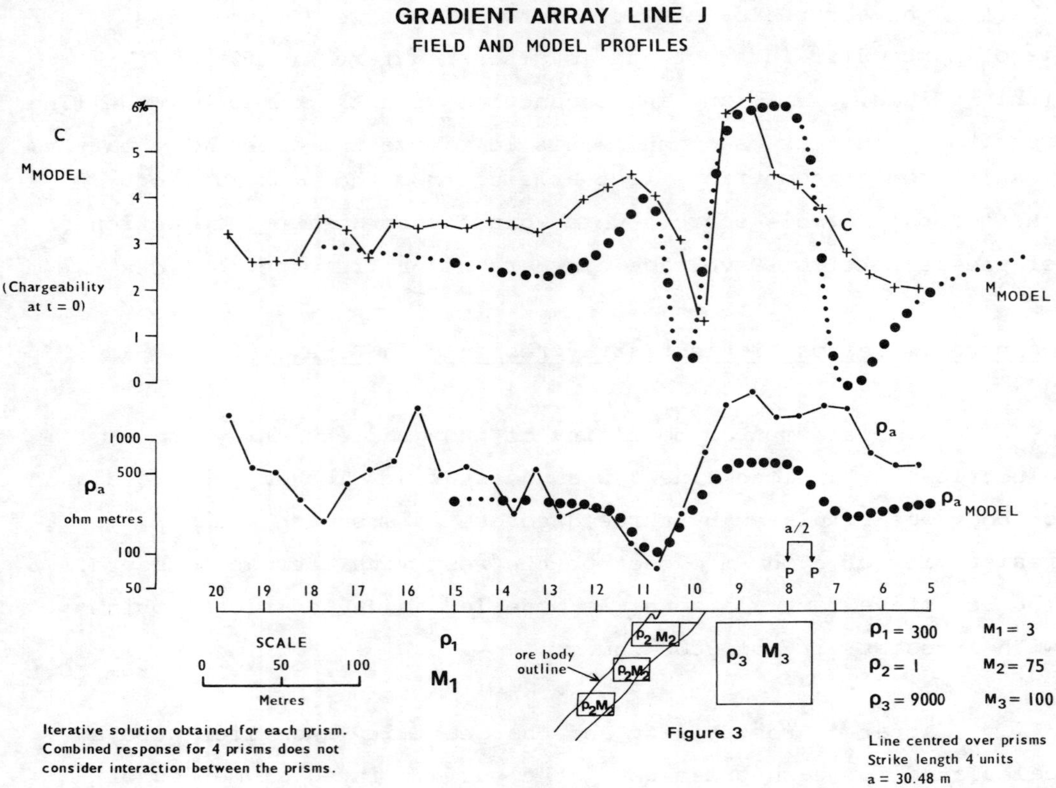

Fig. 3 Gradient array model and field profiles, Line J

gradient array along line J. It should be noted that the apparent chargeabilities shown are actually V_s/V_p expressed as a percentage where V_s = peak secondary IP voltage and V_p = primary voltage at the potential electrodes at the time of cessation of the primary current (i.e. at t=o). For the field profile these apparent chargeabilities were derived from analysis of the transient voltage decay curves as described by Tyne (this volume, p375).

Since the model used to represent the Woodlawn situation was fairly simple, the relative shape of the computed electrical profiles with respect to the field profiles was considered of more importance than an exact amplitude fit.

As shown in Fig. 3 both the computed and field apparent resistivity profiles show a similar shape, however, a reasonable amplitude fit was only obtained in the region of the resistivity low above the orebody. It is of interest to note that the computed resistivity anomaly reaches a minimum of only 100 ohm m despite the very low resistivity for the orebody prisms of 1 ohm m and their proximity to the surface.

It is conceivable that had the resistive prism been shallower and extended further to the east a better fit could have been obtained over the footwall region.

The computed IP profile is quite similar to the field profile although the lows on either side of the larger eastern anomaly are more prominent. These abrupt lows are characteristic of IP anomalies over discrete, resistive, polarizable bodies.

The adjacent conductive and resistive polarizable regions give rise to the abrupt low near J10 above the orebody on both profiles. The low is formed by the subtractive interaction of IP voltage discharges.

The fit between the computed and field IP profiles to the east of J7.5 could probably have been improved by adding another shallow, resistive, polarizable prism in this region.

The simple nature of the IP model is sufficient to affirm that the double peaked IP field profile arises from two discrete polarizable bodies of very different electrical properties.

Surface Profiling, Line G

Figs. 4 to 8 show the electrical profiles obtained with the different electrode arrays on line G (Fig. 1) near the northern

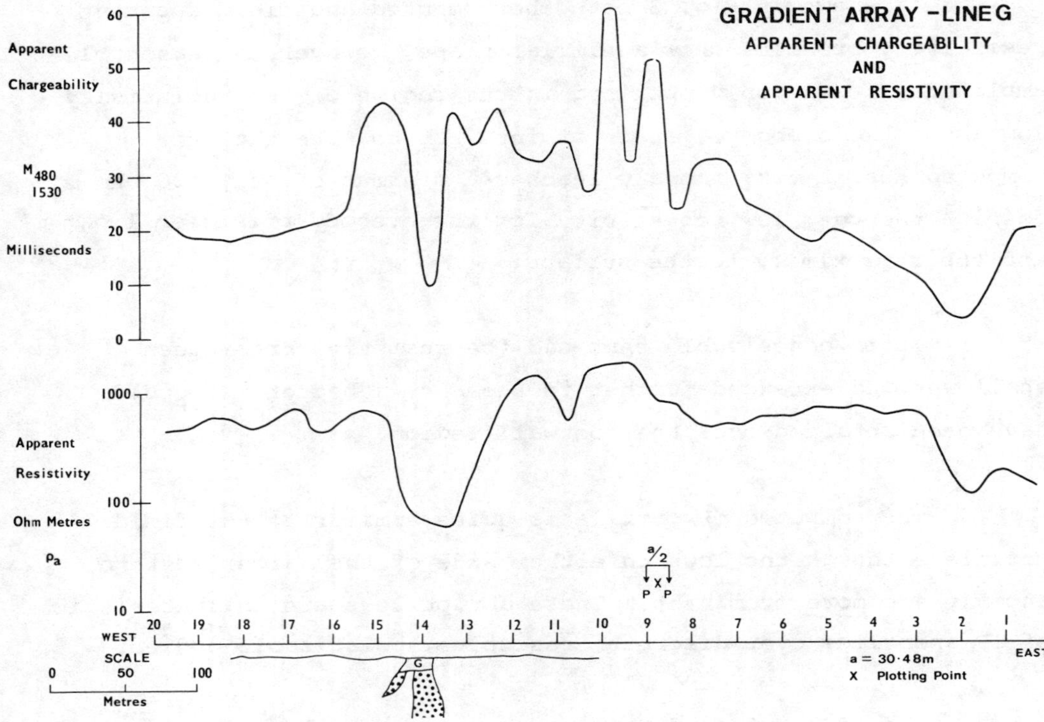

Fig. 4 Gradient array electrical profiles, line G

end of the Woodlawn orebody. At this location the orebody is essentially vertical, very shallow and strikes obliquely to line G.

Gradient Array

The current electrode position and spacing was the same as for the profiling along line J discussed earlier. A 15.2m potential electrode and station spacing was again used.

Fig. 4 shows the apparent resistivity and chargeability profiles along line G. The apparent resistivity data show a pronounced low near the orebody at about G14 with a minimum resistivity of about 60 ohm m below a general background of more than 500 ohm m. This gives an anomaly amplitude to background ratio of at least 1 to 9. The almost symmetrical nature of this anomaly indicates the near vertical attitude of the massive sulphide orebody although the anomaly is quite broad relative to the width of the orebody. This is due to the obliqueness of the orebody strike to the profile line. High apparent resistivities of more than 1000 ohm m from about G9 to G12.5 correlate with the resistive footwall rocks. The hanging wall rocks have apparent resistivities of about 500 ohm m close to the general background level. These are consistent with the in situ resistivities for the hanging wall rocks obtained from drill hole logging (Tyne, this volume, p538).

The gradient array IP profile on line G shows similar complex behaviour to that on line J (Fig. 2). The anomalous apparent chargeability peak of about 43ms at G14.75 is about twice the background level. This is probably related to the orebody and offset to the west because of the obliqueness of the line to the strike of the orebody. This discrete peak is again separated from a broad polarizable region associated with the footwall rocks by an abrupt low at G13.75.

Further to the east and superimposed on the polarizable region are high amplitude narrow peaks with apparent chargeabilities up to 60ms. These are believed to be caused by shallow highly polarizable tuffs or sediments.

The gradient array has good lateral resolution of anomalies on this line because of the small potential electrode and station spacing relative to the width of many of the anomalous sources.

Three array

The equispaced three array profile data for line G are shown in Fig. 5. The lateral resolution of this array is

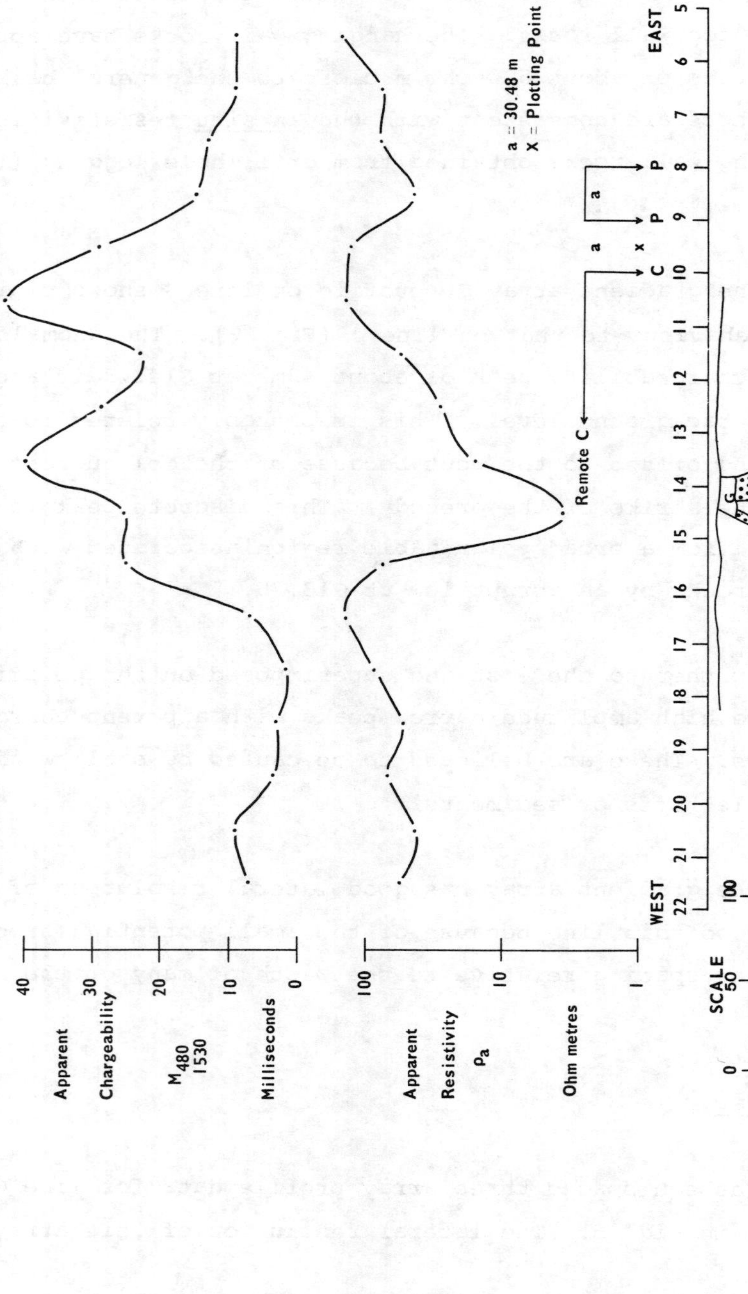

Fig. 5 Three array electrical profiles, line G

decreased relative to the gradient array because of the larger electrode and station spacing of 30.5m.

The apparent resistivity profile shows quite a different character to the gradient array profile in Fig. 4 with a broad asymmetric resistivity low whose minimum is displaced slightly to the west of the orebody. The shape of this anomaly is typical of that over a shallow vertical conductive tabular body with a three array whose spacing is greater than the width and depth of the body (Pratt and Whiteley, 1974).

The well defined apparent resistivity minimum over the orebody has values down to 3.6 ohm m which is closer to the true resistivity of the massive sulphides. The anomaly amplitude to background resistivity has increased to 1 to 16. The resistive footwall region noted on the gradient array profile is not as obvious on the three array profile. Also, in general, the apparent resistivities measured with the three array are about an order of magnitude less than those for the gradient array. This is due to generally lower resistivities in surficial layers and weathered rocks relative to fresh rocks and the fact that the three array (and the other arrays) samples a much smaller and shallower volume of rock material than does the gradient array.

The three array IP profile shows a narrow plateau in the vicinity of the orebody with an amplitude of 25ms followed by an abrupt peak at G13.5 with an amplitude of about 40ms and an additional peak of similar amplitude to the east at G10.5.

It is difficult to interpret the three array anomalies in the vicinity of the orebody because of poor lateral resolution and distortion due to proximity of sources. It is likely, however, that the plateau region is produced by the orebody while the following peak is enhanced by the nearby polarizable footwall rocks. This would suggest, for the three array IP anomaly over the orebody, an amplitude to background ratio of about 5 to 1 compared to about 2 to 1 for the gradient array.

For the eastern IP anomaly at about G10.5 the poorer lateral resolution of the three array can be seen by comparing this anomaly with the double peaked gradient array anomaly in the same area.

Dipole-Dipole array

Fig. 6 shows electrical profiles obtained with the dipole-dipole array using the same electrode spacing of 30.5m as for the three array and dipole separations (n) of 1, 2 and 3.

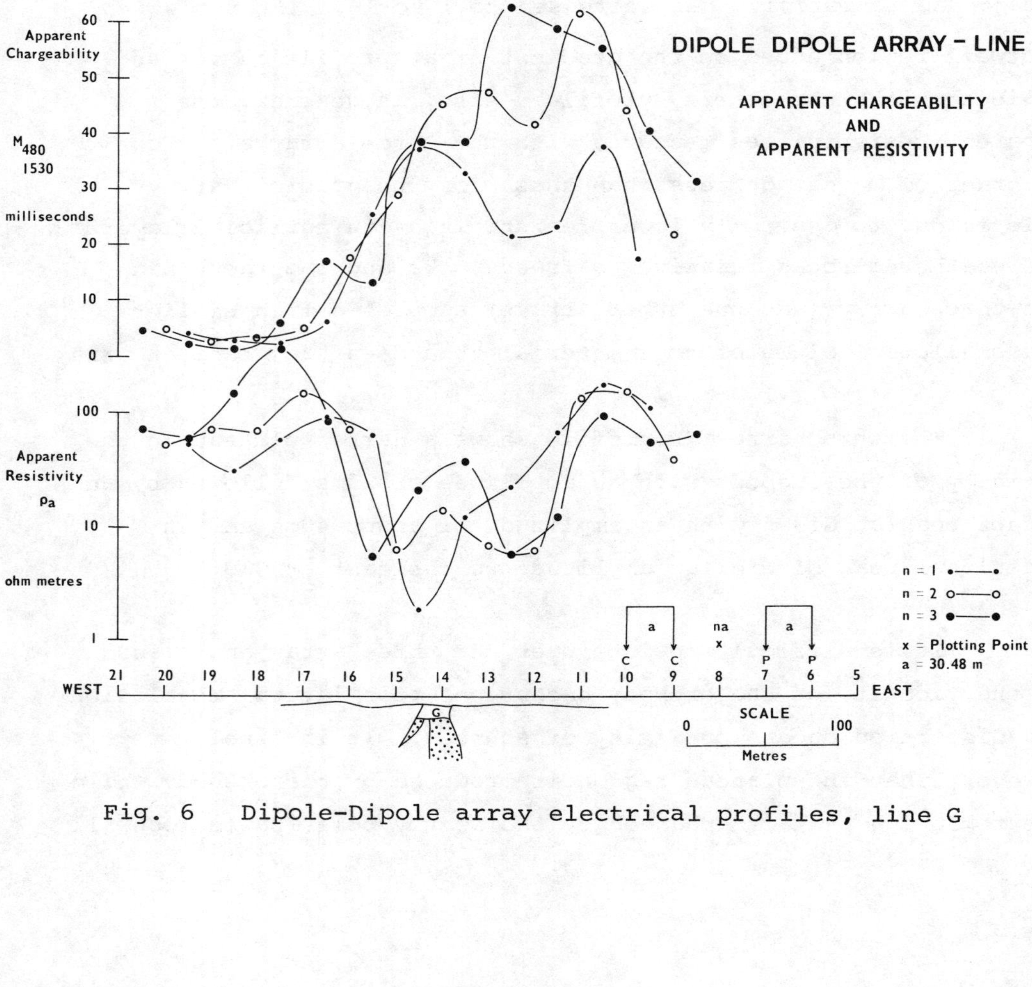

Fig. 6 Dipole-Dipole array electrical profiles, line G

The dipole-dipole apparent resistivity profiles show a fairly symmetrical, complex low over the massive orebody. At n = 1 the shape of the anomaly is remarkably similar to the three array anomaly with a minimum of about 2 ohm m compared to 3.6 ohm m for the three array giving an anomaly amplitude to background of about 1 to 25. At the larger dipole separations of n = 2 and 3 the responses are typical of a shallow, vertical, conductive body whose width and depth are less than the electrode separation (Pratt and Whiteley, 1974, Fig. 4).

The dipole-dipole resistivity profiles do not show an anomaly from the resistive footwall rocks as obtained with the gradient array (Fig. 4). These rocks must therefore have comparable resistivities to the hanging wall rocks at shallow depth.

The dipole-dipole IP profile for n = 1 shows two prominent peaks and is similar in shape and amplitude to the three array IP profile but is displaced one station to the west. The western peak correlates with the resistivity low over the orebody and has an IP anomaly amplitude to background ratio of about 15 to 1.

At the larger dipole separations of n = 2 and 3 the IP response increases significantly. The two peaks on the n = 1 data merge into a single large anomaly which is displaced to the east of the orebody. This shows that the greater volume of polarizable material on the eastern side of the orebody swamps the response of the conductive orebody at larger dipole separations.

Tripotential arrays

Carpenter (1955) has shown that there are only three unique arrangements of current and potential pairs in any four electrode array. These arrangements are known as the alpha, beta and gamma arrays and provide apparent resistivities whose signs are dependent on the direction of current flow and measurement conventions. Together, these arrangements are called the tripotential arrays.

For the equispaced tripotential arrays the alpha array is equivalent to the Wenner array and the beta array to the dipole-dipole array for n = 1. Apparent resistivities obtained with the tripotential arrays are related by a simple expression which enables the third apparent resistivity to be calculated from measurements with the other two arrays. Similar expressions also exist for the IP measurements, alternatively, if all three apparent resistivities are measured, then instrumental accuracy may be checked and indirectly the principle of reciprocity may be verified (Habberjam, 1967).

Electrical profiles obtained with the equispaced tripotential arrays using an electrode spacing of 30.5m are shown in Fig. 7. For these arrays, four non-polarizing electrodes were used and measurements were made by successively switching to each array at every station.

Electrical profiles obtained with the alpha, beta and gamma arrays will intersect wherever a lateral electrical contrast, within the depth of investigation of the arrays, is closely approached or is within the array (Pratt and Whiteley, 1974, Fig. 6). When this occurs, for the arrangement used at Woodlawn, the alpha and gamma profiles will generally move in the same direction while the beta profile will move in the opposite direction. These effects can be seen on the apparent resistivity profiles in Fig. 7 which also show a small symmetrical resistivity low with the alpha and gamma arrays over the orebody centred at G13.75. The anomaly amplitude to average background resistivity for the alpha or Wenner array is about 1 to 7.

TRIPOTENTIAL ARRAY-LINE G

APPARENT CHARGEABILITY AND APPARENT RESISTIVITY

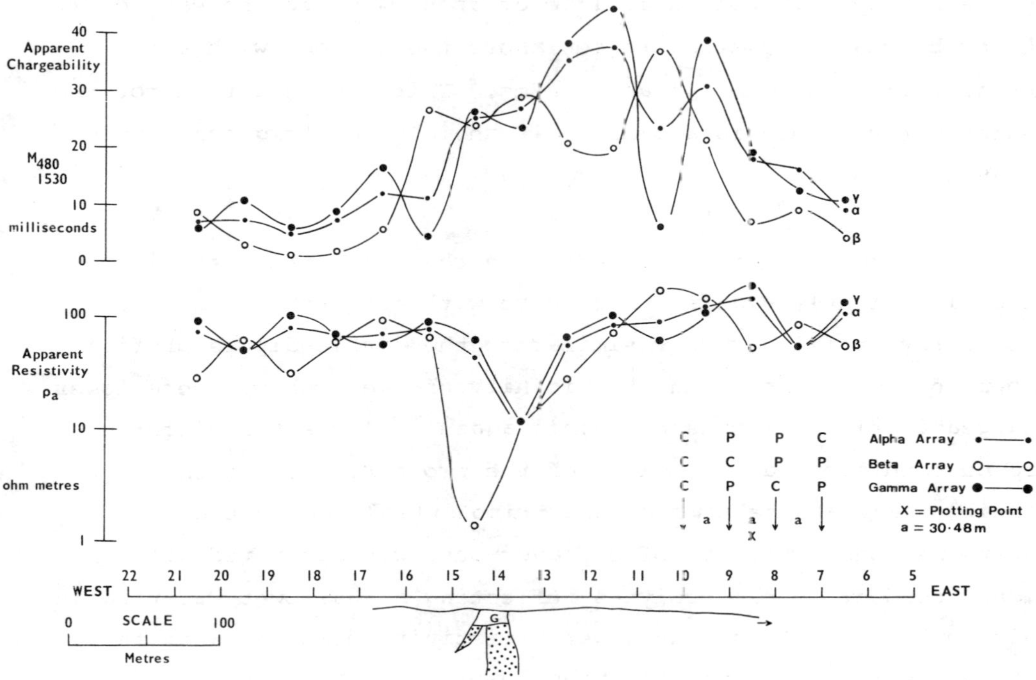

Fig. 7 Tripotential arrays electrical profiles, line G

The resistivity low obtained with the beta array is much more prominent than that obtained with the alpha and gamma arrays and is displaced approximately one station to the west. This anomaly is almost identical to that for the n = 1 dipole-dipole array in Fig. 6, however, minimum resistivity is somewhat lower indicating more conductive conditions in the vicinity of the orebody at the time of survey.

Other minor resistivity anomalies are also indicated where the tripotential profiles intersect on either side of the orebody.

The apparent chargeability profiles obtained with the tripotential arrays are quite complex and intersect frequently indicating a number of polarizable sources particularly on the footwall side of the orebody. The alpha array shows one broad chargeability high with a maximum of about 40ms to the east of the orebody and a plateau region above the orebody with a chargeability maximum of about 27ms. This suggests that for the Wenner array, there is a maximum IP anomaly to background ratio of about 4 to 1.

Chargeabilities obtained with the beta array are, in general, slightly smaller than those with the alpha array. The beta array profile is very similar to the dipole-dipole profile shown in Fig. 6 except in the vicinity of the orebody where lower chargeabilities are present. This again reflects the differing ground conditions at the times of the two surveys. It is possible that at the time of the tripotential survey the supergene zone, which is often cavernous, was saturated with highly conductive electrolytes reflecting a shallower water table. This would account for the lower resistivity measured with the beta array and, consequently, the lower chargeability. The adjacent polarizable footwall rocks, particularly black shales, would not be greatly affected.

Unipole array

The unipole array is rarely used in base metal exploration. This array has four mobile electrodes and a fixed current electrode some distance away. Equal current strengths of the same polarity are transmitted through the outer electrodes of the profiling array. This was done at Woodlawn by means of variable resistors and specially designed switching circuitry (Emerson et al, 1969) which allowed measurements with the Wenner and the other tripotential arrays at each station without electrode rearrangement. The equispaced unipole or unipole-Wenner array with an electrode spacing of 30.5m was used at Woodlawn.

The unipole array is a special type of field compensation array called a pure anomaly array (Whiteley, 1973), in which, over a homogeneous and horizontally layered earth zero voltages differences are obtained. Non-zero voltages will only be measured when lateral electrical variations are present. As a result, apparent resistivities cannot be calculated and the primary potentials must be normalized. At Woodlawn, voltages were normalized with a theoretical homogeneous ground potential and a zero level was arbitrarily chosen.

The normalized primary unipole and Wenner voltage ratios and IP profiles are shown on Fig. 8.

The Wenner array voltage ratio profile shows a prominent minimum over the conductive orebody and maximum over the resistive host rocks to the east. This profile is simply a linear version of the alpha array logarithmic apparent resistivity profile on Fig. 7. The unipole profile shows a definite crossover anomaly in the vicinity of the orebody with the crossover correlating with the minimum of the Wenner array. This anomaly is typical of a highly conductive body of limited extent (Gupta and Bhattacharya, 1963).

The anomaly evident on the eastern side of the orebody with the Wenner and other conventional arrays and associated with resistive rocks does not appear to be present on the unipole profile.

The unipole IP profile shown on Fig. 8 is interesting and fairly complex relative to the Wenner profile with large negative and positive crossover anomalies. The negative anomalies indicate a reversal in the IP current directions. These large negative anomalies are difficult to interpret although they correlate closely with the edges of the broad polarizable region over the Woodlawn orebody and to the east of it as shown for the Wenner and other arrays. Within this region unipole chargeabilities are similar in amplitude to Wenner chargeabilities.

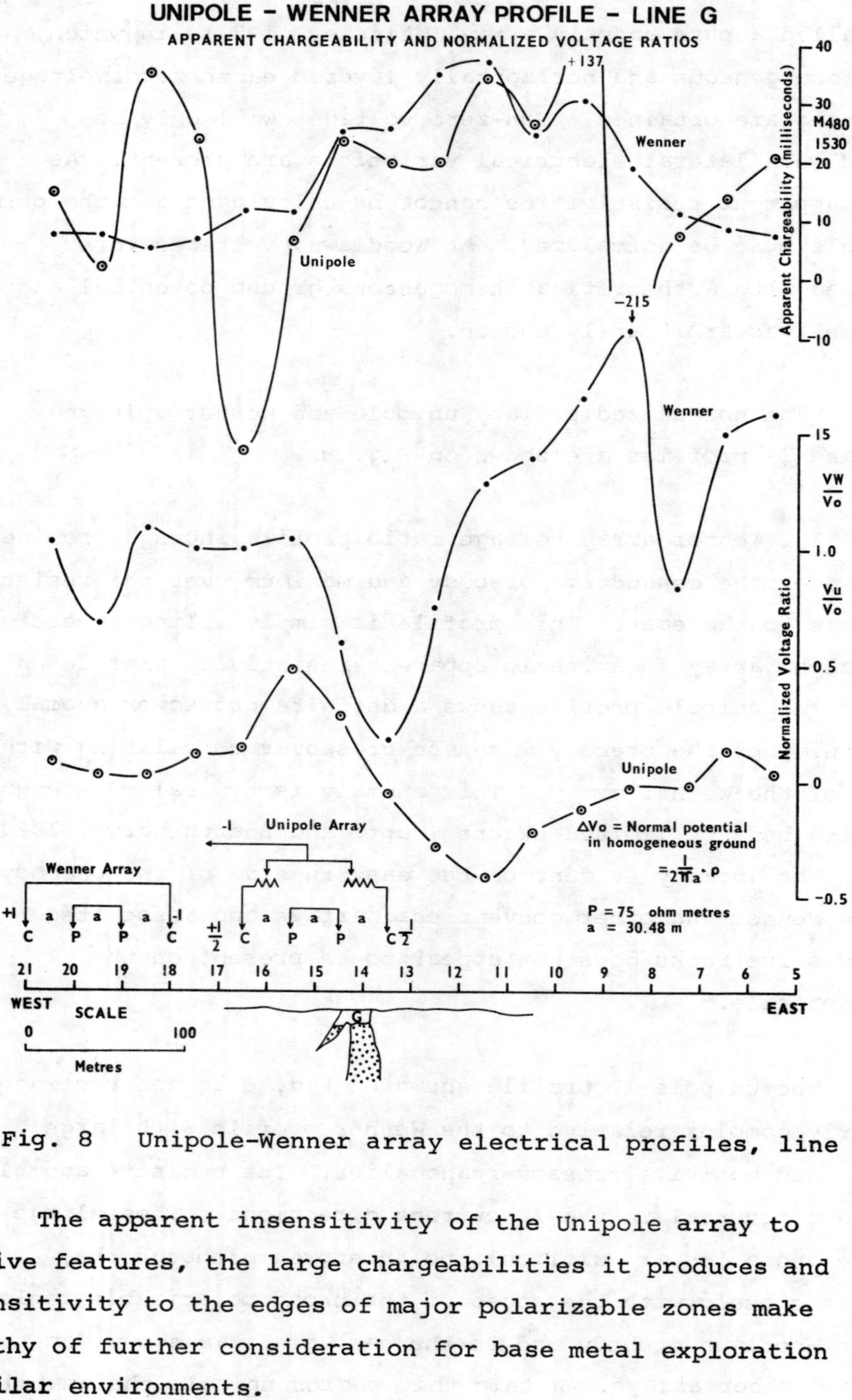

Fig. 8 Unipole-Wenner array electrical profiles, line G

The apparent insensitivity of the Unipole array to resistive features, the large chargeabilities it produces and its sensitivity to the edges of major polarizable zones make it worthy of further consideration for base metal exploration in similar environments.

Downhole to surface array

Electrical surveys are commonly completed in drill holes following the intersection of massive sulphides, to determine the dimensions of the sulphides and to locate nearby sulphides not encountered in the drilling. Generally a current electrode is placed within the sulphide intersection and applied potential (mise-a-la-masse) surveys are carried out with measuring electrodes on the surface. Such surveys have been completed at Woodlawn (Templeton et al, this volume, p 509). These generally rely on a substantial resistivity contrast between the sulphides and the host rocks, however, the IP contrast is not always considered.

A test resistivity and IP survey at Woodlawn was completed with a downhole current electrode and remote current return electrode on the surface.

The current electrode was placed within the massive sulphides in drill hole W254 (Fig. 1). Resistivity and IP measurements were then made along line G with a potential dipole and station spacing of 30.5m.

Fig. 9 shows the electrical profiles obtained with this array. The location of the downhole current electrode in W254 is about 30m south of G16 and has been projected onto the section shown in Fig. 9.

Apparent resistivities were calculated using geometric factors to account for the earth's surface (Hauck, 1970). The profile shows a resistivity low associated with the orebody and a resistivity high immediately to the west over a shallow dolerite sill. Evidence of this resistive dolerite is seen as a small resistivity peak on Figs. 5 and 6. This array configuration emphasizes the dolerite at G16.5 because the current electrode lies directly beneath it. The resistivity low over the orebody is no more distinct than the low on any of the surface resistivity profiles.

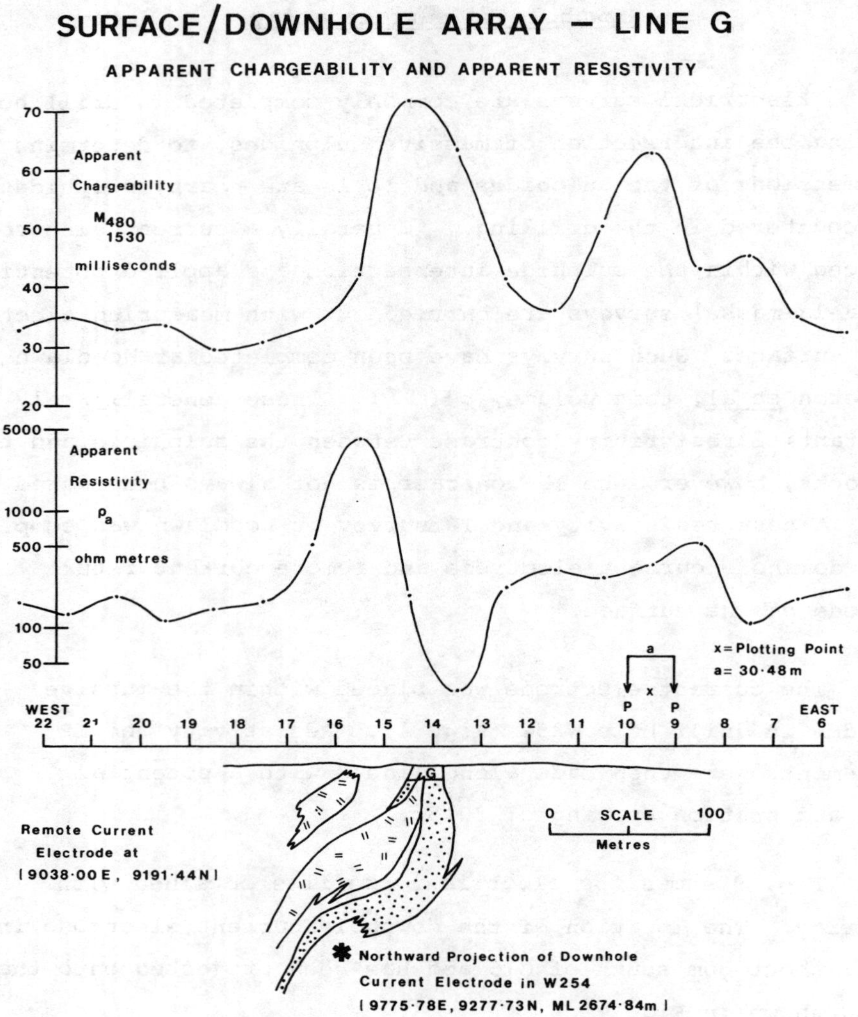

Fig. 9 Downhole to surface array electrical profiles, line G

The IP profile shows two anomalous peaks similar in form to those obtained with the three and dipole-dipole arrays (Figs. 5, 6). The profile confirms that the orebody is polarizable. The emplacement of the electrode within the massive sulphides serves to enhance the IP anomaly of the orebody to more than 70 ms, and clearly differentiates this anomaly from the eastern footwall anomaly. The amplitude of the orebody anomaly is almost twice that obtained with the surface profiling arrays. However the background IP effects have also increased such that IP anomaly amplitude minus IP background is about the same as for the surface dipole arrays.

It appears that the surface to downhole IP profile shows some enhancement of the orebody anomaly over the conventional surface array profiles. Therefore, the above results imply that this type of follow up surveying would be a most useful adjunct to mise-a-la-masse surveys over a polarizable, conductive body of the Woodlawn type.

Summary and Conclusions

All the arrays tested at Woodlawn produced a distinct and simple resistivity anomaly consistent with the shallow highly conductive massive sulphide orebody. Only the gradient array, with its larger depth of investigation, produced strong resistivity anomalies over the resistive host rocks to the east of the orebody.

The IP profiles for all arrays were complex due to polarizable sources in close proximity. These were principally the massive sulphide orebody and the polarizable sources within footwall rocks. The latter sources produced the largest IP anomalies which tended to dominate the IP profiles. This dominance increased with increasing depth of investigation as shown by the dipole-dipole n = 2 and 3 IP profiles and the gradient array data.

The Woodlawn orebody was not easily resolved in the IP data possibly because the 30.5m electrode and station spacing used with most of the arrays was too large relative to the depth and width of the orebody. However, the gradient array profile at station spacing of 15.2m, although having greater resolution, served only to emphasise the complexities in the IP response.

Time and frequency domain gradient array profiles show consistent IP anomaly characteristics. A computer generated gradient array IP profile indicated that a simple electrical model for the dipping orebody and resistive footwall zone is sufficient to explain the complicated gradient array IP field response.

Some of the features of the electrical responses of the Woodlawn orebody to the conventional arrays are presented on Table 1. From this table it can be seen that the three and

Table 1
Summary of electrical response of the Woodlawn orebody.

Features of Response over the Woodlawn orebody	Array			
	Gradient	Three	Dipole-Dipole (n=1)	Wenner
Min. app. resistivity (ohm m)	60	3.6	2	10
Minimum resistivity anomaly to background	1:9	1:16	1:25	1:7
Max. IP anomaly (ms)	43	40	38	27
Max. IP anomaly to background	2:1	5:1	15:1	4:1

dipole-dipole arrays produce the most discernable resistivity anomaly over the massive sulphides. The dipole-dipole array gives the lowest apparent resistivities which are closest to the true resistivity of the massive sulphides. The maximum IP response attributed to the orebody was similar for the gradient, three and dipole-dipole (n = 1) arrays with the most discernable IP anomaly produced with the dipole-dipole array.

The broad Wenner array IP anomaly showed a maximum amplitude of about twice that for the gradient array.

As with the conventional arrays the equispaced unipole array produced a well defined resistivity anomaly over the orebody. The IP response obtained with this array was also complex and of comparable amplitude to that of the Wenner array with very large anomalies defining the edges of the major polarizable region which included the orebody and the eastern host rocks.

The downhole to surface array using a buried current electrode within the orebody produced electrical profiles similar to those obtained with the three array, however this array appeared to be useful for confirming an IP source.

References

Carpenter, E.W., 1955. Some notes concerning the Wenner configuration. Geophys. Prosp. Vol. 3 pp 388-402.

Carpenter, E.W. and Habberjam, G.M., 1956. A Tripotential Method of Resistivity Prospecting. Geophysics, V.21, 455-469.

Emerson, D.W., Whiteley, R.J. and Henderson, R.J., 1968. A field test of the unipole method of resistivity profiling. Aust. Jour. Sc. Vol. 31, No. 4.

Gupta, R.N. and Bhattacharya, P.K., 1963. Unipole Method of Electrical Profiling. Geophysics, V. 28, 608-616.

Habberjam, G.M., 1967. On the application of the reciprocity theorem in resistivity prospecting. Geophysics, Vol. 32, No. 5, p 918.

Habberjam, G.M. and Jackson, A.A., 1974. Approximate Rules for the Composition of Apparent Resistivity Sections. Geophysical Prospecting, V. 22, 393-420.

Hauck, A.M., 1970. A Reconnaissance Downhole Induced Polarization and Resistivity Survey Method. Publication of McPhar Geophysics Ltd., Ontario, Canada.

Pratt, D.A., 1972. The Surface Integral Approach to Solution of the 3D Resistivity Problem. Bull. Aust. Soc. Explor. Geophys., V. 3, No. 4, 33-50.

Pratt, D.A. and Whiteley, R.J., 1974. Computer Simulation and Evaluation of Electrode Array Responses in Resistivity and IP Prospecting. Bull. Aust. Soc. Explor. Geophys., V.5, No. 2, 65-87.

Sumner, J.S., 1976. Principles of Induced Polarization for Geophysical Exploration. Developments in Economic Geology, 5. 277 pp. Elsevier Scientific Publishing Company, New York.

Templeton, R.J., 1975. A Fortran Program System for Electrical Modelling of Prism-Shaped Bodies. Rep. geol. Surv. N.S.W., GS 1975/224 (unpubl.).

Whiteley, R.J., 1973. Electrode Arrays in Resistivity and IP Prospecting: A Review. Bull. Aust. Soc. Explor. Geophys., V.4, No. 1, 1-29.

Acknowledgements

This paper is published with the permission of the Under Secretary, New South Wales Department of Mineral Resources and Development, Sydney.

8.4. *A Field Study of First and Second Order Transient Induced Polarization Effects and Electromagnetic Coupling at Woodlawn*

E. D. TYNE

Contents

Abstract

Analysis of transient voltage decay curves for several electrode arrays over the deposit provides an insight into electromagnetic (EM) inductive coupling and the more subtle expressions of the induced polarization (IP) phenomenon. These field decay curves can be accurately described by a minimum of three exponential functions.

The initial rapid decay represents an EM coupling transient which may require as long as 0.4s to dissipate to negligible levels. Coupling appears to be mainly caused by lateral variations in bedrock resistivity rather than variations in overburden thickness. It is therefore desirable to monitor and to quantify the coupling component of the transient voltage decay curve in new survey areas so that the chargeability integration period can be optimized.

The longer period decay component substantially represents the IP transient. Analysis of this decay component over both the economic massive sulphides and uneconomic pyrite mineralization suggests that the shape of the IP decay curve may yield some geological information which is not apparent from the first order IP effect, chargeability.

The results suggest that, although it is difficult, it may be possible to identify the massive and disseminated sections of a polarizable source, as well as differentiating between the conductive massive sulphide zone and Black Shale containing abundant disseminated pyrite.

Introduction

The Geological Survey of N.S.W. has conducted an extensive geophysical study of the Woodlawn orebody over a period from 1974 to 1976. This included surface electrical surveying, an important phase of which was the recording of transient voltage decay curves for three different electrode arrays over both economic massive sulphides and uneconomic pyrite mineralization within black shales. The aims of the field study were to determine if the various components or the shape of decay curves provide a possible means of:

1. Recognizing and quantifying inductive (EM) coupling from the decay curves for the three electrode array, dipole-dipole array and gradient array in this typical eastern Australian acid volcanic environment.

2. Differentiating the polarization response of the massive sulphide Woodlawn orebody from that of zones within the footwall rocks containing abundant disseminated pyrite,and weakly mineralized host rocks.

3. Differentiating the polarization response of the Woodlawn orebody from that of conductive, pyritic shales which are common in the Woodlawn region and which present major exploration problems because their electrical responses are similar to those of massive sulphide ore targets.

Review of Recent Research

The measured magnitude of the IP response can be categorized into first order and second order effects (Sumner, 1976, p106). The first order effect is expressed as the amplitude of the IP effect or chargeability. The shape of the IP decay curve is the expression of the second order effect.

Recent studies of the IP effect reveal that the shape of the transient voltage decay curve or its frequency spectrum may convey useful geological and geophysical information that is otherwise discarded by recording the IP signal over only a short time period or frequency range.

Zonge and Wynn (1975) have recently suggested that complex resistivity spectral techniques show promise for mineral discrimination in the field (especially between pyrite and chalcopyrite) as well as the ability to remove EM coupling and identify cultural coupling effects. Pelton et al. (1976) used multifrequency IP over a number of North American sulphide, graphite and magnetite deposits. Their results indicate the possibility that massive sulphides and graphite can be distinguished in the field on the basis of a difference in spectral response. Despite the evidence, which appears contrary to that on Zonge and Wynn, that there is no significant difference between the response of chalcopyrite and pyrite mineralization of the same textural form the authors claim that sulphide mineralization of quite different form, such as stockwork veins and discrete disseminations, shows different spectral response characteristics.

Current north American research is mainly directed at recording data in the frequency domain which offers some instrumental advantages over time domain recording. This bias towards frequency domain may also have arisen as a result of an assumption that subtle variations in transient decay rate related to sulphide and host rock responses may be more readily observed in the frequency domain data when plotted on complex amplitude or frequency versus phase diagrams. However, for very long current charging times, measurement of the entire transient voltage decay curve contains as much information as multifrequency measurements. The problem is to extract the information in the most diagnostic form.

Grant (1971) examined the possibilities of using the IP transient voltage discharge for discrimination of sulphides having gross textural differences such as massive or disseminated deposits, and for identification and removal of undesirable EM coupling. Grant (ibid) asserts that the aim of quantitative IP interpretation should be to initially estimate the first order IP effect or apparent chargeability i.e. ratio of secondary voltage at switch off, Vs, to the primary voltage, Vp, after subtracting the effect of inductive coupling from the data. However, apparent chargeability estimates cannot be made effectively from IP measurements collected over a narrow time and frequency range. A recording of the entire decay curve is required for effective quantitative interpretation.

Theoretical examples presented by Grant (ibid) indicate that the shape of the decay curve i.e. the second order IP effect, is not constant and may vary by subtle but measurable amounts over given anomaly profiles.

Phillips and Richards (1974) and Bertin and Loeb (1974) analysed or factored transient IP decay curves into a number of exponential functions. There is no published investigation using the IP decay rate and amplitude of the resulting exponential functions for distinguishing mineral content and textural form of polarizable sources and evaluation and removal of inductive coupling effects prevalent in regions of highly conductive overburden.

Phillips and Richards (ibid) concluded that there was no detectable difference in the pattern of decay over the surveyed mineralized zone and over unmineralized rocks. The results show the possible influence that current electrode polarization for metal stake electrodes may have on transient decay curve measurements. It has been the author's experience that the discharge of current electrode polarization is not significant if the time domain IP transmitter can be considered to totally disconnect itself from the ground circuit during the current off-time. There is no published study on the influence of electrode polarization on complex resistivity measurements.

Despite the fact that modern time domain IP receivers (eg. Huntec MkIII and Scintrex IPR8) allow the recording of a substantial section of the transient voltage decay curve, there appears to have been very little serious interest in attempting to obtain additional field information from the decay curve regarding the nature of polarizable sources and levels of EM coupling for various electrode arrays and spacings.

Relevant Geological and Petrophysical Factors

Woodlawn Deposit

Depth of oxidation at Woodlawn varies from about 6m at the northern end of the body to about 15m at the southern end. Overburden resistivities are generally much greater than 30 ohm m. Estimates of EM coupling and masking effects from the weathered layer indicate that they are not significant for the electrode arrays and spacings used at Woodlawn.

IP traverses were completed along lines G and J over the orebody (Fig.1). These lines were chosen because they cross the extremes in geometric disposition of the orebody. Line G, (9310mN) is near the northernmost tip of the orebody which is essentially vertical and strikes obliquely to the line. About 90m to the south on line J (9220mN) the orebody dips approximately 45° to the west and is orthogonal to the line.

The massive sulphides on the drilled sections in Fig.2 show the very fine grained complex ore of the Woodlawn orebody which consists of about 75% sulphides, mainly pyrite, sphalerite, galena and chalcopyrite, in order of decreasing abundance. The minimum resistivity of the massive sulphides has been determined *in situ* by drill hole logging (Tyne, this volume, p 538). This is less than 0.1 ohm m.

The hanging wall rocks consist of acid tuffs, tuffaceous shales and dolerite sills in thick tabular units (Fig.2). There is generally a sharp

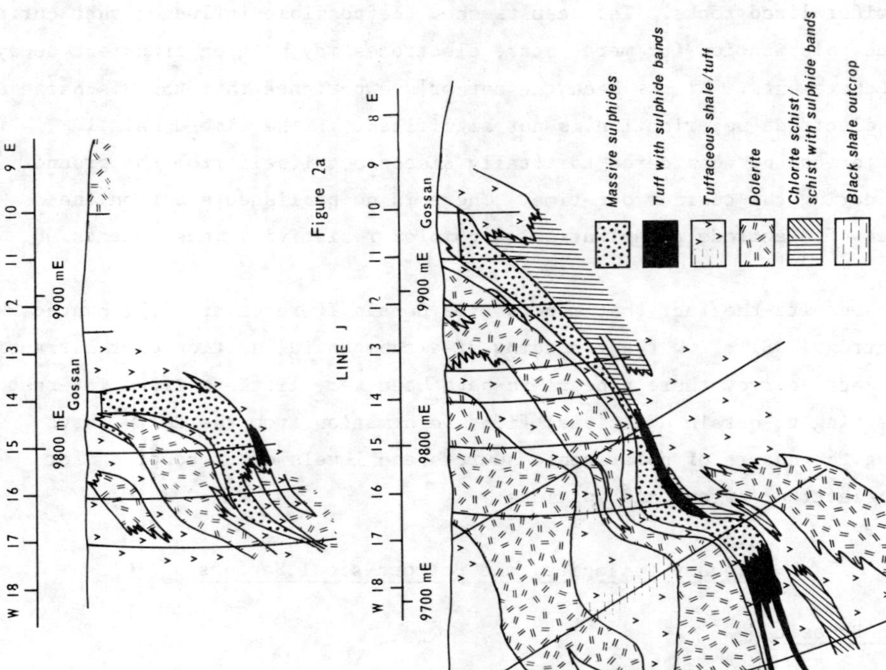

Figure 2 Woodlawn Geological Cross Sections

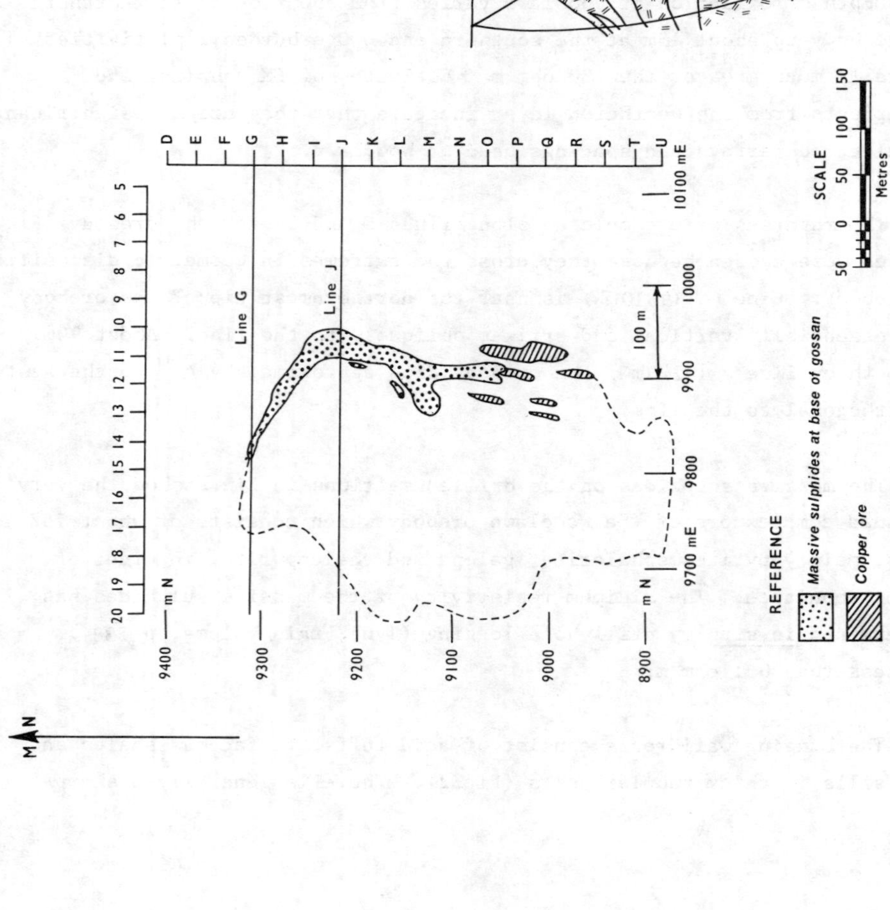

Figure 1 Woodlawn Orebody and Test Grid

geological contact between the massive sulphides and the immediate hanging wall rock, a 10m thick tuffaceous shale unit, which contains local quartz veining and sulphide bands. Rock resistivities vary from several hundred ohm m for tuffs and tuffaceous shales to 7,000 ohm m for the dolerite sills (Tyne, this volume,p538) thus providing an extremely strong and abrupt electrical contrast with the massive sulphides.

The footwall sequence is chloritized and silicified to varying degrees and consists principally of schist, tuffaceous shale, welded tuff and graphitic shale (Malone, this volume, p 49). According to Malone et al (1975) the footwall boundary of the massive sulphides is generally in sharp contact with non-mineralized footwall and is marked by an abrupt drop in total sulphide content where chlorite schist ore underlies the lens. It is misleading to say that the footwall is unmineralized because it contains common to abundant pyrite with some minor chalcopyrite in places. The sulphides appear to be more common within black shale and take the form of thin bands, disseminated small crystals, and smearing on foliation planes. No in situ resistivity or chargeability measurements have been made within the footwall rocks.

Common sulphides throughout the footwall rocks suggest that this zone will produce strong IP effects because the abundant pyrite disseminations and thin bands offer a large polarizable surface area. The massive sulphide orebody would be expected to polarize at its upper and lower boundaries, where electrode/electrolyte interfaces are present. Also there should be very little internal polarization within the fresh ore because the fine-grained and non-porous nature of the massive sulphides would largely preclude the development of internal polarization interfaces. However the supergene zone and supergene contaminated ore may be more polarizable.

It is difficult to estimate and attribute an intrinsic chargeability to the massive sulphide body from drill hole logging and especially from core measurements because both these techniques rely on measuring polarization effects distributed throughout a rock unit (as for the footwall rocks) rather than at the discrete boundaries of a rock unit.

These considerations suggest that the shallow IP currents arising from the massive sulphide body and the polarizable footwall zones may lead to transient voltage decays of a sufficiently different character at the surface.

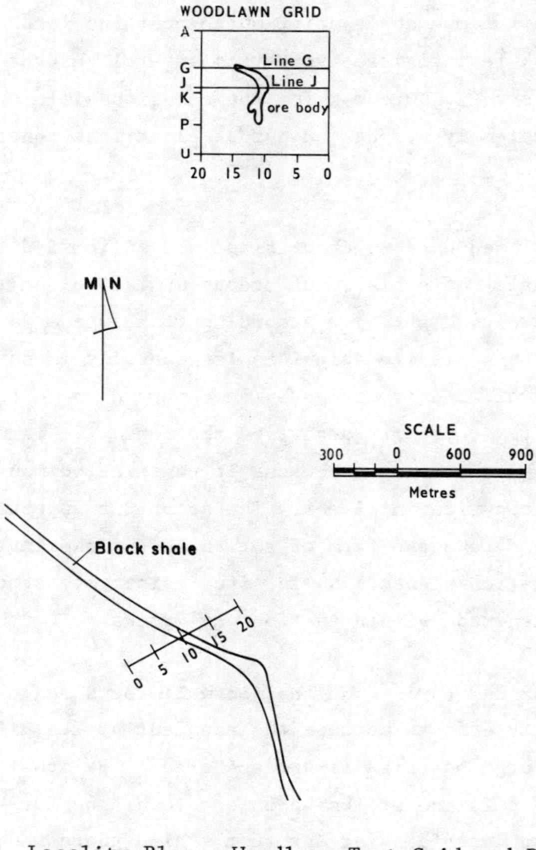

Figure 3 Locality Plan - Woodlawn Test Grid and Black Shale

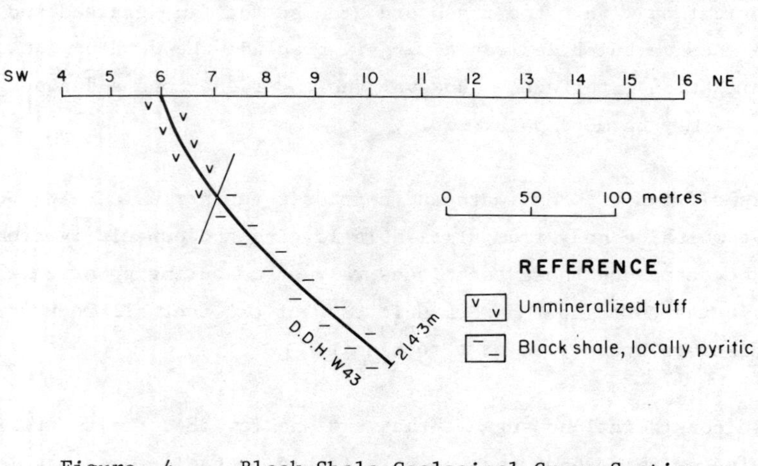

Figure 4 Black Shale Geological Cross Section

Black Shale

The Upper Ordovician Black Shale located about 2.5km south of the orebody and selected for comparative study (Fig.3) has a strike length in excess of several kilometres. Fig.4 shows a section across the Shale where a strong self potential (SP) anomaly (Cifali and Whiteley, this volume,p 321) led to the drilling of DDHW43. This hole encountered a graphitic slaty tuff containing common pyrite and trace base metals as disseminations and stringers. The surrounding rocks are siliceous welded tuffs which are essentially unmineralized.

This type of conductive, polarizable shale is common throughout the Woodlawn region and is responsible for numerous electrical and EM anomalies. If there is poor geological control in the vicinity of these anomalies it is often difficult to discriminate between a possible massive sulphide body and a graphitic shale.

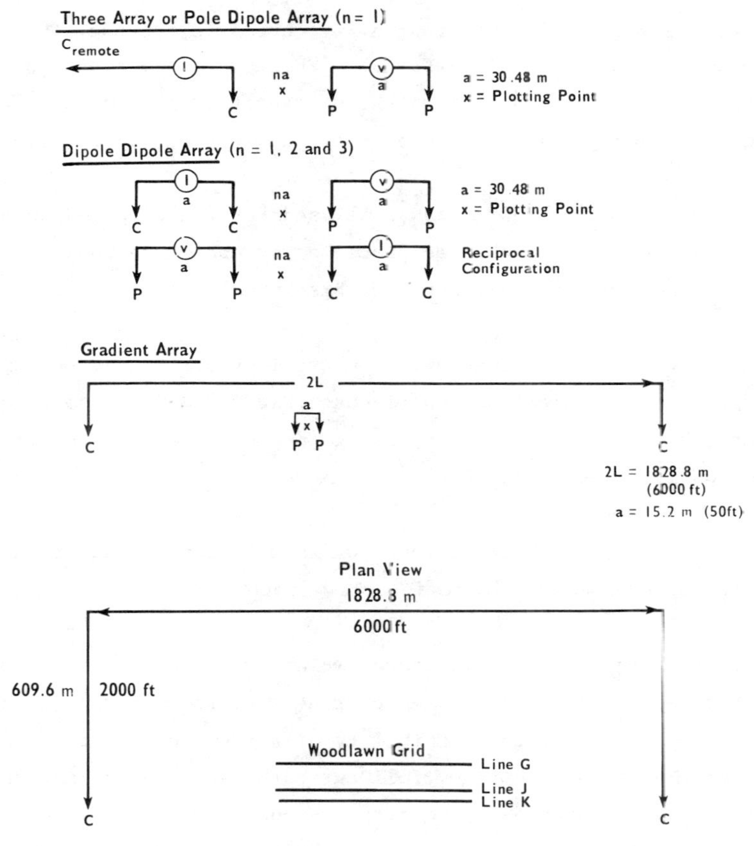

Figure 5 Electrode Arrays used for Transient Voltage Decay Curve Measurements

Transient Decay Curve Measurements at Woodlawn

Electrode Arrays and Current Electrodes

The three different electrode arrays used at Woodlawn are shown in Fig.5, and represent the most commonly used electrode arrays in IP prospecting. Each of the arrays is suitable for mapping steeply-dipping or vertical structures (Whiteley, 1973) and is therefore applicable to exploration in the Woodlawn area.

The three array or pole-dipole array (n=1) was the only array employed on all three survey lines. The interelectrode spacing was chosen at 30.5m (100ft) This is the optimum spacing, both to resolve the steeply dipping formations of interest which are generally of the order of 30m thick or more and to allow effective investigation below the 6m to 15m of overburden above the orebody.

The remote current electrode for this array on lines G and J is coincident with the westernmost gradient array current electrode, that is, at least 600m to the west of each line (Fig.5). The remote electrode for the line over the Black Shale is located 300m to the northeast of this line. The orientation of this asymmetrical array is shown on each of the relevant figures that follow.

A spacing of 30.5m (100ft) was also used for the dipole-dipole array. Decay curves were recorded for the reciprocal array configurations for n=1, 2 and 3 (Fig.5) on line G but only for n=2 over the Black Shale.

Decay curves were obtained with the gradient array only on lines G and J. The fixed current electrodes were spaced 1829m (6000ft) apart and located symmetrically about the orebody along an east to west line coincident with line K of the Woodlawn Test Grid (Fig.5).

The current return line was laid out in a U-shaped loop to the north of the Grid (Fig.5) so as to minimize inductive coupling.

Non-polarizing Cu-$CuSO_4$ electrodes which are normally used for potential measurements were also used as current electrodes for each array configuration so as to reduce the possibility of current electrode polarization. An exception was the use of aluminium foil current electrodes during an orientation study of the dipole-dipole array response over the Black Shale. No problems associated with current electrode polarization were encountered.

Field Recording of Transient Decays

Transient IP methods use substantially different current charging times and IP signal recording times. In theory, a long charging time or an effectively direct current is ideal. Bertin and Loeb (1974) and Komarov (pers. comm. 1978) employ charging times as long as tens to hundreds of seconds and recording times of the same order. This allows the polarization process to reach a near saturated condition, but greatly slows field operations. Bertin and Loeb (1974) show the increase in amplitude of the transient voltage decay curve which results from an increase in current charging time from 5s to 300s. Their results confirm that for certain mineralized environments, polarization does not reach saturation levels until the charging current is applied for hundreds of seconds. Decay constants derived from the voltage transients may also vary with current charging time. IP field crews operating in the USSR appear to be the only ones to routinely use very long charging and recording times to study first and second order IP effects (Komarov, pers.comm. 1978).

At Woodlawn, an Austral T842 time/frequency domain transmitter was used with an 8 second cycle time current waveform together with a Huntec Mark III time domain receiver which simultaneously measured the primary voltage and four segments of the transient IP decay curve. This equipment was designed for rapid surveying and was incapable of providing the desirable long charging times.

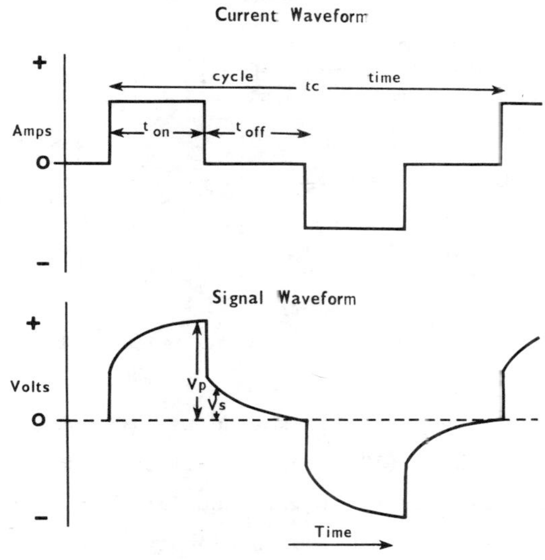

Figure 6 Time Domain Current and Signal Waveforms

However, it has been found in eastern Australia that these current and recording times are generally long enough to allow polarization to reach a near steady state condition and for conventional IP measurements to be delayed sufficiently to avoid EM coupling decay. Therefore a study of the first and second order IP effects using these short charging and recording times was still considered worthwhile.

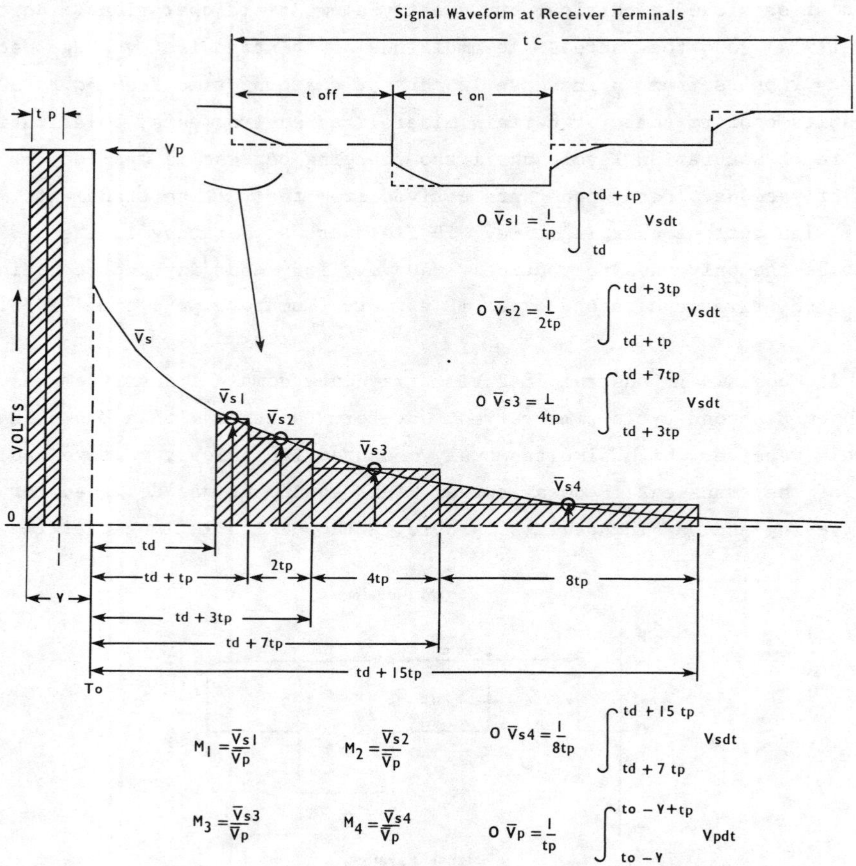

Figure 7 Parameters of IP Signal Measured by MK111 Receiver

The principles of operation of the Huntec IP receiver have been described in detail by Hutchins (1971) and Phillips and Richards (1974), and are illustrated diagrammatically in Fig.7. The instrument integrates and averages the primary voltage, Vp and four secondary voltages, Vs over as many cycles as required until repeatable readings are obtained. The secondary voltages are normalized by Vp and the integration period and are presented as percentages or M factors for each segment of the decay curve as shown on Fig.7. The integration window is doubled for each successive M factor, so that as the transient voltage drops to low amplitude levels at longer times, a larger integration period allows effective averaging of small IP signals and cancellation of random noise. The transient voltage decay curve is obtained by progressively increasing the delay time (td), and the period of integration (tp) (Fig.7) for each successive measurement set of four M factors until a sufficient number of points are recorded to fully define the shape of the transient decay curve.

Six sets of readings were made at each recording station giving 24 discrete points on the transient decay curve. Table 1 lists the td and tp settings which were selected for each of the six sets of readings, as well as the central plotting position for each of the 24 curve points.

Since the Huntec Mark III receiver does not actually measure the secondary voltage Vs at a particular point in time but rather as the average over an integration time period, there is some question as to whether the 24 M factors accurately portray the amplitude and rate of decay of the curve, especially at short time delays where the decay is rapid.

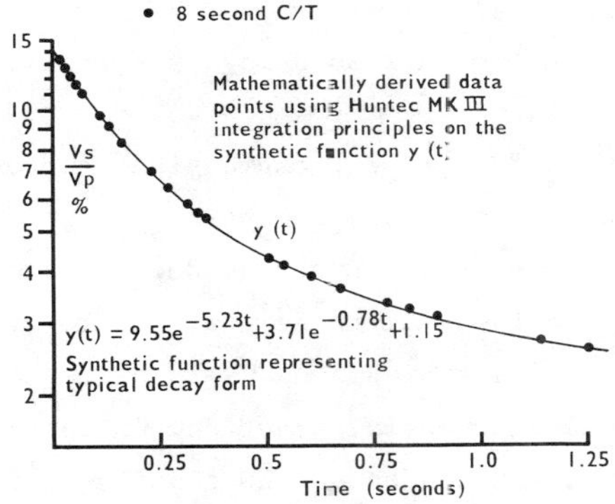

Figure 8 Fidelity of Huntec Measurements to Actual Decay Form

Fig.8 shows a synthetic exponential type decay curve typical of that obtained at Woodlawn. If this decay curve is mathematically integrated using the same principles and parameters (Table 1) as for the Huntec receiver, it is found that the resultant 24 discrete data points for the 8 second cycle time faithfully represent the true amplitude and shape of the synthetic function. The td and tp parameters for the 8 second cycle time (Table 1) are therefore suitably chosen for this study.

Table 1. Decay Curve Recording Parameters

8 SECOND CYCLE TIME

Setting of Time Delay and Period (milliseconds)	Plotting Point for M Factors (milliseconds)			
Td/Tp	M1	M2	M3	M4
15/20	25	55	115	235
15/30	30	75	165	345
15/60	45	135	315	675
240/60	270	360	540	900
480/60	510	600	780	1140
480/70	515	620	830	1250

Analysis of Decay Curves

A simple, practical means of quantifying the amplitude and decay rate of the decay curve components is to factor the smoothed field curve into a number of exponential functions. There is a sound basis for regarding the inductive and IP decay components as being essentially exponential.

Dey and Morrison (1973) present expressions for an exponential-like inductive voltage decay over a multilayered earth. Also the assumption of an exponential-like decay form for the IP voltage decay is consistent with the equivalent circuit model and the electrochemical nature of the IP phenomenon presented by Lynam (1972). The expressions by these authors are of course much more involved than a single exponential function. Although a simple exponential decay may not exactly represent the actual EM coupling and IP behaviour within the earth, it does provide a meaningful basis for comparing the characteristics of measured transient decays.

The six groups of four M factors obtained at each station are plotted on a log-normal scale for graphical analysis. Fig.9 shows the groups of M factors at a noisy station on Line J. Each group of M factors shows a decay rate which is consistent with its neighbours, however, there is some discrepancy in amplitude level which is particularly noticeable between the last two M factor groups in Fig.9. The discrepancy usually arises at stations where there is a very low absolute signal level for Vs and a moderate to high self potential (SP) gradient. Because only four decay curve points can be recorded simultaneously

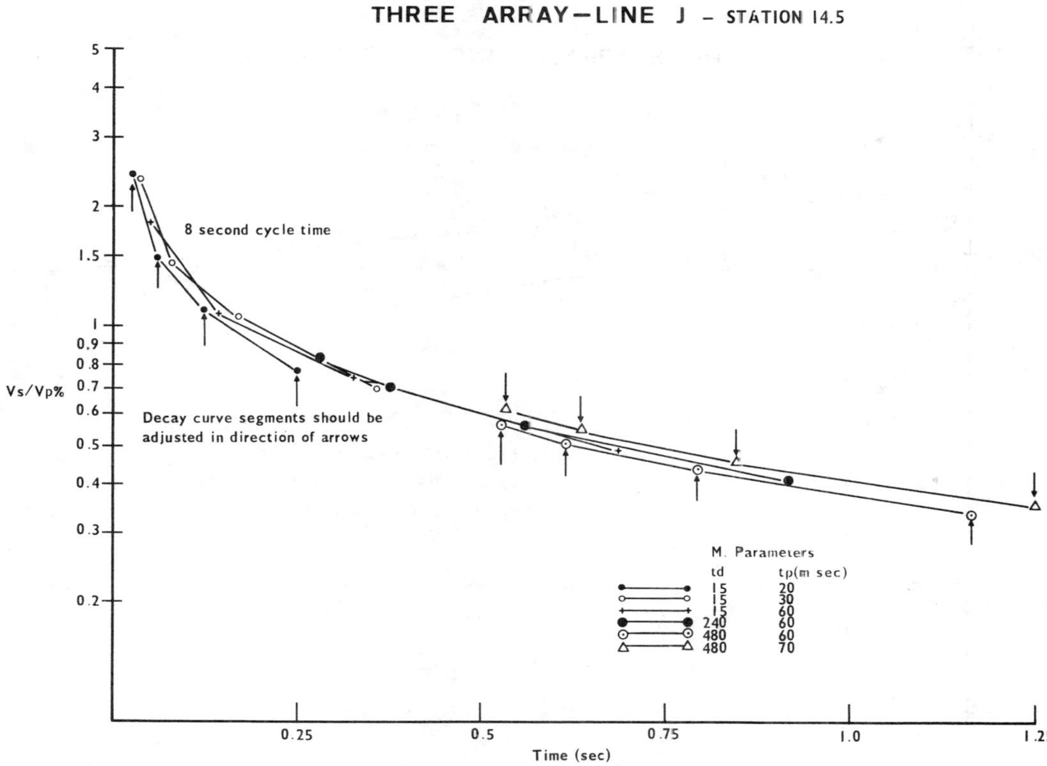

Figure 9 Adjustment of Transient Voltage Decay Curve Segments to Allow for Variations in SP Base Level at Each Recording of Four M Factors

with the IP receiver it is possible in this situation for a momentary drift in the SP base level, which is determined by the Mark III in the automatic SP cancellation loop, to cause a constant datum shift to the amplitude level of each of the four recorded M factors.

It is apparent from Fig.9 that smoothing the recorded points may give an incorrect representation of the decay curve shape. It is therefore necessary to remove this SP datum shift before producing a hand-smoothed decay curve. The curve segments for each of the M groups must be adjusted vertically by a constant numerical amount until they match the remaining segments of the curve. The segments are adjusted in a sense parallel to the ordinate axis (Fig.9).

Of the 170 decay curves measured at Woodlawn, it was possible to smooth 70% of the decay curves from the recorded M factors alone. 20% of the curves required very minor adjustments of the M factors to account for drifts in the

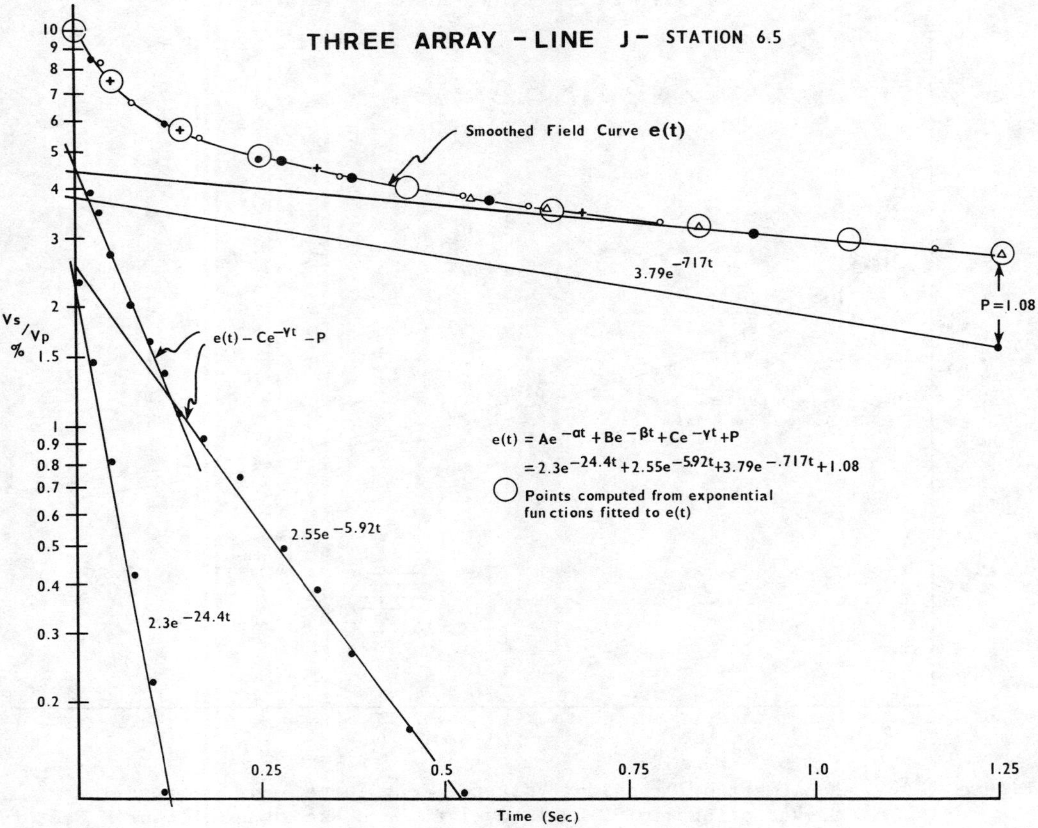

Figure 10 Decay Curve Factoring Procedure

SP base level and less than 10% of the curves required significant adjustment or were too noisy to factor. Almost all of the very noisy curves occurred away from the important areas.

Fig.10 shows a typical field curve which required no segment adjustment. The smoothed curve shows a rapid decay at early times which is related to the dissipation of EM coupling voltages in the ground. An almost steady state decay rate occurs at the longest times. This may be considered to represent the longer-lived IP voltage discharge through the ground.

Fig.10 shows the graphical factoring procedure used to analyse the decay curve forms from Woodlawn into a number of exponential functions. It appears that at least three functions are required to give a good representation of the recorded field curves. The graphical factoring procedure for three exponential functions and the parameter notation is discussed by Phillips and Richards (1974). An exponential function is represented by a straight line on a log-linear plot. The factoring procedure involves fitting a straight line segment to the tail of this curve and graphically subtracting this straight line (or exponential function) from the smoothed line and so on until the decay curve is analysed into three straight line segments or exponential functions. The exponential parameters obtained from the factoring process are then used to compute a transient voltage decay curve which is compared with the smoothed field curve. The exponential parameters are adjusted until a close visual fit to the field curve is obtained. Fig.10 shows the typical close coincidence required between computed curve and field curve.

The fitted curve function takes the form:

$$Ae^{-\alpha t} + Be^{-\beta t} + Ce^{-\gamma t} + P. \qquad (1)$$

where P is the residual secondary voltage at the SP cancellation time of the instrument. P is essentially an instrumental datum shift and is subtracted from all decay curve points to obtain the true amplitude value of the transient decay curve. The first exponential is related to the short lived EM coupling component of the decay curve, A being the amplitude of the coupling and α its decay rate. The last exponential is considered to substantially represent the IP component of the decay curve, C being the IP amplitude and γ its decay rate. C is essentially an apparent chargeability or first order IP effect measured at zero time with EM coupling removed. This is the parameter required for quantitative interpretation (Grant, 1971). C is simply obtained by extrapolating this section of the decay curve back to zero, a very simple empirical procedure for substantially removing EM coupling from the decay curve.

The need for an intermediate exponential function, $Be^{-\beta t}$, to fit the central portion of the field curve indicates that $Ae^{-\alpha t}$ and $Ce^{-\gamma t}$ do not totally define the EM coupling and IP components of the decay curve. This hybrid function constitutes part coupling and part IP component. However, the function appears to be more closely related to variations in $Ce^{-\gamma t}$, than to the rapid EM decay, $Ae^{-\alpha t}$.

The variations in the first and last exponential functions may be used to compare the relative changes in the coupling and IP components of the transient field decay curves with geological changes along a traverse line.

The exponential decay constants for the curve in Fig. 10, α = 24.4, β = 5.92 and γ = 0.717 are typical of the groupings of decay constants obtained at Woodlawn. Variations in the decay constant γ, may be considered to constitute a second order IP effect. Figures 11 to 22 present the decay rate as the relaxation time, which is the time (seconds) for each transient voltage component to decrease to 1/e of its value at zero time. It is in fact the inverse of the decay constants α, β and γ. Using the notation of Phillips and Richards (1974), the amplitude components A, B, C and the relaxation times r_α, r_β, r_γ are presented in profile form as a basis for comparison of variations in decay form with the geology.

It should be emphasized, that the actual magnitudes of voltage components and relaxation times presented here do not uniquely define the total transient decay curve since they refer only to the section of decay curve from 0.025s to 1.25s (8s cycle time) and are determined by the method of factoring as well as the current charging time. It is the relative variation in the amplitudes of the voltage components and relaxation times that is pertinent.

Equispaced Three Array

Line G (Fig. 11)

The massive sulphides of the Woodlawn orebody are defined by a strong resistivity low which is broader than would be expected for the narrow orebody section shown on Fig. 11. This is because the body strikes at an acute angle to the line.

The C component profile essentially shows the variations in first order IP effect or chargeability along the line. There are two adjacent anomalies of similar magnitude and extent centred at G14 and G10.5, and coincident with two dissimilar polarizable sources; the massive sulphides and the pyritic shales and tuffs to the east. The C anomaly over the massive sulphide orebody is coincident with the apparent resistivity low. The eastern C peak occurs over the resistive pyritic black shale (Fig.2).

The chargeability profile indicates that the type of polarizable material causing the adjacent chargeability anomalies can not be differentiated on the basis of the first order IP effects alone.

The relaxation time ($r\gamma$) profile essentially reflects variations in second order IP effects along line G. In contrast to the similar amplitude of the two C component peaks, the IP relaxation time is measurably greater in the

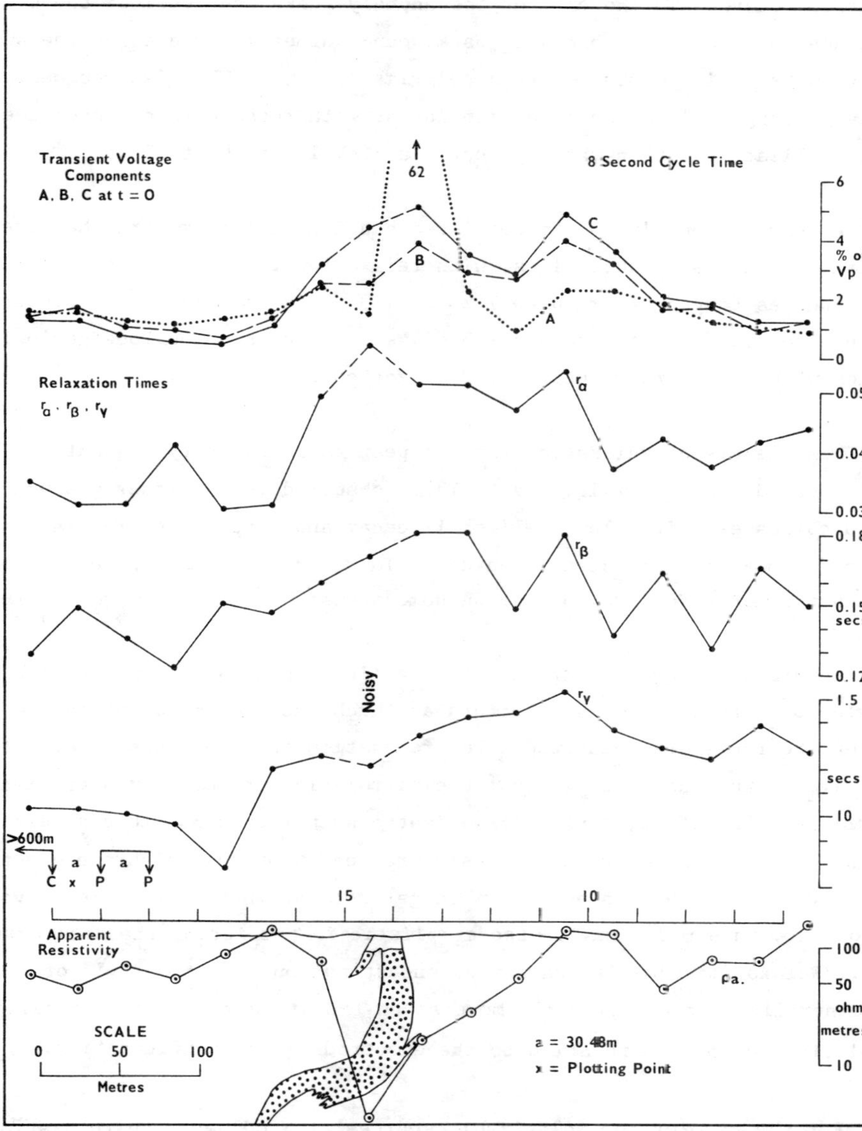

Figure 11 Transient Decay Curve Parameters, Three Array - Line G

vicinity of the eastern anomaly peak than over the western anomaly peak. That is, the IP voltage transient decays more slowly over the eastern resistive zone than over the massive sulphide orebody.

To the west of the orebody within unmineralized hanging wall rocks, background C component levels are accompanied by short relaxation times or rapid voltage decays. To the east of the anomaly peaks the level of the C component does not fall as rapidly to background values as it does to the west, indicating an extension of the eastern polarizable zone. This is accompanied by relatively large relaxation times consistent with those obtained over the pyrite mineralization within the adjacent black shale at about G10.5.

In summary, on line G unmineralized hanging wall rocks are characterized by IP relaxation times of about 1.0s, with relaxation times over the conductive massive sulphides ranging from about 1.2s to 1.35-1.4s and over the resistive footwall pyrite zone by greater than 1.35-1.4s. There is no obvious geological correlation with the sharp reduction in relaxation time at G17.5.

These levels of relaxation times appear to be contrary to what would be expected. For example, Grant (1971) observed that coarser grained massive sulphides exhibit a longer-lived IP decay and hence large relaxation times than for more disseminated sulphides. To reconcile this apparent discrepancy the model for overvoltage phenomena must be examined more carefully.

In this model grain surfaces of metallic sulphides in contact with electrolyte constitute potential barriers at which the phenomenon of overvoltage occurs when the rocks are polarized. The relaxation time for these overvoltages is indicative of the ease with which the electrolytic current flow overcomes these barriers. Therefore, finely disseminated sulphides constitute a less substantial potential barrier than coarser grained sulphides which are electrically well connected and which thus exhibit a larger relaxation time for the IP voltage decay. However, this model may be too simplistic in predicting the real earth behaviour. Relaxation time is heavily dependent not only on the style of sulphide mineralization but probably more so on the gross shape and dimensions of the polarizable body in relation to the electrode array configuration.

For the case of the polarizable sources at Woodlawn, the relaxation time ($r\gamma$) profile on line G implies that the polarizable footwall zone presents a greater overvoltage barrier than the massive sulphides in relation to the three electrode array. It may therefore be concluded that despite the style of mineralization of the footwall zone it exhibits longer relaxation times than the massive sulphides because it is considerably larger in areal extent and overall geometry than the massive sulphides. The pyrite zone may be as much as 100m wide within the footwall region, whereas the massive sulphide orebody only attains 30m at its widest point. In terms of gross structure, the pyrite zone probably constitutes a more substantial overvoltage barrier to electrolytic current flow than does the narrower massive sulphide orebody. The resistive nature of the footwall zone probably also tends to increase its relaxation time.

It is of interest to compare the C and rγ parameters with curve parameters defining the rapid decay behaviour. The A component profile on line G, is dominated by a very strong anomaly with an amplitude of 62% of V_p. This occurs almost directly over the very conductive orebody where the most favourable EM coupling geometry between current cable return, orebody and potential dipole is present. Apart from this one station anomaly there is a general trend of increased A component values, up to about 2% over the two major polarizable zones, compared with background levels of about 1% over surrounding unmineralized rocks. The corresponding relaxation time profile, r_α also shows a definite increase over the major polarizable zones, indicating that EM decay is longer-lived over the mineralized area.

It is noteworthy that the 62% anomaly takes as long as 0.3s to dissipate to less than 5% of the level of the C component. Beyond 0.3s the A component contributes a negligible proportion of the measured chargeability response.

In summary, the three electrode array profile on line G shows strong changes in coupling level corresponding to lateral variations in earth resistivity. Background levels of EM coupling over resistive unmineralized host rocks are fairly constant at about 1% to 1.5% of Vp. These background levels decay to less than 5% of background chargeabilities within about 0.1s, therefore indicating that coupling due to the weathered overburden material is negligible for the three array when compared to that caused by lateral variations in bedrock resistivity.

The final profiles on line G to be considered are those of the parameters of the intermediate exponential function i.e. B and rβ.

Fig. 11 Shows that the amplitude component B shows strong similarities in shape to the C profile. Two anomalous peaks of equivalent amplitude are again evident. This profile appears to bear no relationship to the A component.

The corresponding relaxation time, rβ profile shows a general increase over the central area, however, variations are not as distinctive as on the rα and rγ profiles.

The variations in the intermediate exponential parameters B and $r\beta$ appear to be closely related to variations in the exponential parameters C and $r\gamma$. It appears that for the three array on line G, the exponential functions $Be^{-\beta t}$ and $Ce^{-\gamma t}$ probably closely represent the IP transient decay. The rapid decay function, $Ae^{-\alpha t}$ has features which are unlike those of the other two functions and can be considered to substantially reflect the EM coupling effects in the vicinity of the Woodlawn orebody.

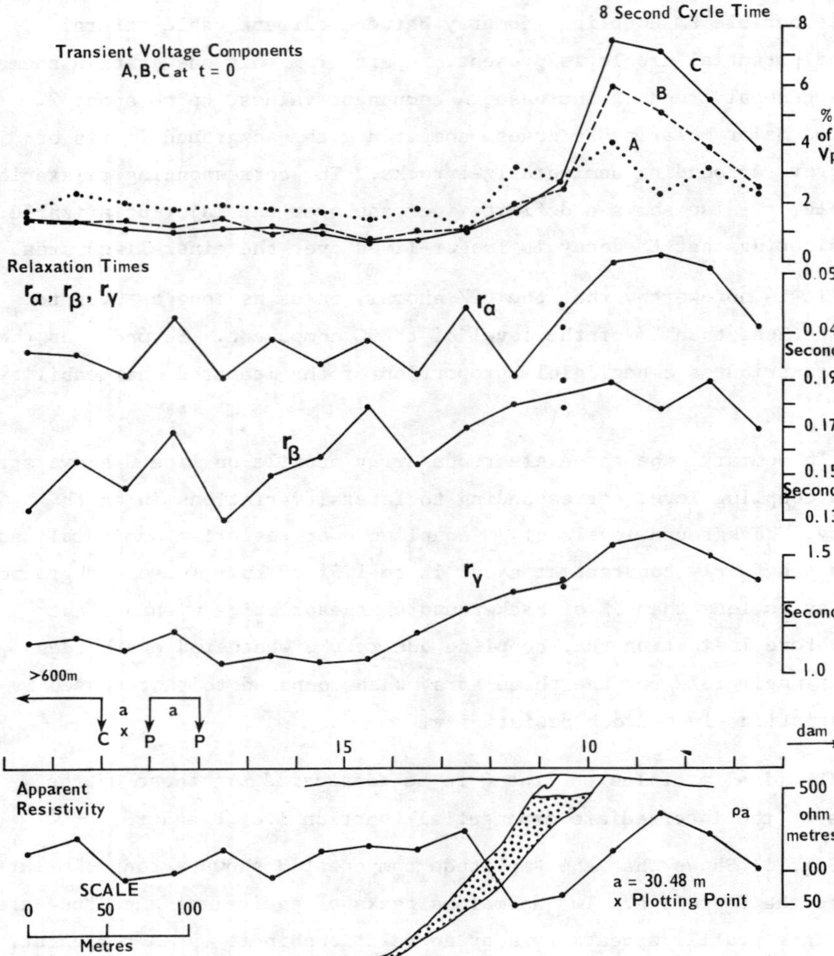

Figure 12 Transient Decay Curve Parameters, Three Array - Line J

Line J (Fig.12)

As shown on Fig.12 the shallow region of the westerly dipping massive sulphide orebody on line J is well defined by the resistivity low anomaly, however, there is no corresponding polarization anomaly peak on the C component profile as for line G.

The polarization response on line J is characterized by a single strong anomaly on both the B and C component profiles, coinciding with the footwall tuffs and shales. Pyrite concentrations occur very close to the footwall boundary of the massive sulphides on this line, which is probably why two separate polarization anomalies are not discernible on the B and C profiles as on line G (Whiteley and Tyne, this volume, p 423).

It is apparent that pyrite concentrations in the footwall are the primary cause of the chargeability anomaly on line J. The only reflection of the massive sulphide IP response may be the change on the B and C component profiles from a steep anomaly gradient east of J10.5 to a gentler gradient west of J10.5 which decreases gradually to J13.5, beyond which background polarization levels occur.

The rγ profile provides additional information about the polarizable zone on line J. The profile shows relaxation time levels of 1.4 to 1.6 seconds, coincident with the largest values on the C component profile. This is consistent with the levels of the relaxation times over the eastern mineralization on line G.

A notable feature of the rγ profile is that the rγ level drops to an intermediate level of 1.2s to 1.4s in the vicinity of the massive sulphides in a similar fashion to the rγ profile over the massive sulphide section on line G. Background relaxation times of 1.05s to 1.15s occur beyond J14.5 which are comparable with those on line G.

The results on line J suggest that the variations in rγ or second order IP effect are more closely related to lateral changes in the gross style of sulphide mineralization than to the variations in the C component i.e. the first order IP effect.

The rβ profile shows similar trends to the rγ profile east of J13.5, however, to the west background levels are irregular.

The A component profile shows background levels at about 1.5% to 2% of Vp to the west of J12.5. There is a definite increase in EM coupling levels on the A component profile of between 2% to 4% over both the massive sulphides and footwall pyrite. The corresponding relaxation time rα profile shows an increase directly over the polarizable zone, similar to that on line G. However, the level of coupling is negligible because it dissipates to less than 5% of the C component level within 0.13s.

Despite the lack of similarity of first order IP effect anomaly forms on both lines G and J, the rγ profile, representative of the second order IP effect, shows consistent relaxation times of greater than 1.35 to 1.4s over the footwall polarizable region. Relaxation times of 1.2s to 1.4s coincide with the low resistivity anomaly above the massive sulphide orebody.

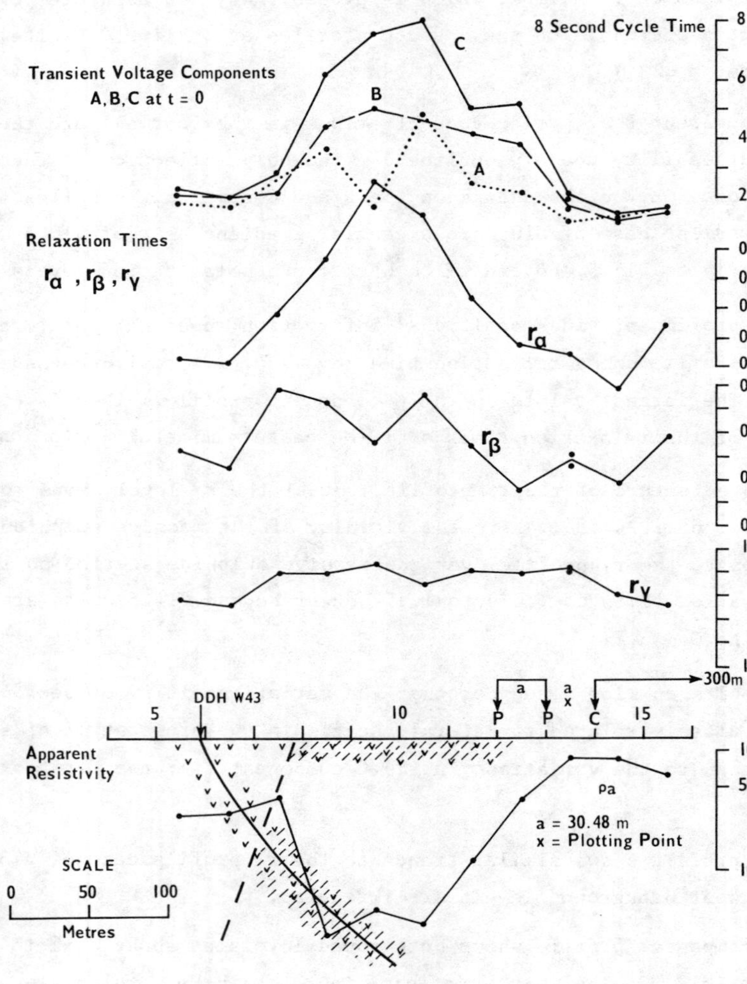

Figure 13 Transient Decay Curve Parameters, Three Array - Black Shale

Black Shale (Fig.13)

Results obtained over the Black Shale are shown in Fig.13. These may be compared with data from lines G and J because the tuffaceous shale exhibits a resistivity anomaly of similar amplitude to that observed over the massive sulphide Woodlawn orebody. However, the pyrite mineralization within the graphitic slaty tuff is of quite a different textural form to the massive sulphides. It is of a similar gross form and extent to the footwall pyrite mineralization to the east of the massive sulphides.

Profiles of the B and C component across the Black Shale show the same anomalous form which is unlike that of the A component profile. The C profile has a similar amplitude to that on lines G and J.

The rγ profile remains almost constant over the polarizable, conductive Shale at times greater than 1.35s. This is about the same level as the second order IP effects recorded over the footwall zone at Woodlawn, despite the fact that the pyritic Black Shale is very conductive and the footwall pyrite zone is resistive.

EM coupling increases from a background level of 1.5% of Vp to as much as 5% over the polarizable conductive zone above the Shale. The shape of the rα profile differs from that of the rβ and rγ profiles and shows a strong anomalous increase over the Shale. The Shale is therefore an EM conductor with a large decay constant. Maximum rα relaxation times are significantly larger than those over the Woodlawn orebody on lines G and J. Despite what appears to be strongly anomalous coupling behaviour over the Shale, the amplitude of component A decays to less than 5% of the C component about 0.16s after current switch-off.

EM Coupling (Fig.14)

Fig. 14 shows EM coupling behaviour for the three array on lines G and J over the Woodlawn orebody and on the line over the Black Shale.

The decay curve A component and rα profiles appear to adequately represent EM coupling behaviour for the three array because they are unlike those for the longer-lived decay components.

On all of the A component profiles the coupling amplitude falls to a negligible level over the unmineralized rocks surrounding the orebody and the Black

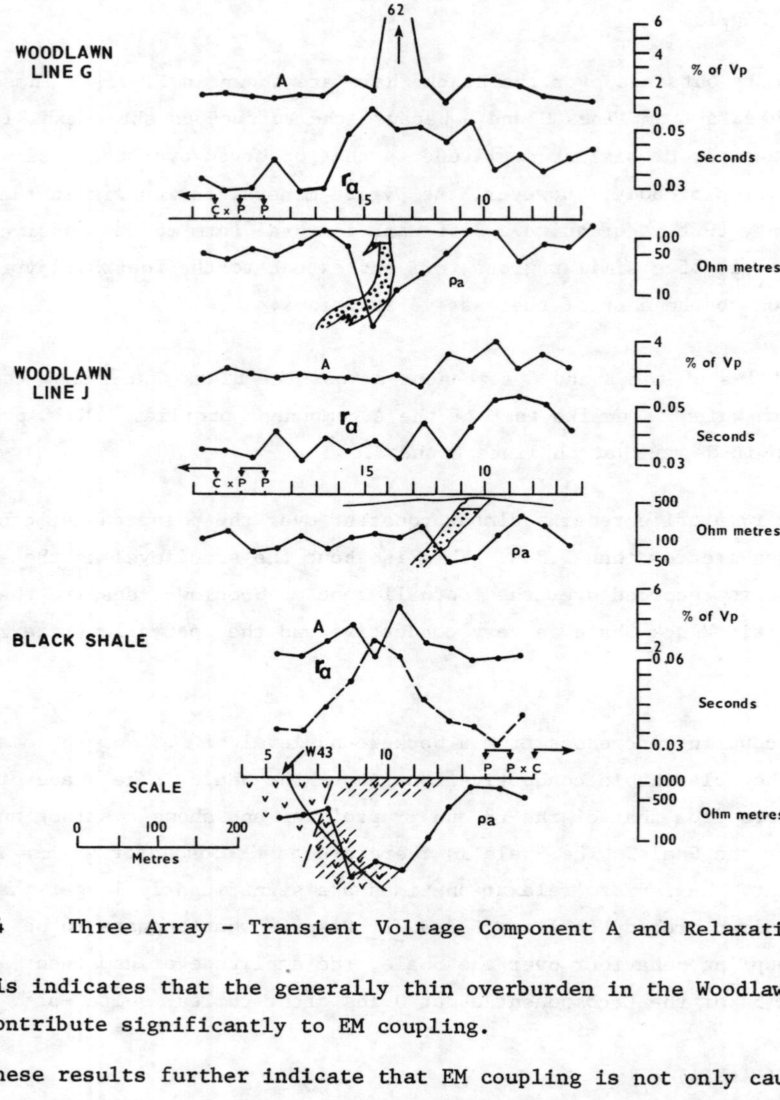

Figure 14 Three Array - Transient Voltage Component A and Relaxation Time rα

Shale. This indicates that the generally thin overburden in the Woodlawn region does not contribute significantly to EM coupling.

These results further indicate that EM coupling is not only caused by the electrical properties of the overburden or weathered layer but also shows considerable variation in magnitude and decay rate with lateral changes in bedrock resistivity.

Each of the six parameter profiles show anomalous behaviour over the mineralized zones. The level of coupling reflected by the A component is almost directly related to the magnitude of the apparent resistivity anomaly over the conductive zones as would be expected. The largest coupling level occurs on line G and is associated with an apparent resistivity value of less than 5 ohm m.

Coupling levels are higher over the Black Shale than on line J. The absolute resistivities over the Black Shale are higher than on line J although the amplitude of the anomaly over the Shale is greater than over the orebody.

The relaxation time profiles also show a distinct increase over the mineralized zones. The Black Shale shows the most dramatic increase in rα which reveals that the wide conductive zone within the shale is energized very effectively by the three electrode array.

The largest anomalous values recorded in coupling amplitude (A) and relaxation time (rα) for the three electrode configuration (a = 30.5m) demonstrate that, in this, and other environments with relatively thin and moderately resistive overburden and weathered layers, conventional transient IP measurements of the decay curve after a time delay of about 0.3s should be substantially free of coupling effects which arise mainly from lateral variations in bedrock resistivities.

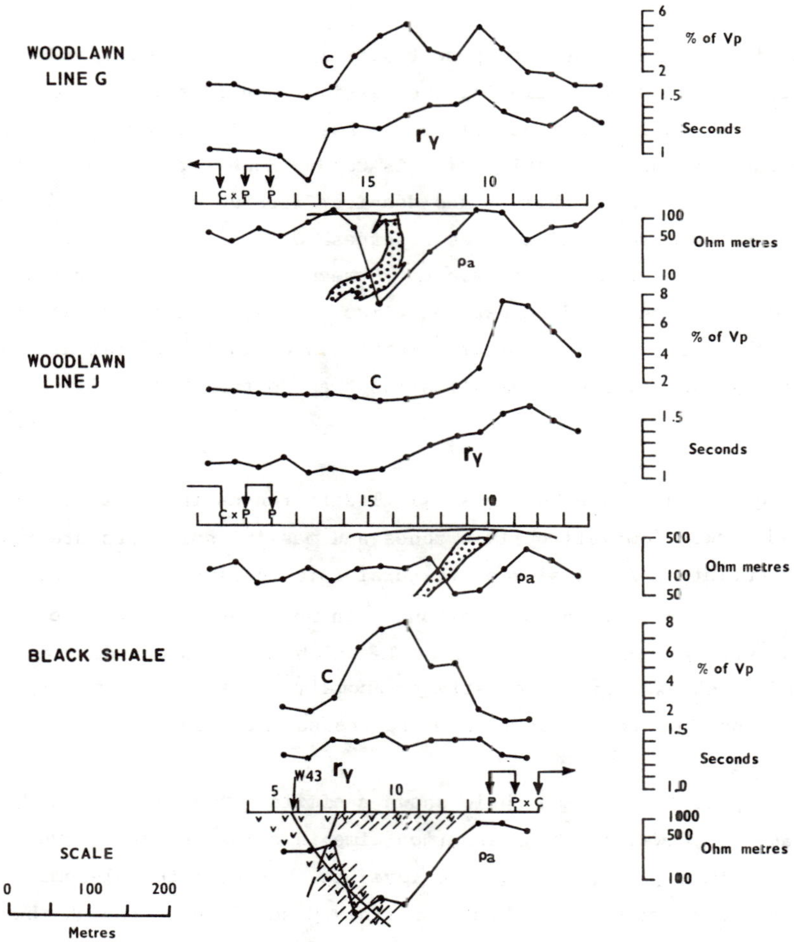

Figure 15 Three Array - Transient Voltage Component C and Relaxation Time rγ

First and Second Order IP Effects (Fig.15)

Fig.15 shows the transient decay parameters C and rγ for the late time section of the decay curve for all three traverses. Although the transient IP behaviour is probably more realistically represented by a combination of the exponential functions, $Be^{-\beta t}$ and $Ce^{-\gamma t}$, the profiles of the C component and rγ are much smoother and therefore more diagnostic than those of the B component and rβ. This is partly due to the graphical method of factoring.

Since the relative variation in the IP decay form along each line rather than a precise mathematical representation of the IP decay is of importance, the C and rγ profiles are considered suitable to display lateral variations in first and second order IP effect.

On lines G and J the C component profiles are dissimilar and do not provide a diagnostic means of differentiating the massive sulphides and disseminated pyrite mineralization in the footwall on the basis of IP response, unless additional information such as apparent resistivity is used. However, the relaxation time (rγ) profiles are similar on both lines despite differences in the C component profiles. Both rγ profiles show relaxation times in excess of 1.4s over the eastern pyritic zone as well as a measureable decrease to between 1.2s to 1.4s over the massive sulphide orebody. Background relaxation times are fairly constant at about 1.0s to 1.1s on both lines. These similar rγ profiles exhibit a closer affinity to changes in the gross style of mineralization than do the chargeability or C component profiles.

The question of whether these small differences in relaxation times between unmineralized rocks, footwall pyritic zones and massive sulphides are related to sulphide concentration and its gross textural form or simply to lateral changes in resistivity along the profile is important. In an attempt to resolve this question, the profiles over the conductive Black Shale which contains abundant disseminated sulphides and which exhibits a resistivity anomaly comparable in magnitude to that over the shallow orebody on line G are reconsidered below.

The C profile over the Shale shows a similar response amplitude to that on lines G and J. However, the relaxation time is almost constant over the polarizabl zone at greater than 1.35s, or the same level as that over the disseminated pyritic zone within the eastern footwall rocks on lines G and J, even though the Black Shale is conductive and the footwall zone is resistive.

From these three array results a number of observations may be made regarding the first and second order IP effects represented in Fig.15. These are:

(a) The pattern of the IP voltage decay, shown as relaxation time is not constant along each profile. Mineralized zones show significantly longer relaxation times than unmineralized host rocks. This is contrary to observations made by Phillips and Richards (1974).

(b) There is no direct correlation between the first order IP effect, i.e., chargeability and the second order IP effect or decay rate. The decay rate does not reflect lateral variations in chargeability. This implies that the first and second order effects do not reflect the same features of the electrochemical processes that are believed to constitute the IP effect and that additional geological information may be obtained from the decay rate of IP transient voltage discharges.

(c) Consistent measurable differences in relaxation time, $r\gamma$, between unmineralized host rocks (1.0s to 1.1s), disseminated sulphide zones (>1.35s to 1.4s) and the Woodlawn massive sulphides (1.2s to 1.4s), i.e., between mineralization of very different textural form and gross geometry appear to be present. These differences in $r\gamma$ seem to be related to the gross shape of the polarizable body as well as to changes in the mineralization style from massive to predominantly disseminated and not to differences in the resistivity over each sulphide concentration. This is suggested because the conductive Black Shale containing abundant disseminated pyrite gives rise to the same order relaxation times as those over the disseminated pyrite within the resistive footwall lithologies at Woodlawn. Both pyritic sources are similar in gross shape.

Dipole-Dipole Array

Line G (Fig.16,17,18)

Transient decay curves were recorded on line G with the dipole-dipole array at three different dipole separations, of n=1, 2 and 3 using an electrode separation of 30.5m. Figs. 16, 17 and 18 show respectively the decay curve parameters for each of the three separations.

The profiles obtained with the smallest separation of n=1 are shown on Fig.16. A strong resistivity anomaly above the massive orebody is present. This is similar in shape but of larger amplitude to the three array anomaly (Fig.11). If the component profile again shows two anomalies, the western anomaly coincides

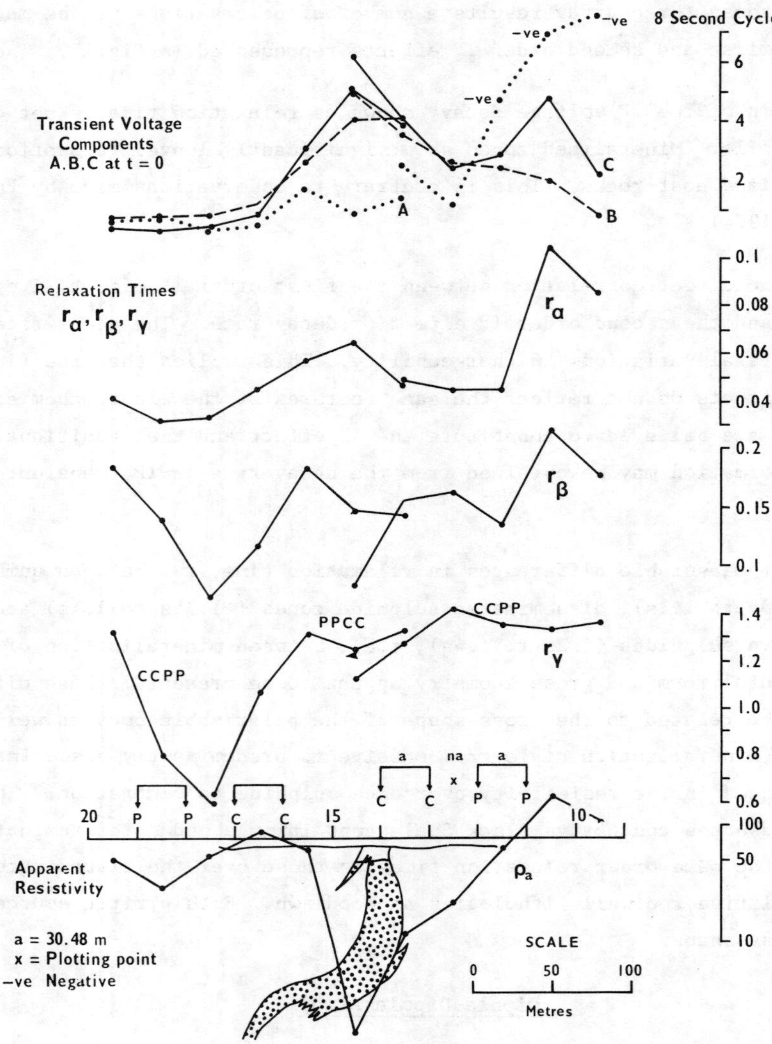

Figure 16 Transient Decay Curve Parameters, Dipole Dipole Array, n=1 - Line G

with the Woodlawn orebody and the eastern anomaly with the footwall pyritic zone. The electrode spacing and separations used are suitable for resolving the two adjacent polarizable sources.

The relaxation time $r\gamma$ profile, shows large variations from 0.6s to 1.4s. Variations over the two polarizable sources are not as obvious as for the three array. However, $r\gamma$ is greater than 1.35s over the resistive polarizable footwall zone and less than 1.35s over the orebody. Relaxation times corresponding to the low resistivity region above the orebody are less than 1.35s and are of the order of

1.15s to 1.25s. The sharp low in $r\gamma$, evident at G17.5 on Fig.11 also occurs on the dipole-dipole array profile. Although this feature appears to have some geological significance an adequate explanation of it has not been found.

A comparison of the reciprocal array configuration (i.e. CCPP and PPCC) at G13.5 and G14.5 for parameters $r\gamma$ and C suggests that reciprocity for these parameters is obeyed only at station G13.5. The differences at G14.5 may imply that either reciprocity of the IP phenomenon does not occur over the conductive body or that the very low signal levels of Vs, resulting in noisy decay curves at this point, do not allow an accurate comparison of the reciprocal parameters. This is considered more likely. Nevertheless such differences in reciprocal values emphasize the need for profiling the whole line, perhaps with the reciprocal electrode arrangements.

The coupling parameters A and $r\alpha$ obtained with the dipole-dipole array show anomalous behaviour. Background levels are low over the hanging wall and both A and $r\alpha$ increase marginally over the massive sulphides. However, A becomes strongly negative and $r\alpha$ increases sharply over the footwall pyritic zone, to the east of G12, to almost twice the levels obtained with the three array. The negative coupling components take as long as 0.385s to dissipate to less than 5% of the C component level which is also longer than the corresponding three array data. These negative coupling effects do not appear to be related to the resistive pyritic zone but are probably due to reflective EM coupling from the massive sulphides resulting from the spatial relationship between array geometry, connecting cables from the central transmitter at G14 to the eastern current electrodes, and the high resistivity of the footwall rocks.

The B component profile shows a strong anomaly over the massive orebody. A corresponding anomaly is not present over the eastern pyritic zone because, in this instance the B component appears to represent an average between the strong positive C component anomaly and the strong negative A component anomaly.

The $r\beta$ profile shows similarities with both the $r\alpha$ and $r\gamma$ profiles and is therefore not wholly related to, the rapid decay associated with EM coupling or to the slow IP decay.

The decay curve parameters for the n=1 data indicate that the second order IP effect shows small measurable differences between the narrow massive orebody and the wider footwall pyritic zone. However, a sharp decrease in $r\gamma$ at G17.5, the cause of which is not known, make it difficult to use the rγ profile.

The chargeability profiles in Figs 17 and 18 for the larger n=2 and n=3, electrode separations shown on Fig. 17 and 18 do not resolve the two polarizable sources beneath line G.

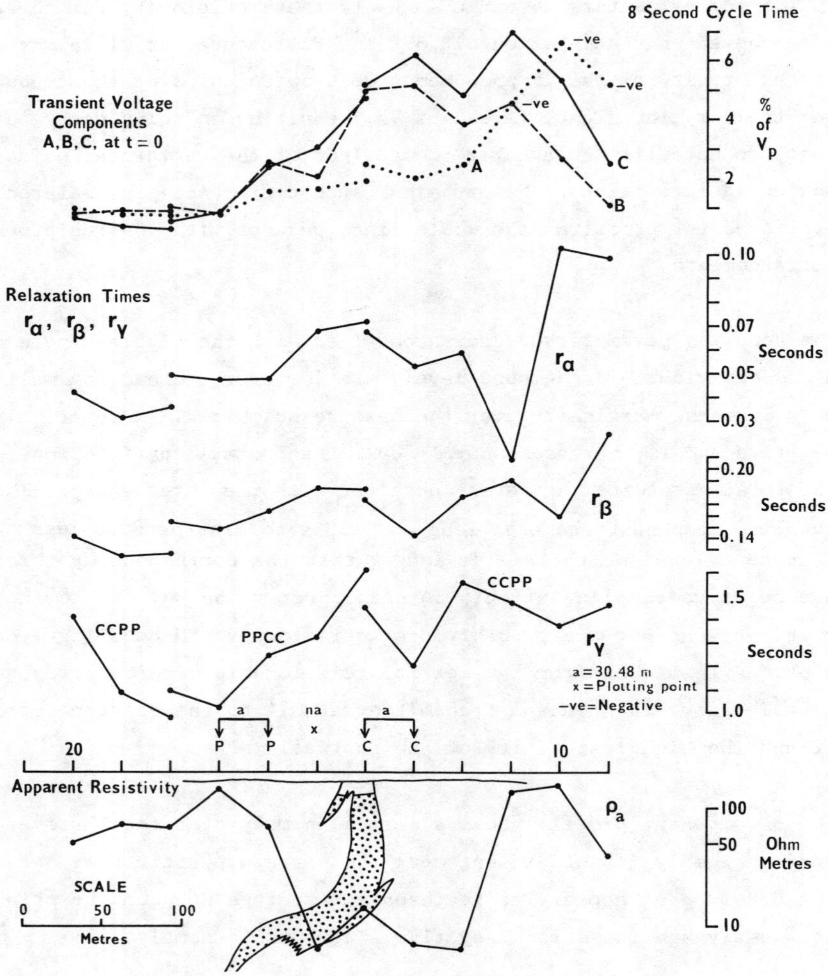

Figure 17 Transient Decay Curve Parameters, Dipole Dipole Array, n=2 - Line G

Fig. 17 shows a broad double minimum resistivity anomaly above the massive orebody.

The C component profile shows a merging of the two anomalies evident at n=1 (Fig.16) with the eastern pyritic anomaly predominating. Also rγ shows a greater increase above background levels over the mineralization than for n=1, however, a difference in rγ levels between both sources cannot be seen. Coupling effects are again negative to the east of the orebody, with the largest A component levels requiring more than 0.355s to dissipate to less than 5% of the C component level.

The parameters B, C, rβ and rγ for the reciprocal arrays at G13.5 and G14.5 show almost equivalent reciprocal values. However, A and rα show large divergences in reciprocal values. This may be explained since EM coupling does not depend on the electrode grounding points but on the mutual inductance of the dipole cables on the ground surface. Despite the fixed relative positions of the grounding points with the reciprocal arrays, the overall receiver and transmitter cable-layout configuration may be quite different for the reciprocal array. When this occurs, a constant mutual inductance between the reciprocal cable configuration is not maintained and so the resulting inductive coupling will not necessarily display reciprocity. It is the precise location of the cable configuration with respect to the buried massive sulphide conductor that is predominant in defining the resultant amplitude and sign of the coupling level and relaxation time at Woodlawn. The lack of reciprocity in A and rα values for n=3 (Fig.18) can be adequately explained by the changes in EM mutual inductance of the connecting cables for the reciprocal arrays.

Phillips and Richards (1974) suggest that a lack of reciprocity in the initial rapid decay of the transient curve may also be caused by polarization of the current electrodes. Although Cu-$CuSO_4$, non-polarizing electrodes were used for current electrodes, they do in fact polarize at high current densities, but to a lesser degree than metal stakes or foil. However, when used in conjunction with a well regulated current source that presents an infinite impedance to the ground circuit during the current off-time, (as in this case) current electrode polarization effects should be negligible. Any depolarization currents arising from the current electrodes during the off-time (Fig.6) will be so localized in scale around the electrodes as to contribute negligible effects at the distant recording dipole electrodes, in comparison with the very much larger scale depolarization currents, arising from the bulk volume of polarizable sulphide grains within the subsurface. EM coupling, will in general have a far greater influence on the initial rapid transient decay than any polarization effects from current electrodes assuming the use of a satisfactory transmitter.

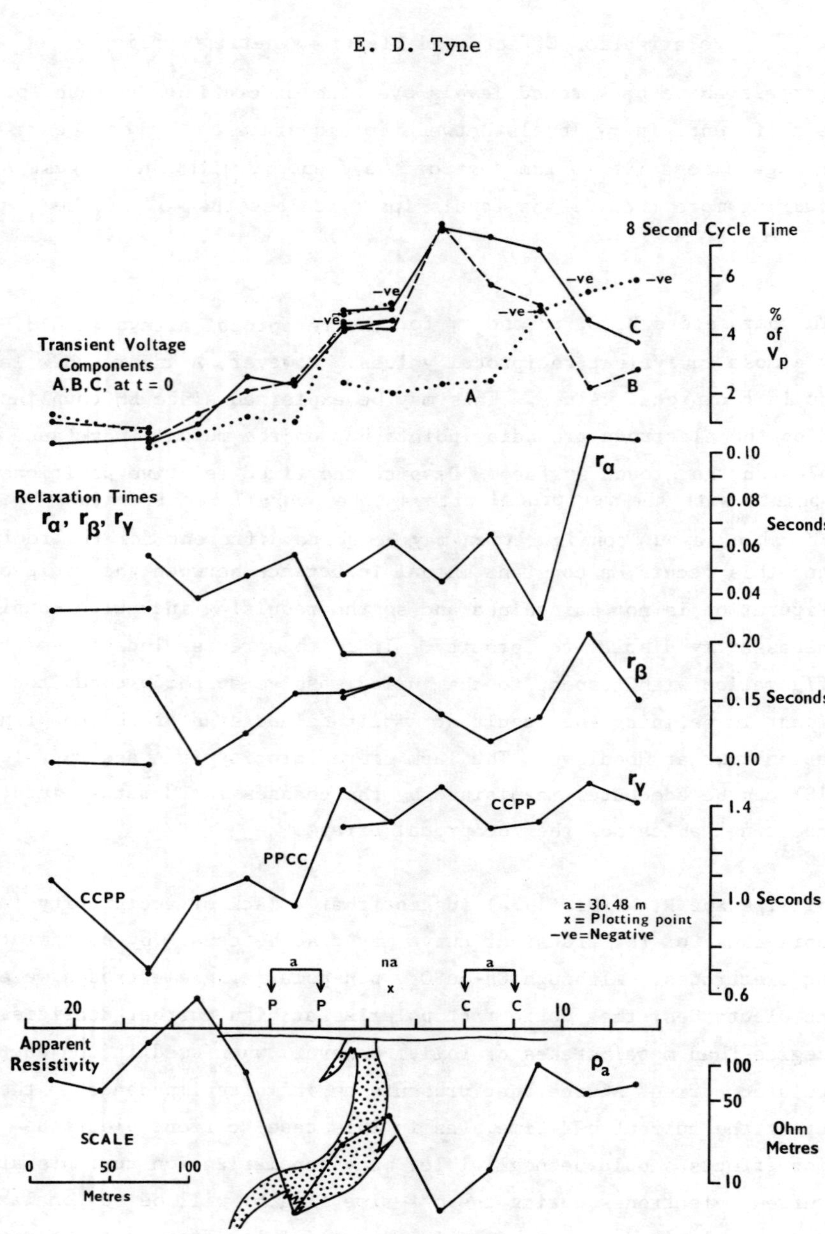

Figure 18 Transient Decay Curve Parameters, Dipole Dipole Array, n=3 - Line G

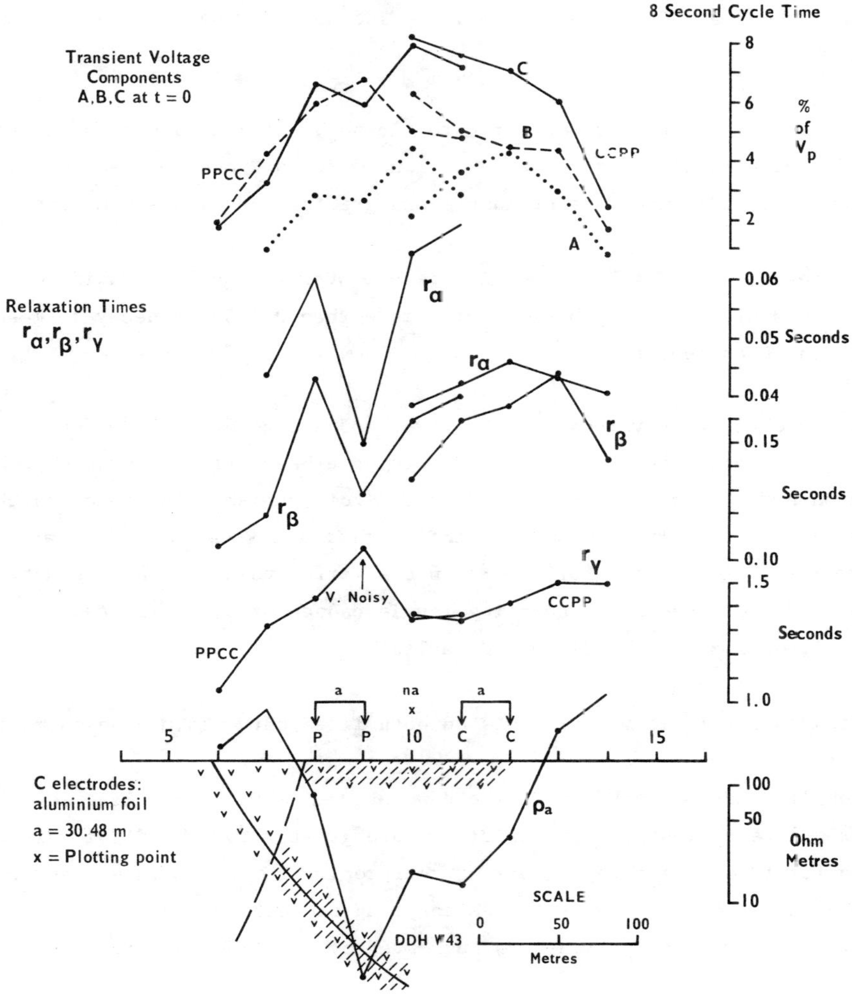

Figure 19 Transient Decay Curve Parameters, Dipole Dipole Array, n=2 - Black Shale

Black Shale (Fig.19)

A dipole-dipole profile at n=2 over the Black Shale (Fig.19) shows a strong resistivity low between stations 8 and 12, over the same region as for the three array (Fig.13).

Both the B and C component profiles also show strong anomalies, above the Shale.

The second order IP effect, $r\gamma$ profile shows similar behaviour to the corresponding dipole-dipole $r\gamma$ profile on line G for n=2 over the eastern pyritic zone (Fig.17) and $r\gamma$ maintains levels greater than 1.35s over the Shale.

Anomalous coupling behaviour is present over the Shale. Maximum coupling levels of the A component require more than 0.175s to decay to less than 5% of the C component.

Reciprocal array values for the first and second order IP effects are almost identical and results do not appear to have been affected by any possible polarization effects associated with the metal foil current electrodes which were only used on this line. However, the initial decay parameters, A and $r\alpha$ again show large divergences for the reciprocal array values. This is most likely to be due to differences in EM coupling levels caused by different cable arrangements for the reciprocal arrays as discussed earlier.

The dipole-dipole measurements on both lines surveyed may be summarized as follows:

(a) EM coupling may be both positive and negative. Maximum coupling levels arising from lateral resistivity variations are greater for the dipole-dipole array than for the three array. Lack of reciprocity for parameters A and $r\alpha$ is probably due to the nature of the EM coupling process rather than the polarization of current electrodes as suggested by Phillips and Richards (1974).

(b) Variations in second order IP effects with the dipole-dipole array do not appear to be as diagnostic as for the three array. The smaller signal levels obtained with the dipole-dipole array as compared with the three array resulted in slightly noisy transient decay curves over the mineralized zones. This may be responsible for reducing any small variations in $r\gamma$ to an insignificant level. Both the C and $r\gamma$ profiles for n=1 show variations that may be correlated with the two different styles of mineralization. The $r\gamma$ times, in particular, are greater than 1.35s over the broad pyritic zone and less than 1.35s over the massive sulphides. These seemingly important variations in C and $r\gamma$ are mostly lost with increases in dipole separation to n=2 and n=3 as a result of a deterioration in resolving power. Relaxation time, $r\gamma$ also increases marginally with increasing dipole separation as postulated by Grant (1971). The pyritic zone east of the Woodlawn orebody appears to dominate the first and second order IP response for the larger dipole separations.

The C component and $r\gamma$ profiles over the Black Shale may be compared to profiles for n=2 on line G over the Woodlawn orebody. Both profiles show similar IP responses arising from broad disseminated pyrite zones with relaxation times, $r\gamma$, greater than 1.35s over both zones.

Gradient Array

Line G (Fig.20)

Fig. 20 shows the gradient array transient decay curve parameters on line G. A very prominent apparent resistivity anomaly occurring over the Woodlawn orebody is broader than would be expected for the narrow vertical body shown in the drill section because the orebody strikes obliquely to line G.

The C component profile is considerably different to that obtained with the other two arrays. The narrow anomalies apparent on this profile show the increased resolution of shallow geological variations for the gradient array as compared with that of the other two arrays. This is due in part to the decreased potential electrode spacing and closer station spacing as compared with the three and dipole-dipole arrays. A prominent anomaly peak at G14.75 appears to be related to the massive sulphides. This is offset to the west of the orebody on the drilled section and is probably due to the oblique strike of the orebody with respect to line G.

Chargeability levels over the resistive pyritic footwall zone from G8 to G13 are about twice those over the hanging wall rocks although higher amplitude fluctuations are superimposed on this zone, for example at G9.75 and G8.75. Chargeability levels fall to background to the east of the pyritic zone.

Between the chargeability anomaly over the orebody and the broad eastern anomaly is a sharp depression which occurs on the immediate footwall boundary of the orebody. The second order effect $r\gamma$ profile also shows a corresponding sharp depression at G13.75.

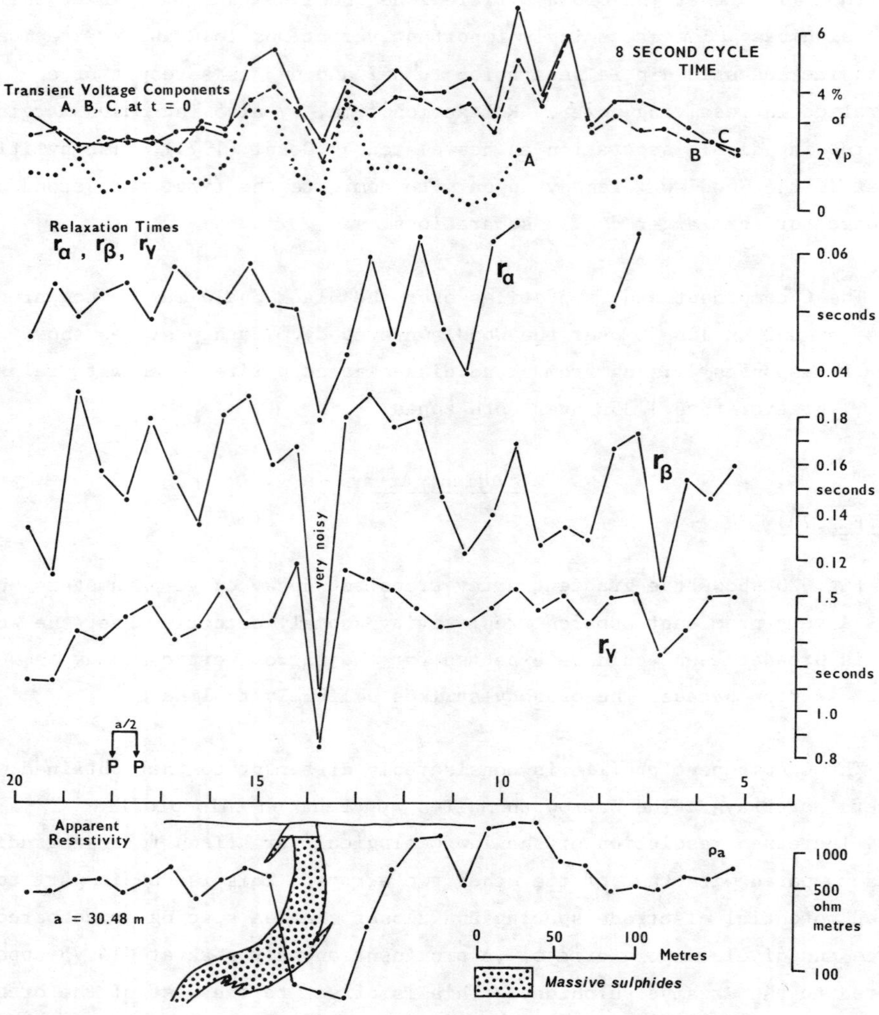

Figure 20 Transient Decay Curve Parameters Gradient Array - Line G

The B and rβ parameters are similar to the C and rγ parameters. Therefore it is inferred that at the footwall boundary of the orebody the IP voltage discharge has a very small amplitude and decays very rapidly; the relaxation time being almost half that of the surrounding region. Although the decay curve points recorded at G13.75 are more scattered than is normal, repeated measurements indicate consistent very rapid decays at the longer times. The feature at G13.75 is not a result of instrumental or measurement errors and may be explained by an interaction of the IP depolarization currents arising from the two adjacent polarizable sources. For very shallow polarizable bodies a negative or small positive chargeability is recorded when the measurement dipole is situated near the edges of the bodies (Sumner, 1976, p194). Very sharp negative troughs are characteristically associated with very resistive, polarizable bodies such as the footwall pyritic zone (Tyne and Whiteley, this volume, p 349). When two polarizable bodies are in close proximity, as on line G, negative depolarizing currents associated with the edges of the bodies will interact in the narrow region separating the bodies. This will give rise to negative or very small positive chargeabilities in this region which will be more pronounced than any low amplitude effects associated with the outer edges of the polarizable bodies. In qualitative terms the above explanation for the depression in the chargeability profile, at G13.75 does not, however, explain the concomitant rapid decay at the same location. To explain this it is necessary to consider in more detail the path taken by depolarizing currents. In the vicinity of the massive sulphide orebody these currents will mainly flow internally through the highly conductive sulphides. The very resistive footwall pyritic zone will probably maintain both internal and external depolarization currents. The external currents will attempt to flow along paths of least resistance around the resistive polarizable regions.

The external depolarizing current flow will therefore be highly distorted by the presence of the adjacent highly conductive orebody. The rapid decay is probably a result of the body shortening the length of the paths taken by the depolarizing currents from the pyritic footwall source.

It is interesting that the feature at G13.75 is not as obvious in the chargeability parameters derived from the three and dipole-dipole array data probably because of the larger electrode and station spacings used with the arrays.

The character of the relaxation time profile ($r\gamma$) obtained with the gradient array is also very different from the profiles obtained with the other two arrays, and shows not obvious variations over the two polarizable sources. The gradient array $r\beta$ profile also shows no obvious differences in decay rate over the two sources.

EM coupling amplitudes on the A component profile are largest in the vicinity of the orebody and require more than 0.15s to dissipate to less than 5% of the C component.

The gradient array is often considered to show poor EM coupling rejection in comparison with the dipole-dipole and three arrays (Sumner, 1976, p179). However, the very low coupling levels recorded for the gradient array on line G attest to the efficacy of using a U-shaped loop of wire connecting the current electrodes (Fig.3) for reduction of EM coupling within the central survey region.

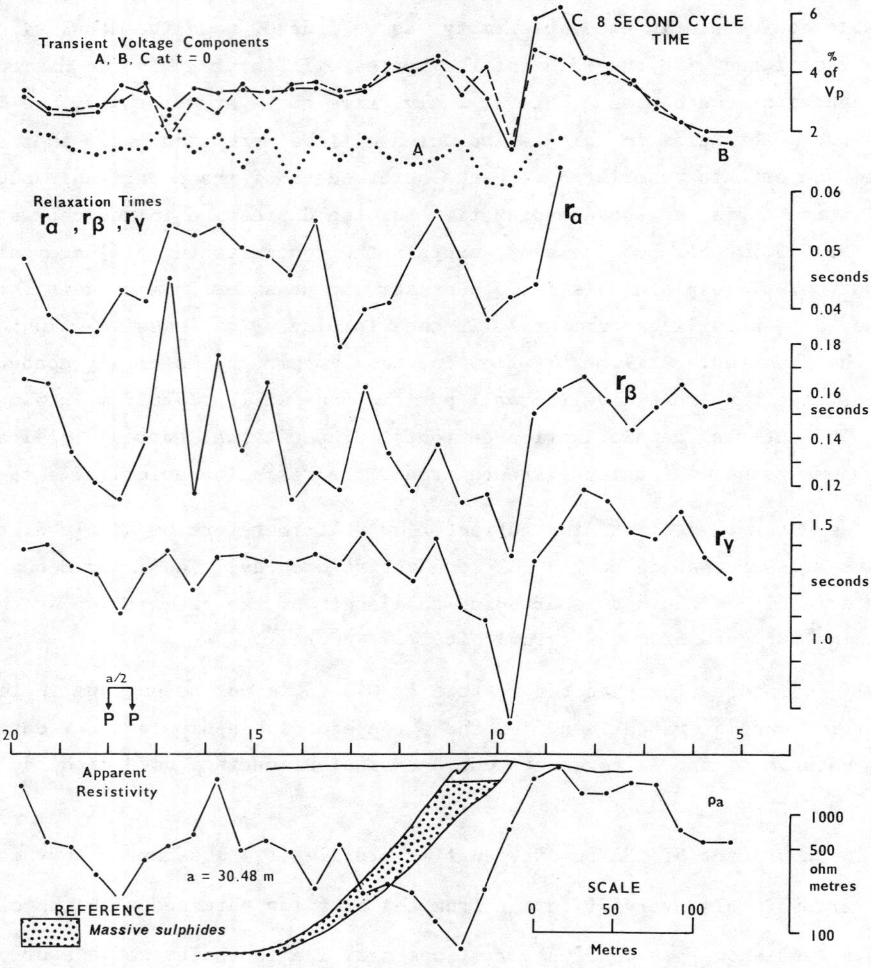

Figure 21 Transient Decay Curve Parameters Gradient Array - Line J

<u>Line J</u> (Fig.21)

Fig.21 shows that the dipping massive sulphide orebody on Line J is again well defined by a prominent gradient array apparent resistivity anomaly. The chargeability profile shows a sharp anomaly over the footwall pyritic zone which is narrower on line J than on line G and a low amplitude anomaly above the dipping orebody. These anomalies are separated by a sharp depression at J9.75, centred near the immediate footwall of the orebody. This feature is similar to that obtained on line G at G13.75. Background chargeability levels occur to the west of J13.

The C component profile emphasizes the excellent lateral resolution of the gradient array for narrow, shallow geological variations and shows two distinct chargeability anomalies. In contrast, the corresponding three electrode array profile does not resolve the two polarizable sources.

The relaxation time rγ, obtained with the gradient array again shows abrupt low at J9.75. Relaxation times in this region are less than half those of the surrounding regions indicating that the footwall zone is the site of a rapid discharge of the depolarization currents from the eastern pyritic source through the orebody. Relaxation times tend to be greater than 1.4s over this resistive pyritic zone and are less on the remainder of the profile. The abrupt low at J9.75 probably masks any subtle second order effects related solely to the orebody.

The B and rβ profiles show certain similarities to the C and rγ profiles although these intermediate parameters are rather erratic and are not easily correlated with the known geology.

EM coupling levels are quite small on line J and show little correspondence to the chargeability profile. Since line J is further from the primary current loop than line G, coupling levels would be expected to be less than for line G. Initial rapid decays do not occur to the east of J9. Maximum coupling levels are again relatively low and require less than 0.115s to dissipate to less than 5% of the C component.

The variations in relaxation time for the gradient array do not appear to be a diagnostic means of discriminating between the two different styles of polarizable sulphide mineralization.

In comparing these results with those for the other two arrays one may conclude that,

1. The chargeability profiles provide better lateral resolution of the boundaries of the polarizable sources (partly a function of smaller electrode spacing), and

2. the second order IP effects appear to show less diagnostic variations over the massive and disseminated styles of mineralization.

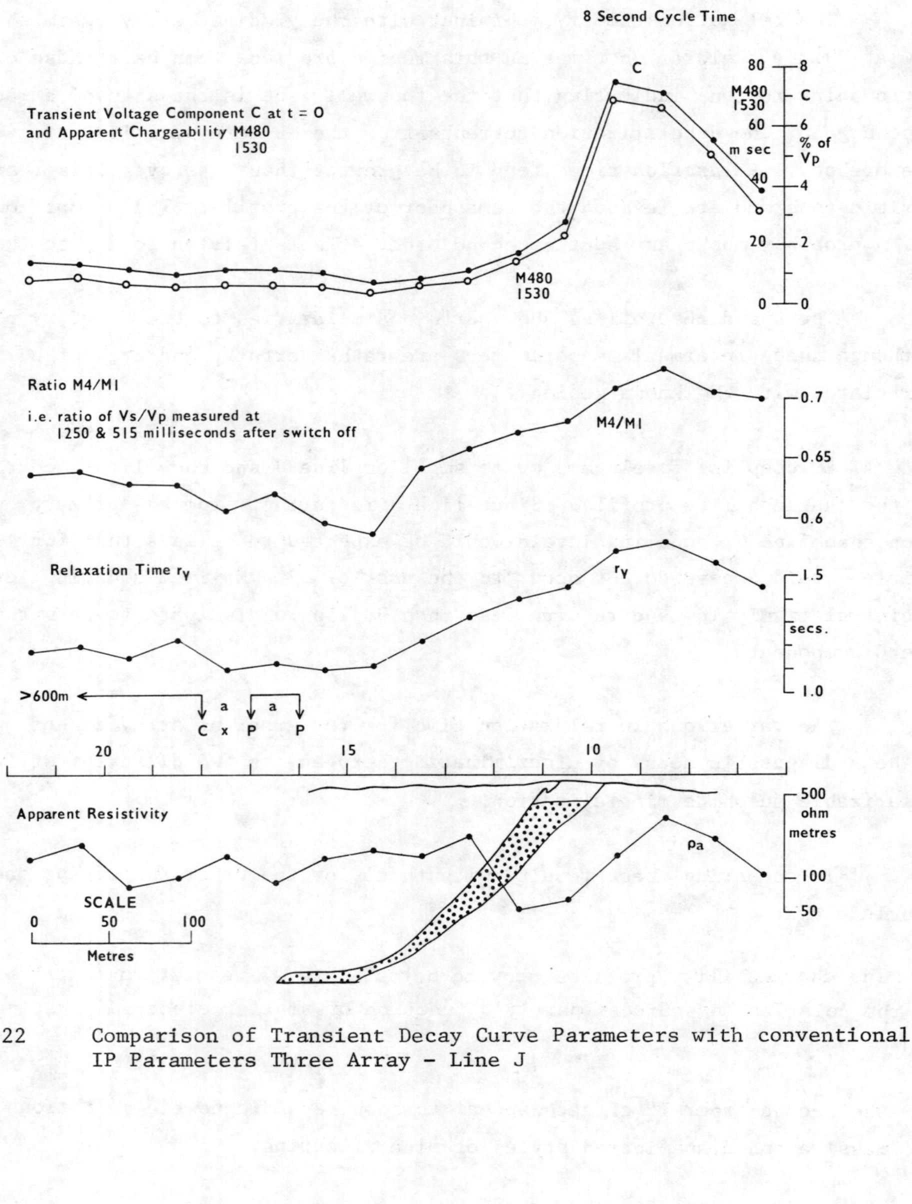

Figure 22 Comparison of Transient Decay Curve Parameters with conventional IP Parameters Three Array - Line J

TABLE 2 - Correlation of M_{1530}^{480} and C Using Linear Regression

ARRAY	LINEAR RELATION	CORRELATION COEFFICIENT
THREE		
Line G	M = 8.07C - 1.04	0.98
Line J	M = 9.53C - 3.35	0.999
Line C	M = 8.65C - 1.74	0.999
DIPOLE DIPOLE		
Line G, n = 1	M = 7.25C + 0.3	0.98
, n = 2	M = 8.71C - 1.04	0.993
, n = 3	M = 8.37C - 1.55	0.993
Line C, n = 2	M = 8.66C - 1.29	0.985
GRADIENT		
Line G	M = 9.26C - 3.24	0.97
Line J	M = 9.26C - 4.59	0.964

Decay Curve Parameters Versus Conventional IP Parameters

Fig.22 illustrates the relationship of parameters C and rγ for line J with the parameters normally recorded in a conventional IP survey. The decay curve has been integrated over the period 480ms to 1530ms and presented as profile M_{1530}^{480} (t_d= 480ms, t = 70ms). The profile correlates closely with the C component profile which represents apparent chargeability as the ratio of secondary IP voltage to primary voltage at switch-off. This indicates that a conventional integration of the IP decay over a limited time interval using a delay time large enough to exclude inductive coupling effects shows a simple relationship and may be used to approximate the apparent chargeability at t=0.

The C component and M_{1530}^{480} profiles for each of the electrode arrays were correlated using linear regression analysis. The correlation coefficients and relationships obtained are presented in Table 2. These are greater than 0.96, indicating a close linear relationship between M_{1530}^{480} and the C component. The different linear relationships obtained indicate that C may be approximated, but not accurately estimated, from the conventional M_{1530}^{480} parameter. However, the close correlation between the two parameters indicates that, apart from absolute amplitudes, variations in apparent chargeability at t=0 are faithfully represented by variations in a conventional IP parameter integrated over a limited time inverval at a delay time sufficiently large to reduce EM coupling to negligible levels.

Discussions and Conclusions

The transient voltage decay curves recorded at Woodlawn with the three different electrode arrays are composed of an initial rapid decay related to EM coupling, and a much slower decay form related to IP effects in the subsurface. Both of these decays have an exponential-like form (see Dey and Morrison, 1973 and Lynam, 1972) and so there is a meaningful basis for factoring the transient voltage decay curves into a number of exponential functions. The graphical factoring procedure applied to the data, revealed that the measured decay curves for each electrode array can be satisfactorily factored into a minimum of three exponential functions (with each function containing an amplitude term (A,B and C) and a decay constant. The inverse of the decay constant is termed a relaxation time ($r\alpha$, $r\beta$, $r\gamma$), and together with the amplitude term is used to compare the transient decay curves.

Profiles of the A and $r\alpha$ parameters for the first exponential function, representing the initial rapid decay are generally dissimilar to profiles of the remaining two other sets of parameters. Anomalous behaviour on the A and $r\alpha$ profiles is associated with regions of low resistivity and can only adequately be explained in terms of EM coupling. Therefore, the A and $r\alpha$ parameters are considered to substantially represent the EM coupling component of the transient voltage decay curves in the Woodlawn environment.

The C and $r\gamma$ parameters, representing the longer time decay, generally show very smooth variations along the profiles for all electrode arrays and are considered to substantially represent the IP component of the transient decay curves. The relaxation times, $r\gamma$, of more than 1.0s are typical of the long-period transient decay of IP secondary voltages for a current charging time of 2s.

Profiles of the B and $r\beta$ parameters, representing the intermediate exponential function, show similarities to profiles of both the first and last exponential functions however, generally they appear more closely related to the C and $r\gamma$ parameter profiles. B and $r\beta$ therefore appear to be hybrid parameters which contain both EM coupling and IP components. Profiles of these parameters are often quite erratic and difficult to interpret in terms of subtle variations on the transient voltage decay curve related to EM coupling and the IP effect. The need to include the intermediate exponential function in the decay curve factoring process indicates that EM coupling decay is not totally defined by the function $Ae^{-t/r\alpha}$ and also that IP transient decay is not totally, defined by the function $Ce^{-t/r\gamma}$. However, the initial exponential function is a good first approximation to the level of EM coupling for each array and the final exponential function is a good approximation to the long period transient decay of subsurface IP effects. The relative variations in the IP decay curve parameters along the survey lines, rather than a precise mathematical representation of the IP behaviour is of prime importance in determining the geological usefulness of measuring the shape of the transient voltage decay curve.

EM Coupling

The parameters A and $r\alpha$ show anomalous behaviour over low resistivity regions which is absent on profiles of the other decay parameters, and substantially represents EM coupling in this environment. The contribution of overburden to EM coupling at Woodlawn is small.

TABLE 3 - Decay Time of Maximum Coupling Amplitudes to Negligible Levels

Electrode Array	Line G (seconds)	Line J (seconds)	Line C (seconds)
Equispaced Three	0.3	0.13	0.16
Dipole Dipole	0.385(n=1) 0.355(n=2) 0.368(n=3)	No Survey	0.175(n=2)
Gradient	0.16	0.115	No Survey

Table 3 lists the times required for the initial coupling amplitude, A to decay to less than an arbitrary level of 5% of the component C level. The table shows that maximum coupling levels vary for each electrode array and also from traverse to traverse depending on lateral resistivity variations. The gradient array shows the most rapid dissipation of EM coupling because of the effective use of the U-shaped current-loop wire. Both the three array and dipole-dipole array show higher coupling levels which require more than 0.3s to dissipate to negligible levels.

The A and $r\alpha$ parameters for the dipole- dipole array do not show reciprocity on line G and the line over the Black Shale. This is caused by variations in EM coupling behaviour due to differences in the reciprocal dipole cable arrangements and not to current electrode polarization effects.

The maximum time of 0.385s for coupling to decay to negligible levels, suggests that conventional transient IP measurements in this type of environment will be substantially free of coupling effects arising from the lateral variations in bedrock resistivities after a time delay of at least 0.4s. Use of a time delay of the order of 0.5s allows an adequate safety margin and would accommodate any small residual EM coupling component which has spilled over into the intermediate exponential decay function $Be^{-t/r\beta}$.

The lateral variability of EM coupling levels at Woodlawn emphasizes that monitoring of the EM coupling component of the transient decay curve in new survey areas is warranted before extensive IP surveying is undertaken so that an appropriate time delay may be chosen to optimize coupling rejection versus chargeability signal to noise levels.

First and Second Order IP Effects

Results obtained for each array at Woodlawn show that there is no consistent relationship between the first order IP effect represented by component C i.e. chargeability, and the second order IP effect represented by $r\gamma$, i.e. relaxation time. Relaxation time does not reflect lateral variations in chargeability which suggests that the chargeability and relaxation time do not represent the same aspects of the electrochemical processes that constitute the IP effect and consequently additional geological information is available from relaxation time.

Relaxation time associated with the IP voltage decay is not constant but is measurably longer over mineralized zones than over unmineralized zones. Relaxation time profiles for the three array appear more useful, principally because of the higher signal levels which reduced the incidence of noisy decay voltages. Both the three and dipole-dipole arrays appear to show consistent differences in relaxation times across the Woodlawn orebody. These range from 1.0s to 1.1s for unmineralized hanging wall rocks, 1.2s to 1.35-1.4s for the massive sulphides and greater than 1.35-1.4s for the disseminated pyrite within the eastern footwall rocks. These differences in relaxation time appear fairly small, but may be expected since the second order IP effect is by nature more subtle than the first order IP effect. These subtle differences in $r\gamma$ seem to be related to the gross shape of each polarizable source as well as to changes in the style of mineralization from massive to predominantly disseminated and not to differences in resistivity over each source. This is inferred because the disseminated pyrite within the Black Shale gives rise to the same order relaxation times as those over the disseminated pyritic zone on lines G and J. Both of these pyritic sources have a similar bulk shape.

In the Woodlawn environment the three array, in particular, exhibits larger relaxation times over the broad disseminated sulphide zones regardless of the resistivity of the host rock than over the narrower conductive massive sulphide orebody. The gradient array relaxation time profile appears to show less diagnostic variations over the two styles of mineralization.

Results indicate that there may be some difficulties in using the relaxation time parameter as a means of discriminating between sulphides of different textural form. For example, the depression in the $r\gamma$ profile at G17.5 for both dipole arrays appears to have some geological significance but this is not immediately obvious because it occurs totally within the unmineralized hanging wall rocks.

Also the abrupt depression on the gradient $r\gamma$ profile at G13.75 which is possibly caused by the interaction of depolarizing currents from the two sources, would probably mask any $r\gamma$ differences between the massive sulphides and disseminated pyrite.

Although the results at Woodlawn are limited by the use of a 2s rather than a very long current charging time, they indicate possible difficulties in recording and interpreting second order IP effects. They are, however, encouraging, and suggest that the shape of IP decay curves may yield useful geological information which is not apparent in the first order IP effect. They also suggest that it may be possible to interpret gross outlines for massive and disseminated sections of a polarizable body, and differentiate between the IP response of a conductive massive sulphide zone and a conductive mineralized zone of very different gross size and textural form, such as the Black Shale containing abundant disseminated pyrite.

Acknowledgements

This paper is published with the permission of the Under Secretary, New South Wales Department of Mineral Resources and Development, Sydney.

References

Bertin, J. and Loeb, J., 1974. Traitement "a la Main" et sur Ordinateur des Transistoires en Polarisation Provoquee. Geophysical Prospecting, V.22, 93-106

Dey, A. and Morrison, H.F., 1973 - Electromagnetic Coupling in Frequency and Time - Domain Induced Polarization Surveys over a Multilayered Earth. Geophysics, V.38, 380-405.

Grant, F.S., 1971 - On the Need for Measuring IP Decay Curves in Situ. Huntec (70) Limited preprint.

Hutchins, R.W., 1971. A New Induced Polarization Instrument. Huntec (70) Limited, 1450, O'Connor Drive, Toronto.

Lynam, J.T., 1972. On the Electrochemical Nature of IP Bull. Aust. Soc. Explor. Geophys., V.3 No.2, 19-36.

Malone, E.J., and others, 1975. - Woodlawn Copper-Lead-Zinc Orebody. Economic Geology of Australia and Papua New Guinea, 1. Metals, AIMM Monograph Series, No. 5.

Pelton, W.H., Ward, S.H., and Hallof, P.G., 1976 - Mineral Discrimination and Removal of Inductive Coupling with Multifrequency IP. Abstracts, V.2, 25th International Geological Congress, Sydney, Australia, August, 1976.

Phillips, W.J. and Richards, W.E., 1974 - A comparison of Transient Voltage Decay Curves Obtained with Different Electrode Arrays and Configurations over a Mineralized Zone. Geophysical Prospecting, V.22, 22-53.

Sumner, J.S., 1976 - Principles of Induced Polarization for Geophysical Exploration Developments in Economic Geology, 5. 277pp. Elsevier Scientific Publishing Company, New York.

Whiteley, R.J., 1973 - Electrode Arrays in REsistivity and IP Prospecting: A Review. Bull. Aust. Soc. Explor. Geophys., V.4, No.1, 1-29

Zonge, K.L. and Wynn, J.C., 1975 - Recent Advances and Applications in Complex Resistivity Measurements. Geophysics, V.40, 851-864.

8.5. *Phase Angle Induced Polarization Survey and Computer Modelling of the Woodlawn Orebody*

R. J. WHITELEY AND E. D. TYNE

Contents

Abstract

A low frequency induced polarization (IP) survey showed that the Woodlawn orebody is highly conductive and polarizable although the IP response is complicated by additional polarizable sources within the footwall rocks. Phase shifts of up to 4° over the frequency range from 0.13 to 1.0 Hz were measured in the vicinity of the orebody. Phase shift anomalies correlated well with PFE anomalies indicating an absence of significant cultural coupling. Inductive coupling was also minimal at the frequencies used.

A linear relationship between phase shift and PFE was evident on all the lines surveyed, however, phase shifts were larger than predicted for disseminated sulphide mineralization.

Two dimensional, finite element modelling substantiated two major polarizable sources, the Woodlawn orebody itself and the eastern footwall rocks. The latter IP sources dominated polarization response. It did not appear possible to distinguish these sources on the basis of phase ratios which were essentially constant in the region of the IP anomalies.

Phase angle IP data from the Black Shale are very similar to data from the Woodlawn orebody. There did not appear to be any way of distinguishing these targets from their PFE or phase shift anomalies.

Phase responses over both the Woodlawn orebody and Black Shale appear to be characterized by a small decrease in phase shift from 0.13 to 0.26 Hz and an increase from 0.26 to 1.0 Hz.

Introduction

The electrical induced polarization (IP) method is based on measurement of variations in the electrical response characteristics of rocks with time or frequency. In the time domain, the IP effect is oberved as a voltage discharge produced in response to a pulsed constant current input. The standard parameter measured is chargeability, the area beneath the discharge curve between two selected times after the current pulse.

In the frequency domain an alternating constant current waveform is used and the IP effect may be measured as a percent frequency effect (PFE), i.e. the difference in voltage amplitude at two frequencies, generally a decade apart. Alternatively, since the voltages are complex, the IP effect may be measured as a difference in phase angle between the received voltage signal and the input current waveform at a single frequency if both are sinusoidal in form. If the input current pulse is a square wave, the phase angle is measured between the first harmonic of the transmitted and received signals.

It has been known for some time that phase shifts are associated with the IP response of massive and disseminated orebodies (Collet and Katsube, 1973; Sumner, 1976; Bertin and Loeb, 1976) although phase measuring systems have not been routinely used in exploration until recently. The belated development of commercial phase measuring field systems is presumably due to the requirement of sophisticated instrumentation in order to accurately measure the small phase angles (generally less than 10°) associated with IP effects from mineralized targets, and the need for an accurate time base reference between the transmitter and receiver. Development of commercial equipment was also hampered by a lack of data on phase shifts due to mineralized rocks prior to the laboratory and field work of Zonge (1972) and Van Voorhis et al. (1973) using complex resistivity spectral measurement systems.

Prior to these studies Newton et al. (1971) also carried out laboratory phase angle measurements on samples of ore and host rocks from the Captains Flat orebody, N.S.W. This deposit is similar to the Woodlawn orebody (Malone and Whiteley, this volume, p23) and has large contrasts in resistivity, phase angle and PFE between the unmineralized host rock and massive ore samples. Samples from mineralized graphitic shales which are in close proximity to the orebody also showed substantial phase shifts.

One of the first commercially available field systems with the capability of making phase measurements at several frequencies was the McPhar Multimode IP system (Smith et al. 1973; Hallof, 1974). This system was used at Woodlawn.

Comparison of IP Parameters

Field measurements (Dolan, 1967; Zonge and Wynn, 1975) as well as laboratory and theoretical studies (Zonge et al. 1972) have shown the similarities between time and frequency domain IP measurements.

There is little doubt that time and frequency domain measurements will detect anomalies with equal effectiveness, provided the equipment is sufficiently sensitive. However, exact comparison is not possible due to inherent differences in the definitions of the various parameters. For example, chargeability is a total polarization parameter while PFE is a differential parameter.

Zonge and Wynn (1975) have also shown that, when the in-phase IP response of a rock is plotted against the out-of-phase response at successive frequencies, PFE and phase angle approximate a quadrature set of data with frequency effect proportional to the in-phase response. Preliminary observations of the laboratory IP response for mineralized samples by these authors indicate that a dual frequency measurement for PFE can be replaced by a single phase measurement. This allows a linear relationship between the two parameters to be calculated. Similar results have been obtained in the field by Georgel (1967) and Scott (1972).

After extensive complex resistivity field measurements over disseminated mineralized targets, Van Voorhis *et al.* (1973) concluded that within the normal one decade measured, the IP effect of rocks can be described as a constant phase response. Situations in which phase responses are not constant were attributed to spectral distortion caused by subsurface geometric effects of electrical discontinuities or to inductive coupling.

Collett and Katsube (1973) have suggested that the differences in phase angle spectra obtained in the laboratory with samples containing different polarizable material may permit mineral discrimination in the field. In later complex resistivity measurements Zonge and Wynn (1975) reported that polarizable mineral discrimination in the field is possible since examples of non-constant phase spectra have been obtained which cannot be attributed to geometry or coupling. In an effort to categorize host rock signatures they distinguished three basic spectral types in rocks (type A, B and C), one of which (type B), was the constant phase response type.

Later discussion of this paper by Nelson and Van Voorhis (1977) has not yet resolved the question as to whether reliable *in situ* mineral discrimination is possible although it was agreed that, to a first approximation at least, the IP effect over the normal range is a constant phase response. This is at variance with the recent results of Pelton *et al.* (1978) over both disseminated and massive mineralization. These authors measured IP spectra over five decades of frequency which show phase responses which are not constant and can be fitted to appropriate Cole-Cole relaxation models. Based on these fits phase angles may not be constant over the normal one decade frequency range used in frequency domain IP exploration.

Additional recent chemical studies on the nature of the metallic IP effect by Angoran and Madden (1977) have led to the conclusion that it is unlikely that polarizable sulphide minerals will be reliably distinguished in IP field measurements, due to the small differences in electrode impedance parameters.

Pelton et al. (1978), however, believe that multispectral IP holds some promise for mineral discrimination on the basis of the sulphide mineral texture which appears to be the factor primarily responsible for characterizing spectral response. These authors also conclude that rudimentary discrimination between graphite and massive sulphides might be possible by observing phase angle variations between 0.1 and 1.0Hz. Phase angle would be expected to increase for massive sulphides and decrease for graphite with a frequency increase from 0.1 to 1.0Hz. This conclusion is based on massive sulphides having a granular texture with poor electrical contact between the grains while graphites are typically fine grained and electrically well connected.

Coupling and Phase Angle

Inductive coupling is present to some extent in all IP field measurements but can be particularly severe in low resistivity environments.

Considerable research has recently taken place on the assessment and removal of inductive coupling. Millet (1967) calculated theoretical coupling curves for a homogeneous earth and the dipole-dipole array. Hohmann (1973) and Dey and Morrison (1973) extended these to multilayered media and both time and frequency measurements. These papers suggest that phase shift due to inductive coupling is directly proportional to applied frequency for a dipole-dipole array, at least approximately, over a homogeneous or layered earth. Based on the assumption that the phase shift due to IP is approximately constant over a small frequency range Hallof (1974) proposed a numerical method of approximately removing inductive coupling by measuring phase angle at two or more frequencies. This method is applied by fitting a straight line or quadratic polynomial to phase angles measured at two or more discrete frequencies. When plotted as a phase spectrum graph the value of these functions where they cross the phase shift axis at zero frequency gives a phase angle supposedly representative of the IP response in the absence of inductive coupling. Wynn and Zonge (1975) report a more sophisticated coupling removal procedure which depends heavily on laboratory and field measurements of complex resistivity spectra over a considerable frequency range and requires interactive computer optimization.

Pelton _et al_. (1978) remove coupling from their spectral measurements by exploiting the much higher frequency dependence of inductive coupling compared to that of polarizable minerals. A Cole-Cole dispersion is fitted to the high frequency data where inductive coupling may be large. This dispersion is subtracted from the observed spectral data to give coupling free IP phase angles.

Extraneous resistivity and IP effects are often produced by a variety of cultural sources such as fences, pipelines and powerlines. The earliest IP surveys at Woodlawn (Smith, this volume, p79) were adversely affected by a grounded fence in the vicinity of the orebody. The fence was subsequently removed. Wynn and Zonge (1975) show examples of IP surveys in areas devoid of cultural sources. In such conditions PFE, phase angle and chargeability anomalies are similar and show excellent correlation. In contaminated areas with both unmineralized and mineralized rocks, the measured PFE values diverge from both the chargeability or phase angle. Examination of correlations between the IP parameters allows an assessment of cultural coupling influences.

Equipment and Field Procedures

A McPhar Multimode IP system (Smith et al., 1973; Hallof, 1974) was used at Woodlawn. This system employs a separate transmitter and receiver and may be programmed to operate in the frequency or time domain, measuring in-phase PFE and phase shift. Timing reference between the transmitter and receiver is maintained by the use of two synchronized crystal clocks which program the complete operating sequence in both instruments.

At Woodlawn, the crystal clocks were synchronized at the commencement of each day prior to operation and checked every half hour. These clocks were found to retain their phase to within about 1 milliradian per hour.

The transmitter used was the McPhar P660 unit operating in the frequency domain mode on a 100 percent duty cycle at frequencies of 0.13, 0.26 and 1 Hz. The transmitting electrodes were aluminium foil in shallow wet pits and were carefully prepared to ensure stability.

Phase shifts at each transmitted frequency were determined at the receiver as the ratio of in-phase to out-of-phase voltages between the receiving electrodes. The entire programmed operation sequence was run over sufficient cycles to achieve a stable meter reading.

The IP survey was carried out over the Woodlawn orebody using the dipole-dipole array with 30m electrode separations along lines G, J, K and M (Fig. 1). An additional traverse was completed over the Black Shale selected for comparative study.

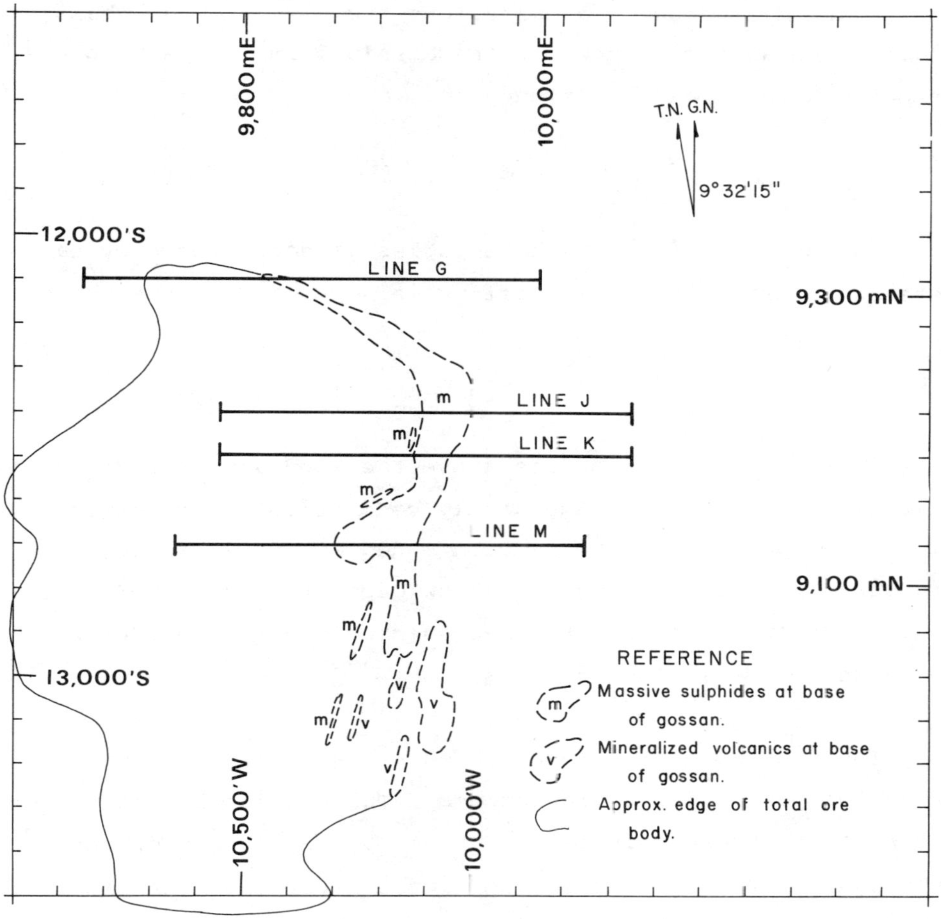

Figure 1. Location of Phase Angle IP Traverses.

Results and Discussion

Phase shift anomalies were obtained on all the surveyed lines to a maximum amplitude of 76 milliradians (about 4°). Apparent resistivity (ρa), PFE between 0.13 and 1 Hz, Metal Factor (MF) and Phase shift (θ) at 1.0Hz, pseudosections for each traverse together with geological sections are shown in the following figures. Phase shifts are presented as negative numbers in milliradians to indicate that the received voltages lag the transmitted current.

Phase difference pseudosections are also shown. These are calculated as differences in the absolute amplitude of phase angle between each pair of frequencies, i.e.

$$|\theta_{0.26}| - |\theta_{0.13}|, \ |\theta_{1.0}| - |\theta_{0.26}|, \ |\theta_{1.0}| - |\theta_{0.13}|$$

If desired, phase ratios and phase angles at each frequency can be determined using the phase difference sections and the phase shifts presented at 1.0Hz.

Line G

A definite ρa low occurs above the Woodlawn orebody on this line (Fig. 2)* illustrating the highly conductive nature of the massive sulphides. This low is open ended at small dipole separations (n=1) indicating a shallow depth to the orebody. There is no indication of any substantial thickness of overburden which drilling indicates is less than 7m.

The PFE anomaly is complicated and appears to result from at least two main polarizable sources. The n=1 dipole data suggest that the top of these sources are at G13.5 and G10.5 respectively. The maximum PFE response at larger dipole separations is displaced to the east of the orebody and appears to be due to a larger target. These features suggest that both the massive sulphide and the eastern host rocks, i.e. tuffs and pyritic black shales are both contributing to the PFE response. These rocks appear to be more highly polarizable than the massive orebody.

The metal factor (MF) anomaly defines the massive orebody with reasonable accuracy and is a useful parameter in this environment. Hallof, (1969), shows comparable behaviour of ρa, PFE and MF anomalies obtained over massive sulphides near Captains Flat, N.S.W. in a similar geological situation.

---

*The folded figs. 2, 4, 5, 9 and 11 can be found in the envelope accompanying the book.

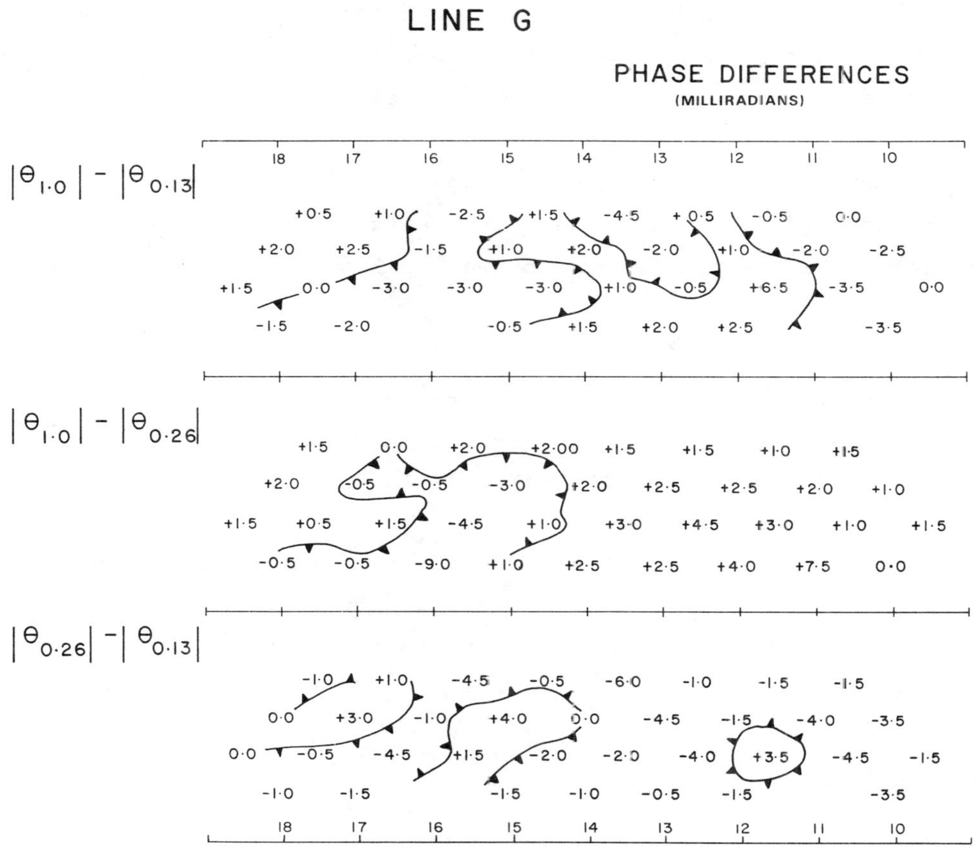

Figure 3. Phase difference pseudosections - Line G.

The phase shift pseudosection shows a high degree of correlation with the PFE section indicating that cultural coupling in the area is minimal. This high degree of correlation of phase shifts with PFE was noted on all the lines. On line G the phase shift in milliradians at 1.0Hz is approximately equal to ten times the PFE between 0.1 and 1.0Hz.

The phase difference pseudosections are shown in Fig. 3. Only relatively small phase differences were observed on this and the other IP lines resulting in phase ratios of about 1.0. They are discussed in general terms as some of these differences may be within the measurement errors. Between 0.13 and 0.26Hz the phase angles generally decrease with increasing frequency over the orebody and polarizable footwall rocks. The contour pattern is more complicated on the hanging wall side west of G14.

Between 0.26 and 1.0Hz the phase angles increase over the orebody and footwall rocks to the east of G14. A region of decreasing phase angles occurs between G15 and G17 on the hanging wall side.

The phase difference pseudosection between 1.0 and 0.13 Hz is very complicated above the orebody and over the polarizable footwall rocks, with no simple systematic behaviour.

It is possible that many of the features on these phase difference diagrams are due to spectral distortion produced by subsurface electrical discontinuities as noted by Van Voorhis _et al._ (1973).

Lines J and K

The pseudosections for these lines are shown in Figs. 4 and 5. Since they show considerable similarity they are discussed together.

The ρa sections show a broad resistivity low over the orebody. The shape of the anomaly suggests a deepening of the source compared to line G and there is no clear indication of dip. The host rocks on the eastern side appear to be at least twice as resistive as those on the western side of the orebody.

The PFE sections show a rapid increase in PFE response in the vicinity of the hanging wall and top of the orebody. Maximum anomalous response occurs near the top of the orebody and a large response, gradually decreasing in magnitude continues to the east of the orebody. This response is too broad to be due to the orebody alone. This region of anomalous PFE values is narrower than that on line G indicating that the polarizable sources are in close proximity. This is also indicated by simple computer modelling of the gradient array IP response on line J (Tyne and Whiteley, this volume, p 349).

The MF sections again provide a good definition of the location of the massive orebody.

The phase difference pseudosections for line J are shown in Fig. 6. Over the majority of this line, phase angles decrease

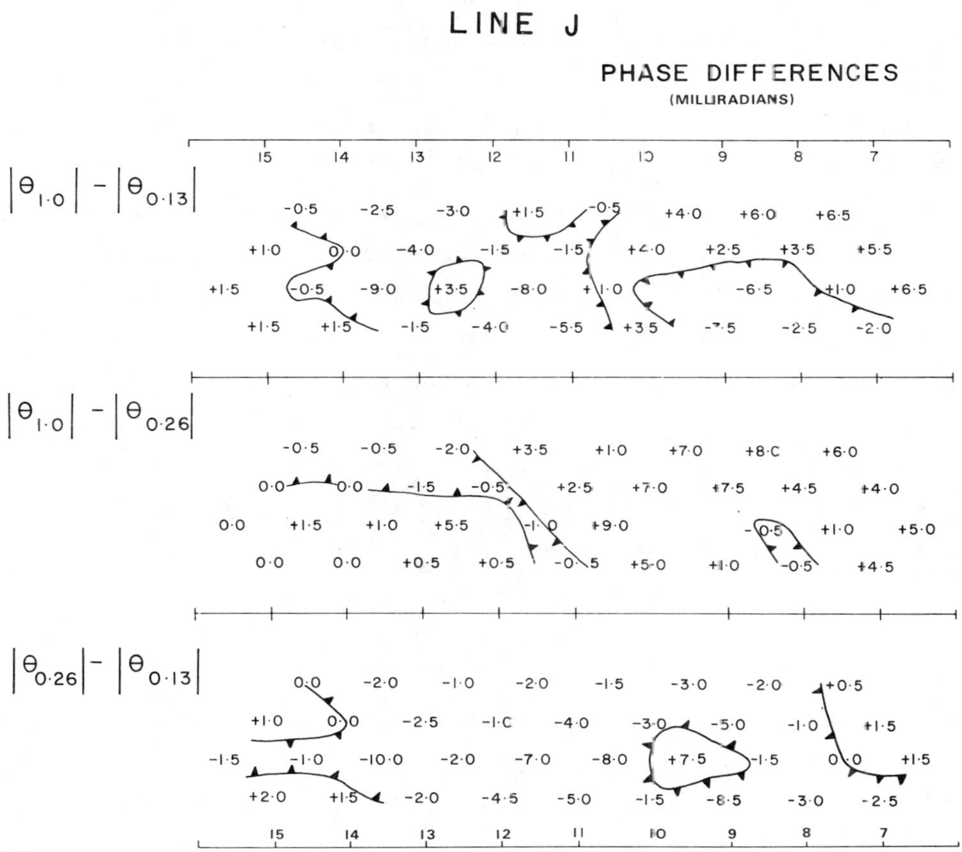

Figure 6. Phase difference pseudosections - Line J.

from 0.13 to 0.26Hz followed by an increase from 0.26 to 1.0Hz, apart from a narrow belt between J12 and J15 where phase angles continue to decrease. The phase difference pseudosection for 1.0 and 0.13Hz is again complicated with generally decreasing phase angles over the hanging wall rocks and increasing phase angles over the shallow region of the orebody and footwall rocks.

Computer Model Comparison - Line J

The geological section beneath line J was simulated as a two dimensional (2D) resistivity and IP model. Since line J is orthogonal to the orebody near its centre, a 2D model was considered adequate to represent the subsurface geological structure. Dipole-dipole apparent resistivity and PFE pseudosections were calculated using a computer program by Rijo (1977) based on the finite element method. An initial model based on the known geology and electrical properties (Tyne et al., this volume, p 99) was adjusted until the calculated pseudosections showed a satisfactory fit with the observed data in Fig. 4. The final model and calculated pseudosections are shown in Fig. 7. The dipping massive sulphide orebody was simulated as a series of prisms with a resistivity of 0.1 ohm m and intrinsic PFE of generally 5%. The steeply dipping, polarizable footwall rocks were modelled as two vertical sided prisms with resistivities of 150 and 250 ohm m and intrinsic PFE of 10% and 2.5% respectively. A variable thickness overburden, known to be present, was modelled with a resistivity of 75 ohm m and intrinsic PFE of 0.5%. Overburden is very thin between J5 and J10.

The host rocks were modelled with a resistivity of 150 ohm m and intrinsic PFE of 1%.

The location, shape and amplitude of the computed ρa, PFE and MF responses on the computed pseudosections closely resemble the observed responses for line J on Fig. 4.

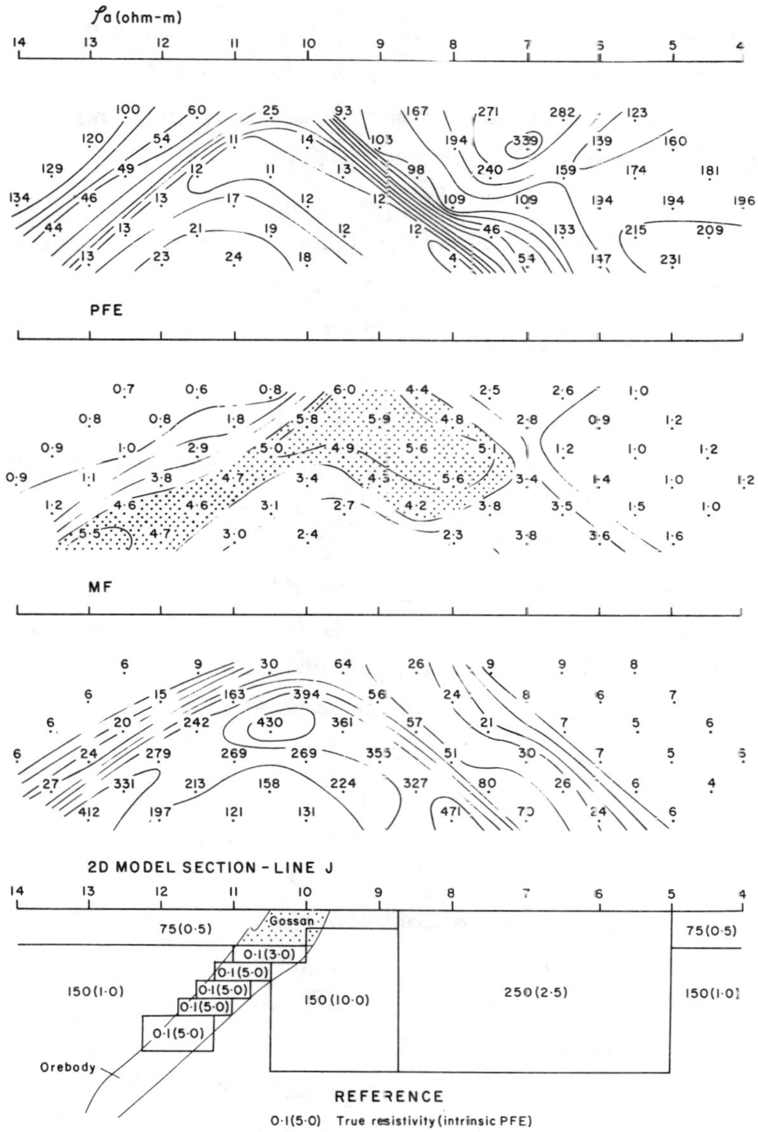

Figure 7. Two Dimensional Finite Element Model - Line J.

The computed ρa anomaly, centred at J10.5 closely fits the observed anomaly, with minimum ρa values of about 10 to 15 ohm m in both cases, although the computed anomaly suggests a slightly shallower conductive source. The resistivity of 0.1 ohm m used for the blocks simulating the massive sulphides is close to the true resistivity obtained from core sample and downhole measurements (Tyne et al., this volume, p 99).

The computed PFE response matches the observed response in overall shape and amplitude although it differs in some details. In the modelling, attempts were made to match the sharp gradient on the western flank of the observed PFE anomaly and the maximum PFE response of about 5 to 6%. This could only be done by attributing an intrinsic PFE to the shallow polarizable footwall region of twice that for the massive sulphide orebody. This low intrinsic PFE for the orebody appears to be somewhat less than obtained by direct measurement (Tyne et al., this volume, p 99). However, intrinsic PFE measurements on core samples and from downhole surveys in the conductive sulphides tended to be very variable and may not represent the polarizable properties of the orebody as a whole. The intrinsic PFE of 5% for the massive sulphides used in the final model is believed to be close to the true bulk intrinsic PFE of the Woodlawn orebody.

On line J the relative contributions of the major IP sources to the model anomaly are shown in Fig.8. Figs.8a and 8b show the observed and modelled PFE pseudosections. Fig.8c and 8d show computed pseudosections for each of the individual sources, i.e. the orebody and polarizable footwall rocks. These were computed by alternately weighting the intrinsic PFE of the orebody and footwall rocks to 0%.

The response due to the polarizable footwall rocks alone (Fig.8c) shows that these are major contributors to the observed PFE anomaly on line J and produce the dominant shallow response.

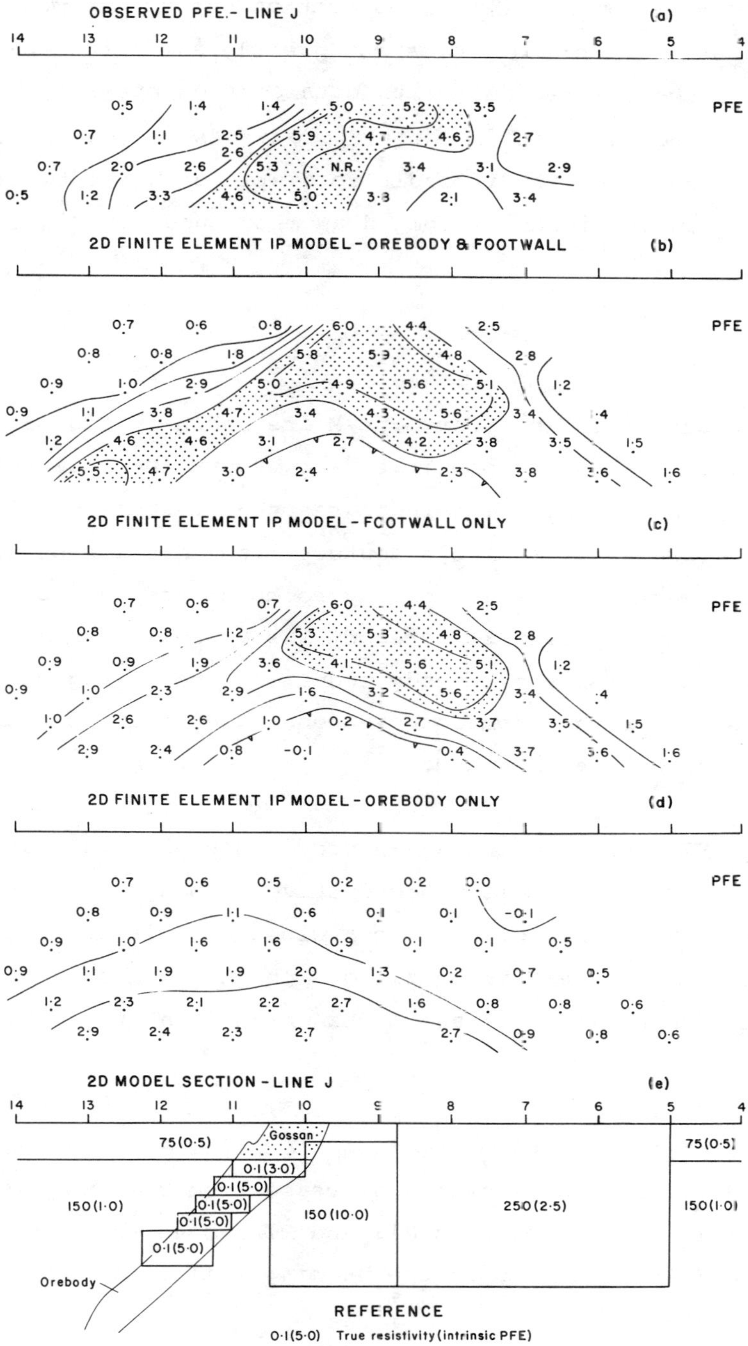

Figure 8. Comparison of Observed and Model PFE Pseudosections - Line J.

Fig.8d shows the PFE pseudosection due to the massive orebody alone. The orebody does not greatly alter the maximum amplitude of the PFE response with both sources present, however, it causes an accentuation of the PFE gradients on the western side of the line and makes it difficult to distinguish the two separate polarizable sources in the combined anomaly shown in Fig.8b.

Line M

The pseudosections for line M are shown in Fig. 9. The ρ_a pseudosection shows a complicated low above the massive orebody. The PFE response on this line indicates the presence of at least two, well separated polarizable sources; the massive orebody beneath M12, and polarizable country rocks to the east of M9. The discrete anomaly over the orebody has a similar shape and amplitude to that for the orebody model on line J (Fig.8d). This suggests that the anomalous response due to the orebody is similar for both lines J and M.

The MF pseudosection again provides a good indication of the location of the massive conductive orebody. The phase angle section at 1.0Hz again shows a close correlation with the PFE section. 1% PFE is equal to a phase shift of about 20 milliradians between 0.1 and 1.0Hz which is about double that observed on line G.

The phase difference sections for line M are shown in Fig. 10. Phase angles generally decrease from 0.13 to 0.26Hz. However, small regions between M11 and M12 above the orebody and between M14 and M15 above the hanging wall rocks show phase angle increases.

Phase angles increase over almost the entire line from 0.26 to 1.0Hz. From 0.13 to 1.0Hz an overall decrease in phase

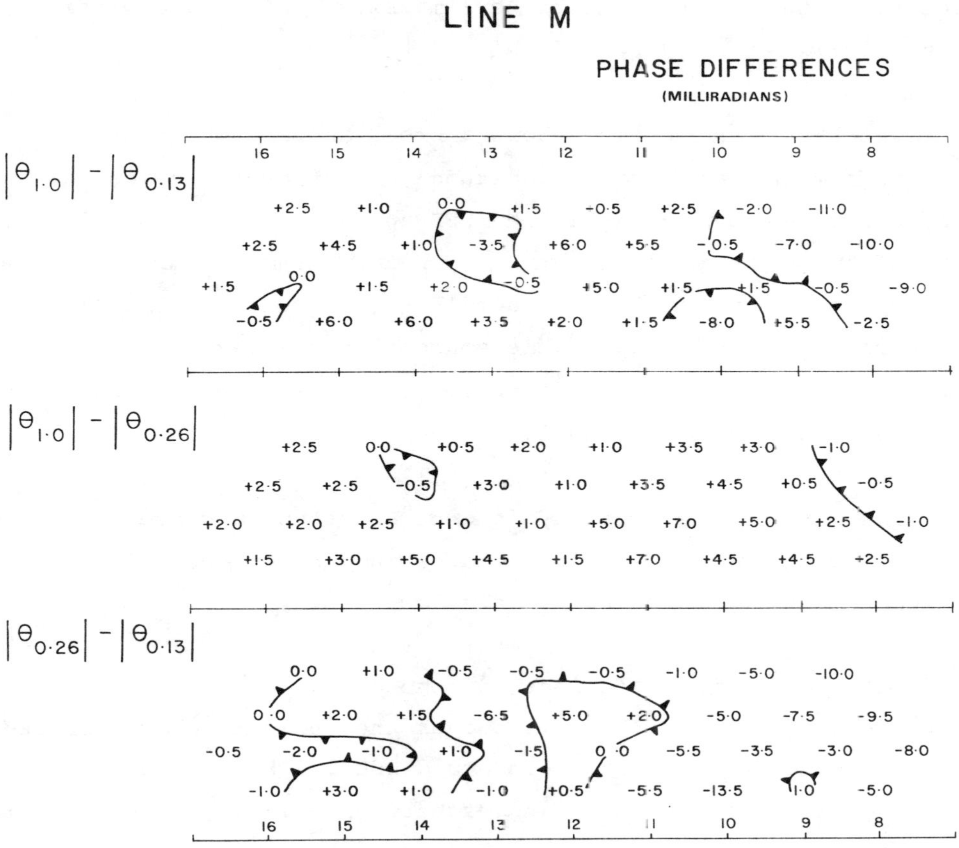

Figure 10. Phase difference pseudosections - Line M.

angle occurs over the footwall rocks while the remainder of the line is characterized by an overall increase in phase angle.

Black Shale

Fig. 11 shows IP results over the Black Shale selected for comparison with those over the Woodlawn orebody. The pseudosection shows a broad resistivity low over the Shale. The resistivity of the Shale appears to be at least twenty times lower than the country rocks. The ρa anomaly is complicated by a more highly conductive region between station 9 and 10. This narrow conductive feature is thought to be a massive pyrite band within the Shale.

The PFE pseudosection shows a broad anomaly which is fairly uniform and similar in magnitude to that over the Woodlawn orebody. The western margin of the PFE anomaly corresponds closely to that of the ρa anomaly. However, the eastern margin extends about 30m beyond the margin of the ρa anomaly. This suggests the presence of additional polarizable sources to the east of the conductive shale. The MF pseudosection emphasises the narrow feature between stations 9 and 10.

The phase angle section, in general, shows a close correlation with the PFE section, indicating the absence of significant cultural coupling at this location.

Phase difference sections for the Black Shale are shown in Fig. 12. Almost the entire IP line over the Shale is characterized by a phase angle decrease from 0.13 to 0.26Hz. This is followed by a general increase in phase angles between 0.26 and 1.0Hz. Overall, there is a phase angle decrease from 0.13 to 1.0Hz over much of the Shale with some phase increase on the eastern side of the Shale.

Computer Model Comparison - Black Shale

The polarizable Black Shale was simulated as a 2D resistivity and IP model using the same procedure as for line J over the Woodlawn orebody. Model results are shown in Fig. 13.

The observed and computed apparent resistivity pseudo-sections over the Black Shale show good agreement. The model results indicate that the country rock has a resistivity of about 1000 ohm m and that the Black Shale is 40 times more conductive. The shallow very conductive unit noted at about station 9 within

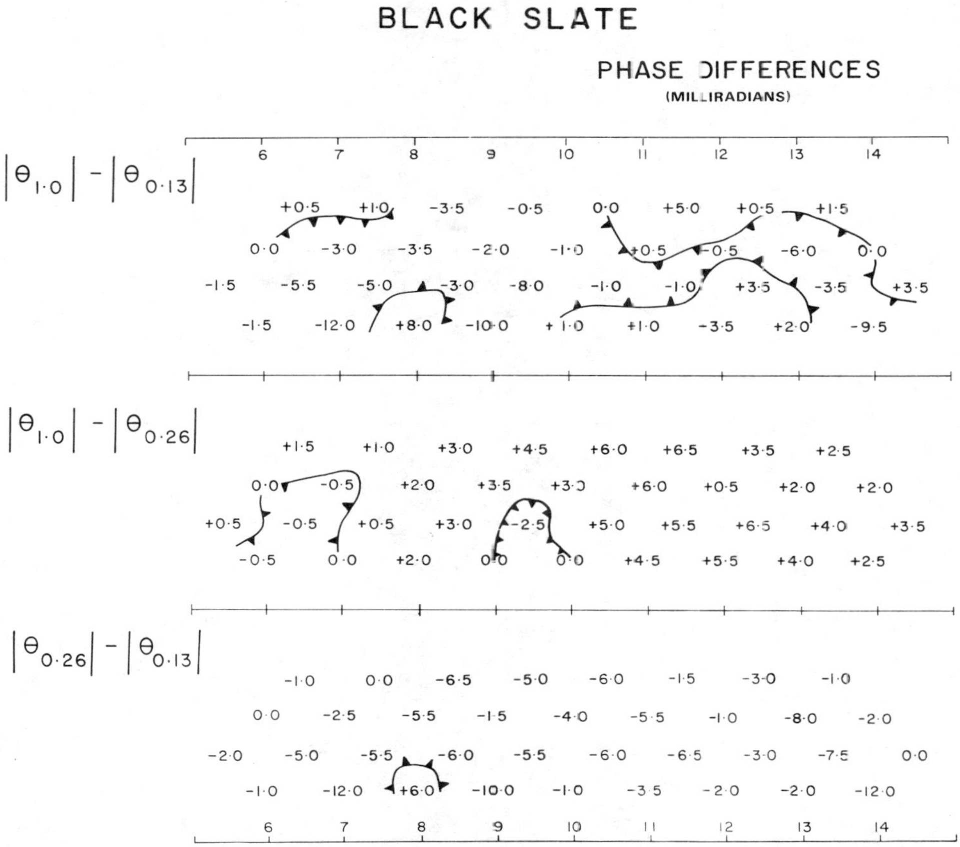

Figure 12. Phase difference pseudosections - Black Shale.

the Shale appears to be satisfactorily represented by a vertical band 2.5m wide having a resistivity of 1 ohm m.

Computed PFE anomalies are also similar in shape and magnitude to observed PFE anomalies. This suggests that an intrinsic PFE for the Shale model of 5.5% is a reasonable estimate of the true PFE of the Shale.

The presence of additional polarizable sources on the eastern side of the Black Shale as indicated by the greater width of the observed PFE anomaly relative to the resistivity anomaly, has not been modelled.

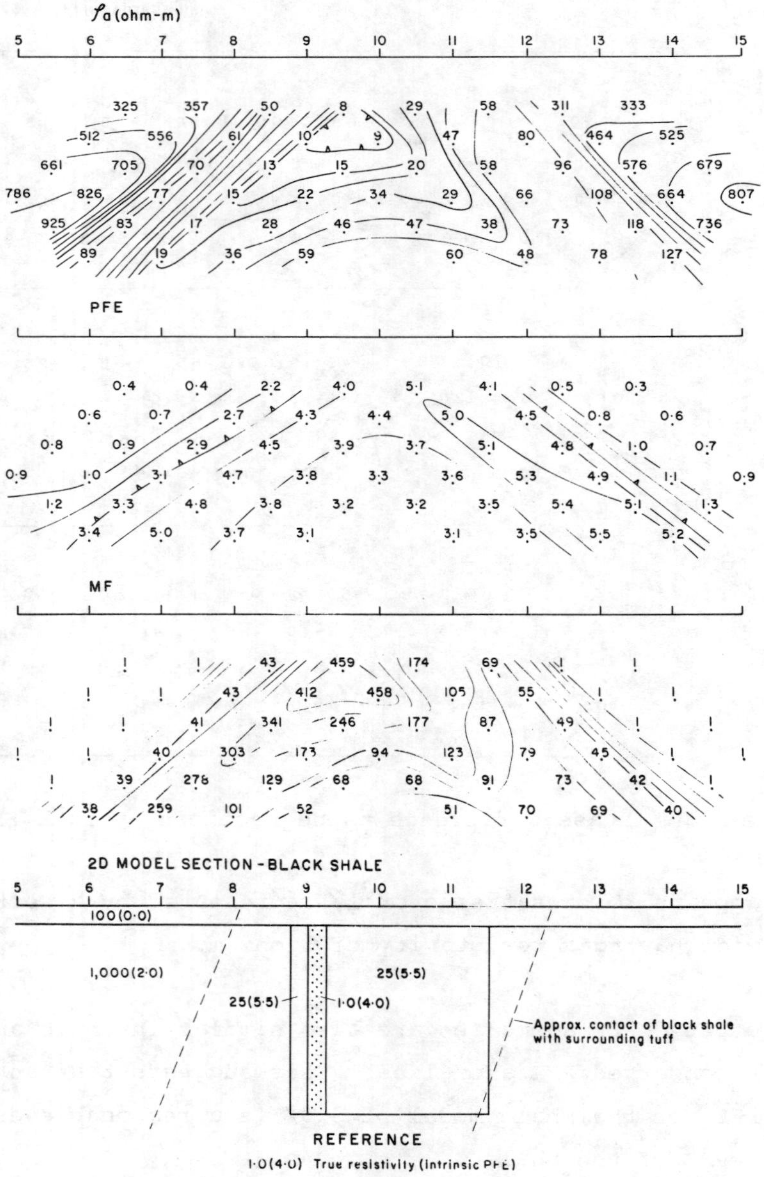

Figure 13. Two Dimensional Finite Element Model - Black Shale.

Comparison of PFE with Phase Angle

Fig. 14 shows phase angle at 1.0Hz versus PFE between 0.13 and 1.0Hz for all the IP lines surveyed. Although somewhat scattered, a linear relationship of PFE to phase angle over this limited frequency range is indicated. The data for lines G, J and K over the Woodlawn orebody and for the line over the Black Shale appear quite similar. Three data points from within the Black Shale anomaly do not appear to follow the general linear trends. These three data points circled in Fig. 14 occur within the Black Shale anomaly in the vicinity of the narrow conductive band near station 9, and probably reflect geometric influences.

It is also of interest to note that data from line M is much less scattered than on the other lines and does not fall within the region of the other data. On line M, larger phase shifts

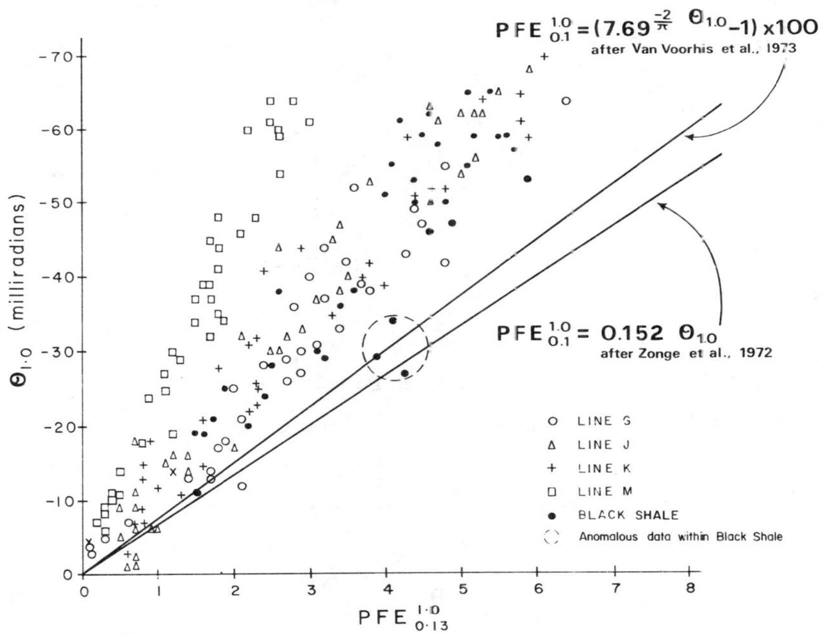

Figure 14. PFE versus Phase Angle.

are present. This could be due to geometric effects especially the greater depth to polarizable sources relative to lines G, J and K.

The linear relationships for PFE versus phase angle derived by Van Voorhis et al. (1973) and Zonge et al. (1972) are also shown. In the Woodlawn area, larger phase shifts are evident than have been predicted by these authors, although too strict a comparison would not be justified considering that the relationships were derived solely for disseminated mineralization. It is of interest to note that PFE appears to decrease more rapidly with increased depth to the orebody than does phase shift as indicated by the data for line M (Fig. 14).

Conclusions

The low frequency IP survey showed that the Woodlawn orebody is highly conductive and polarizable. The IP response of the orebody is complicated by additional polarizable sources in the eastern footwall rocks. Over the frequency range 0.13 to 1.0Hz, large phase shifts of up to 4° occurred over the orebody. The character of these anomalies correlates well with the PFE anomalies indicating an absence of cultural coupling at these frequencies. The lack of consistent phase angle increases with increasing frequency from 0.13 to 1.0Hz suggests that EM coupling is minimal.

Phase angle IP data from the Black Shale show many similarities with those from the Woodlawn orebody. There did not appear to be any simple way of distinguishing these sources on the basis of their PFE or phase shift anomalies.

Phase differences were generally small on all the lines surveyed and phase ratios were essentially constant to within $\pm 15\%$ in the region of the IP anomalies.

Both the Woodlawn orebody and Black Shale appear to be characterized by a small phase shift decrease from 0.13 to 0.26Hz and by an increase from 0.26 to 1.0Hz, although distortions probably due to subsurface geometry are present at some locations.

Phase differences calculated between 0.13 and 1.0Hz do not show a consistent pattern and there appears to be no simple way of distinguishing an orebody such as Woodlawn from the Black Shale on the basis of phase differences between these frequencies. This might be expected because of the fine grained nature and electrical continuity of both the orebody and the shales.

A linear relationship between phase shift and PFE is evident on all the IP lines surveyed. This, however, shows a larger phase shift for a given PFE than predicted for disseminated sulphide mineralization.

Two dimensional computer modelling of the resistivity and PFE data over the Woodlawn orebody closely agrees with the observed data and substantiates two major polarizable sources in close proximity i.e. the massive sulphide orebody itself and polarizable sources in the eastern, footwall rocks. Modelling of each of these sources separately showed that the eastern host rocks are the major contributors to the dominant PFE response in the vicinity of the Woodlawn orebody.

References

Angoran, Y. and Madden, T.R. 1977. Induced polarization: A preliminary study of its chemical basis. Geophysics Vol. 42 No. 4, pp. 788-803.

Bertin, J. and Loeb, J., 1976. Experimental and Theoretical Aspects of Induced Polarization. Vol. 1. Gebruder Borntraeger.

Collett, L.S. and Katsube, T.J., 1973. Electrical parameters of rocks in developing geophysical techniques. Geophysics Vol. 38, No. 1 pp. 92-105.

Dey, A. and Morrison, H.F., 1973. Electromagnetic coupling in frequency and time domain induced polarization surveys over a multilayered earth. Geophysics Vol. 38, p. 380.

Dolan, W.M., 1967. Considerations concerning measurement standards and design of pulsed IP equipment Part 1. Proc. Symp. on Induced Electrical Polarization, Univ. of Calif. Berkeley 1967, pp. 2-16.

Georgel, J.M. (1967). Polarization Provoquee et Recherche Miniere - Bull de Bureau de Rech. Geol. et Mineres Vol. 5 pp 47-65.

Hallof, R.G., 1974. The IP phase measurement and inductive coupling. Geophysics Vol. 39 No. 5, p. 650-655.

Holmann, G.W. 1973. Electromagnetic coupling between grounded wires at the surface of a two layer earth. Geophysics, Vol. 38, p. 854.

Millet, F.B., 1967. Electromagnetic coupling of collinear dipoles on a uniform half-space. Mining Geophysics Vol. 12, SEG, Tulsa, p. 401-422.

Nelson, P.H. and Van Voorhis, C.D. 1977. Discussion on: Recent Advances and applications in complex resistivity measurements by K.L. Zonge and J.C. Wynn, with reply by authors. Geophysics Vol. 24 No. 4, pp 120-124.

Newton, A.W., Smith, M.J., McSharry, P.J. and Emerson, D.W., 1971. Laboratory measurements of electrical resistivity, induced polarization and phase angle spectra from Captain's Flat, N.S.W. Bull. Aust. Soc. of Expl. Geoph. Vol. 2 No. 4 pp 49-69.

Pelton, W.H., Ward, S.H., Hallof, P.G., Sill, W.R. and Nelson, P.H. 1978. Mineral discrimination and removal of inductive coupling with multifrequency IP 1978, Geophysics Vol. 43, No. 3, pp. 588-609.

Rijo, L., 1977. Electromagnetic modelling by the finite element method: PhD Thesis, Univ. of Utah.

Scott, W.J., 1972. Phase-angle measurements in the induced polarization method of geophysical prospecting. Unpub. PhD thesis McGill University Montreal, Quebec 288p.

Smith, R.J., Hallof, P.G. and Sevenhuysen, J., 1973. The McPhar Multi-Mode IP System. McPhar Geophysics Ltd. Technical Paper.

Sumner, J.S., 1976. Principles of Induced Polarization for Geophysical Exploration. Elsevier.

Van Voorhis, G.D., Nelson, P.H., and Drake, T.L., 1973. Complex resistivity spectra of porphyry copper mineralization. Geophysics Vol. 38, No. 1, pp. 49-60.

Wynn, J.C. and Zonge, K.L., 1975. EM coupling, its intrinsic value, its removal and the cultural coupling problem. Geophysics Vol. 40, No. 5, pp. 851-864.

Zonge, K.L., 1972. In situ Mineral Discrimination using a complex resistivity method present at the 42nd Annual International SEG Meeting, Anaheim, California, November 1972.

Zonge, K.L. and Wynn, J.C., 1975. Recent advances and applications in complex resistivity measurements. Geophysics Vol.40, No. 5, pp. 851-864.

8.6. Magnetic Induced Polarization Survey at Woodlawn

D. F. ROBSON

Contents

Abstract

A magnetic induced polarization (MIP) survey of the Woodlawn orebody shows that the normalized magnetic field anomalies clearly delineate the upper part of the orebody indicating it to be highly conductive. Negative chargeabilities over the Woodlawn orebody are similar to chargeability anomalies over black shales and weakly mineralized dolerites and volcanics surrounding the orebody. Results of secondary magnetic field measurements are also similar to the chargeability results. Chargeability decay ratios provide no information on the nature of the chargeable sources detected.

Introduction

During March and April, 1975, the Bureau of Mineral Resources (BMR) conducted a magnetic induced polarization (MIP) survey over the Woodlawn ore deposit. For comparative purposes a traverse was also made over the Black Shale.

The Magnetic Induced Polarization Method

Principles of operation

The MIP method uses a magnetic sensor to measure the magnetic fields associated with current flow in the ground.

Current flow is generated in a conventional manner using a pulse IP transmitter and widely spaced grounded electrodes.

Contrary to conventional electrical IP (EIP) practice the current electrodes are generally placed along the geological strike, and measurements of horizontal magnetic field are made at stations along traverses perpendicular to strike with a sensitive magnetometer whose axis is also oriented perpendicular to strike. With this arrangement, when the primary current is transmitted, the equipment records the primary horizontal magnetic field. After the primary current ceases, the secondary field created by discharging sources in the subsurface is measured.

Seigel (1974) has developed a model to explain MIP response. This model recognises that the current flow due to polarized sources is composed of an internal current flow in the source, and an oppositely directed external current flow around the source. Seigel's model predicts that for highly conductive bodies the internal current density will be greater than the current density around the source and negative chargeabilities may result. Howland-Rose (1976) has reported negative MIP chargeabilities over sulphide deposits in Australia.

Equipment

The MIP survey was carried out using a Scintrex MFM3 fluxgate magnetometer coupled to a Scintrex IPR8 IP receiver. The transmitter was a 2.5kw time domain unit employing a standard 8 second cycle on a 50 per cent duty cycle.

The specifications of the MFM3 magnetometer indicate a sensitivity of about 100 mV per nT, and a noise level of about 10 pT. The IPR8 receiver measures the decaying IP voltage as the average value over 1, 3, or 6 time intervals in the period 130 ms to 1690 ms after the primary current is switched off. Chargeability values recorded with the IPR8 are normalized with respect to a standard induced polarization curve (Dolan and

McLaughlin, 1967). Variations in the value of normalized chargeabilities at different time intervals indicate departure of the measured decay transient from this standard induced polarization decay curve. The time interval over which the chargeabilities are measured are commonly called slices and, depending on the mode of measurement used, chargeability slices are sequentially labelled M_1 to M_6. Under most survey conditions six chargeability slices are recorded.

Parameters Recorded

Apparent chargeability (M)

This dimensionless parameter provides an indication of the chargeability of subsurface sources. It is the ratio of the secondary decay field to the primary horizontal magnetic field and is expressed in mT/T.

Primary horizontal magnetic field (Hp)

This parameter is the measured primary magnetic field and is expressed in units of nT.

Normalized horizontal magnetic field (Hn)

This is a dimensionless, derived parameter which indicates conductivity changes in the ground. The parameter is derived by normalizing Hp with respect to current and geometry as shown in equation 1. I is the current expressed in amps and K is a geometric factor with units of nT/amp.

$$Hn\% = \frac{Hp}{KxI} \times 100 \qquad (1)$$

High normalized fields indicate zones of high conductance, and low normalized fields indicate zones of low conductance. If the ground is homogeneous with respect to conductivity the value of Hn will be 100 per cent at all locations.

Secondary magnetic field (Hs)

This derived parameter has units of pT/amp, and is calculated from Hp, M and I as shown in equation 2.

$$Hs = \frac{Hp \times M}{I} \qquad (2)$$

If Hm is the value of the magnetic field caused by polarization currents, then $M = \frac{Hm}{Hp}$ and $Hs = \frac{Hm}{I}$. Hence Hm is proportional to Hp, and accordingly varies with array geometry, as well as the chargeability of subsurface sources.

Chargeability decay ratio

If chargeability slices have the same sign, the shape of the MIP decay curve is indicated by the ratio of early to late chargeability slices. When six chargeability slices are recorded it is normal to use the parameter M_6/M_1 as the chargeability decay ratio.

Survey Details

Woodlawn survey

The survey was completed along the lines shown in Fig. 1. A number of current electrode pairs with separations ranging from 240m to 300m were placed parallel to the orebody. The location of current electrodes and traverse end points are given in Table 1.

Station spacing was generally 30.5m (100 ft.), with a 15.2m (50 ft.) spacing over anomalous zones. All six chargeability slices (M1 to M6) were measured at each station.

Black Shale survey

A single traverse was made over the selected body of pyritic black shale located about 2.5km southwest of the orebody.

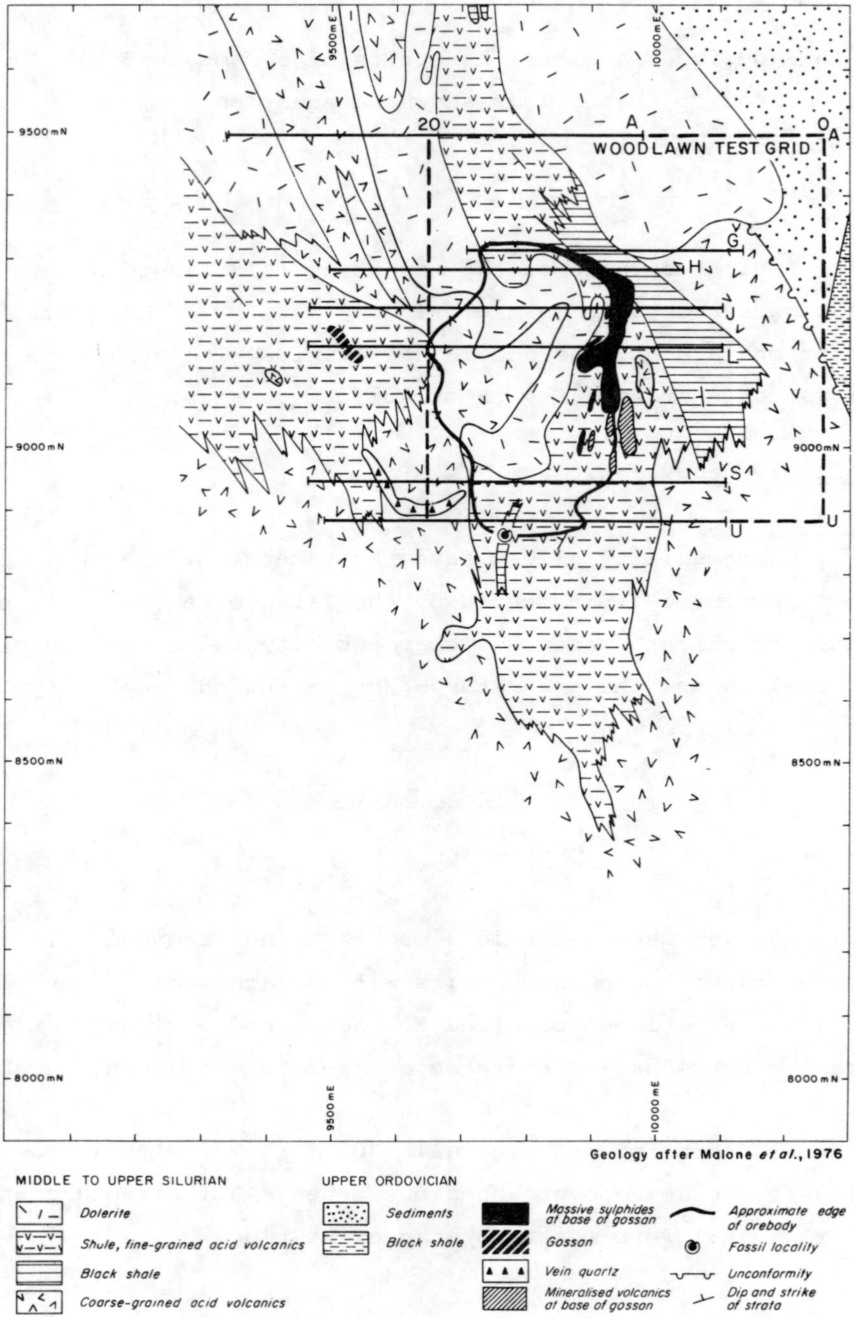

Figure 1. Geology and traverse locations

TABLE 1

Location of current dipoles and traverses

Traverse	Metric grid location*	Current dipoles*		MIP stations*	
		North	South	From	To
A	9500N	9620N/9490E	9370N/9490E	9320E	9700E
		9620N/9790E	9370N/9790E	9670E	9970E
G	9310N	9370N/9910E	9070N/9910E	9720E	10120E
H	9280N	9370N/9640E	9070N/9640E	9460E	9820E
		9370N/9910E	9070N/9910E	9820E	9990E
J	9220N	9370N/9640E	9070N/9640E	9450E	9760E
		9370N/9910E	9070N/9910E	9751E	10080E
L	9160N	9370N/9640E	9070N/9640E	9460E	9760E
		9370N/9910E	9070N/9910E	9760E	10120E
S	8950N	9070N/9640E	8830N/9640E	9460E	9820E
		9070N/9910E	8830N/9910E	9730E	10100E
U	8890N	9070N/9640E	8830N/9640E	9460E	9820E
		9070N/9910E	8830N/9910E	9730E	10080E
Black Shale	6600N	6720N/9670E	6480N/9670E	9590E	9850E

*Locations are shown as approximate positions on Jododex metric grid.

This traverse was 360m long, and employed a 240m current array parallel to strike. A station spacing of 30m was used.

An average of 50 MIP readings were recorded daily during both surveys.

The primary magnetic field between stations varied from 1 to 15 nT and chargeability readings were usually repeatable to within 3mT/T. As a comparison, EIP readings with the IPR8 equipment are generally repeatable to within 0.5 mV/V.

Results from Woodlawn

Profiles of the chargeability and normalized horizontal magnetic field results are shown with geological cross sections in Figs. 2, 3 and 4. Contour plans of the chargeability, normalized horizontal magnetic field, secondary horizontal magnetic field, and chargeability decay ratio are shown in Figs. 5, 6, 7 and 8 respectively. Note that only the M_3 chargeability is shown, this being a measure of the average chargeability between 650 ms and 910 ms after current cut off.

The MIP chargeability results (Fig. 5) show six distinct negative chargeability zones (A to F), of which the largest is zone A. Each of the zones has a characteristic chargeability and normalized magnetic field. The chargeability decay ratios do not correlate with the chargeable zones or with any geological units.

Zone A

This zone occurs over the massive ore immediately west of the subcrop. Chargeabilities of greater than -8mT/T are recorded in this zone. Updip of this chargeable zone, and directly over the conductive subcropping mineralization, normalized magnetic fields (Fig. 6) are in excess of 400 percent. Secondary magnetic fields (Fig. 7) have a similar form to the chargeability results.

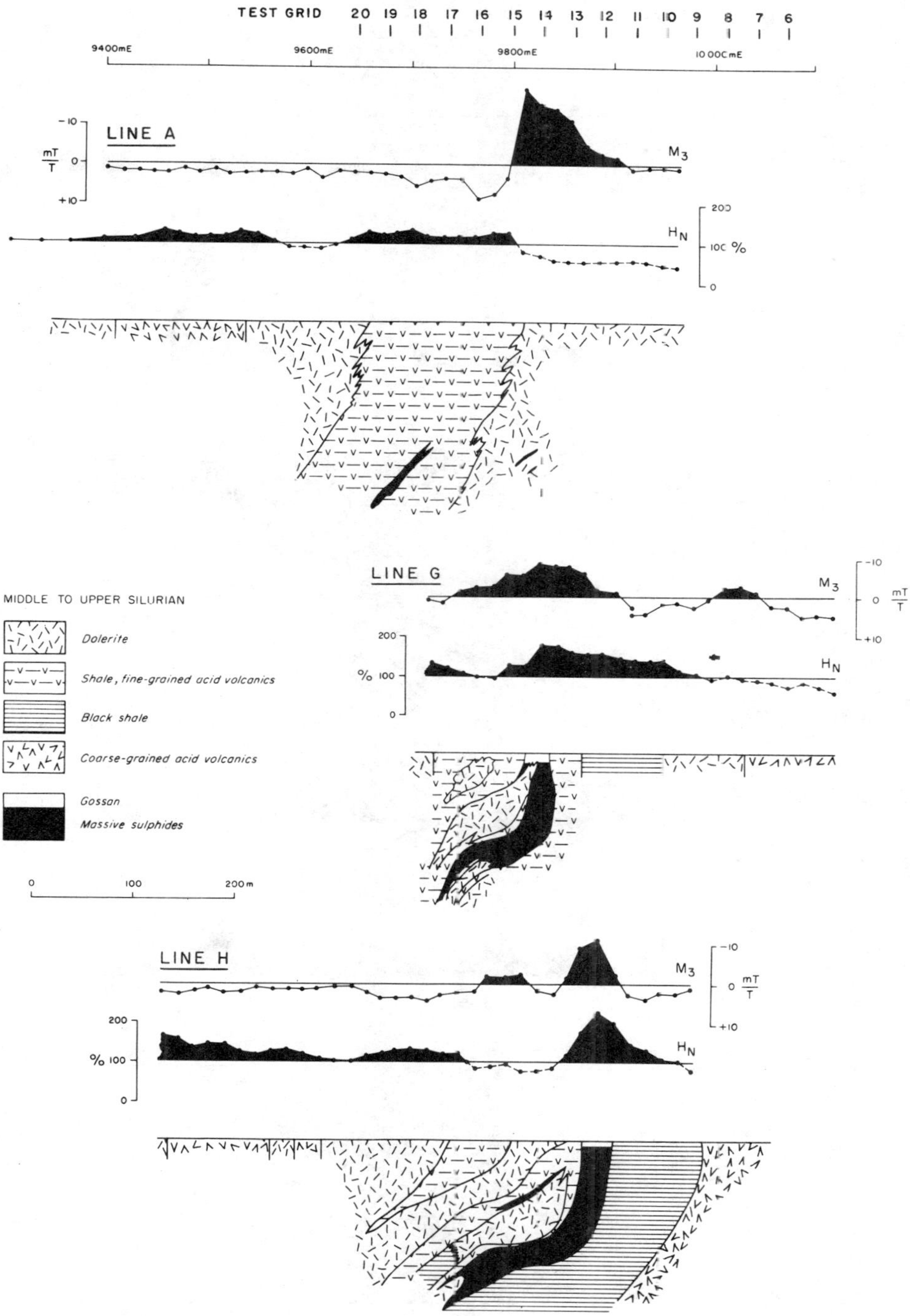

Figure 2. MIP results and geological cross sections, lines A, G and H.

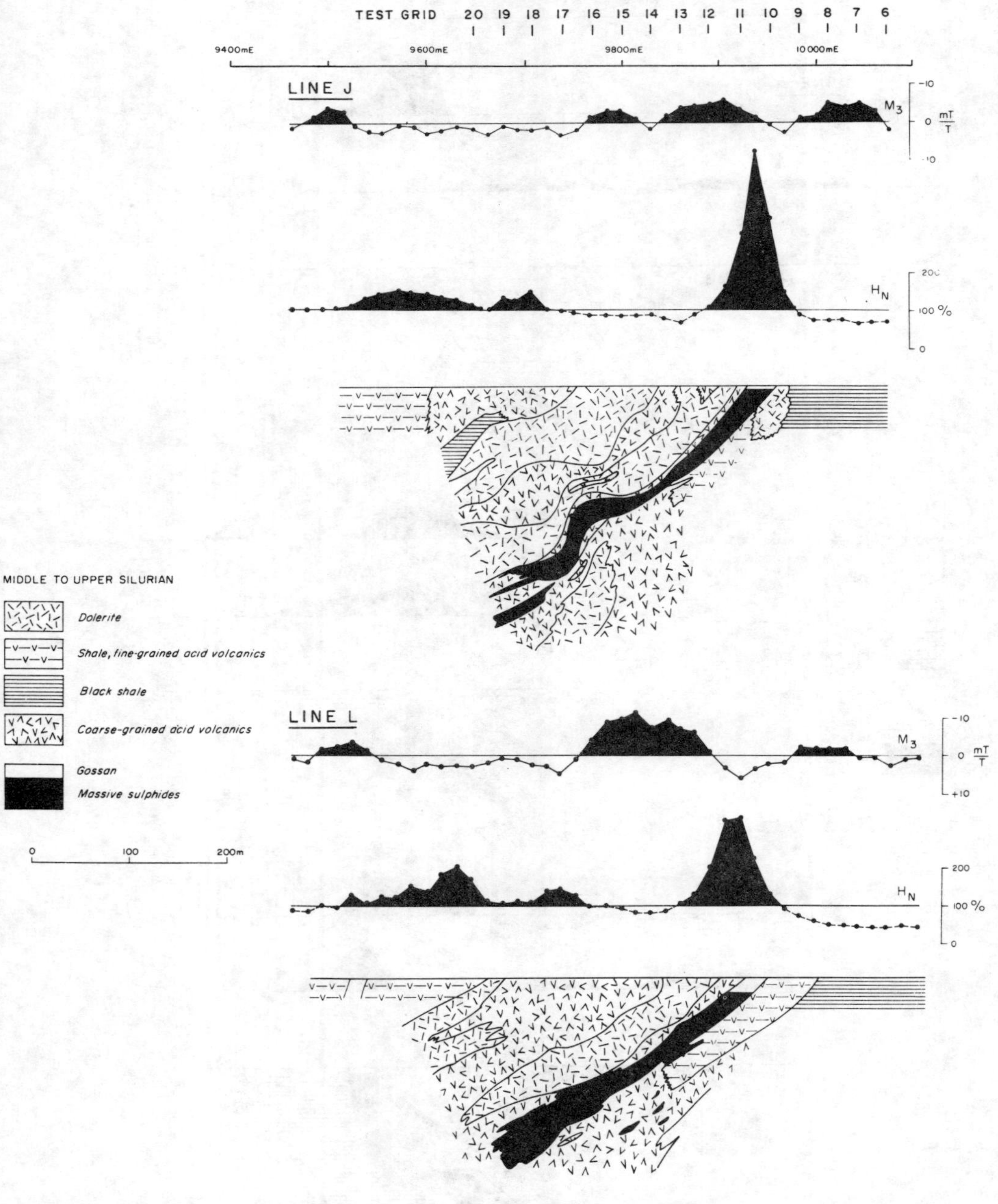

Figure 3. MIP results and geological cross sections, lines J and L.

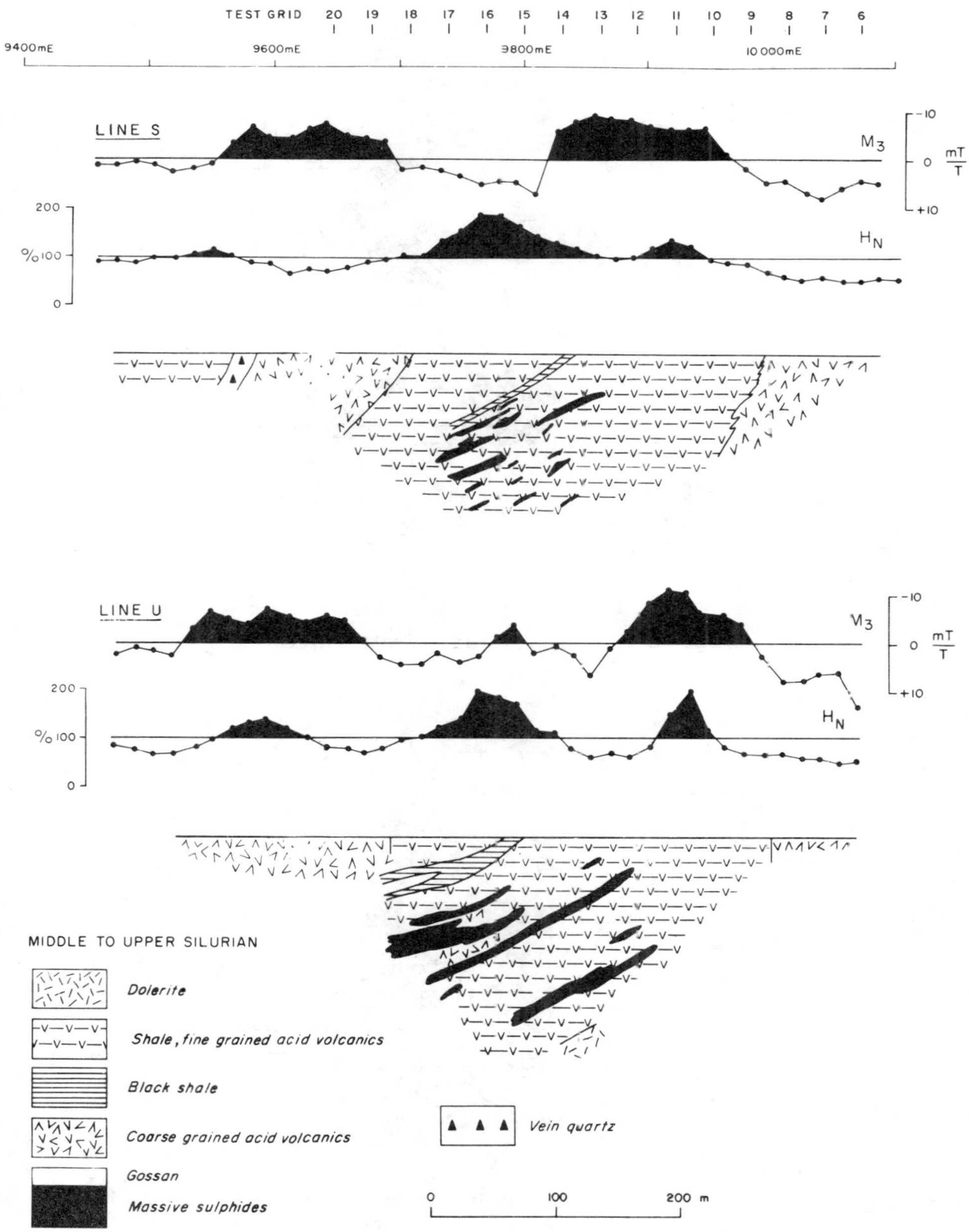

Figure 4. MIP results and geological cross sections, lines S and U.

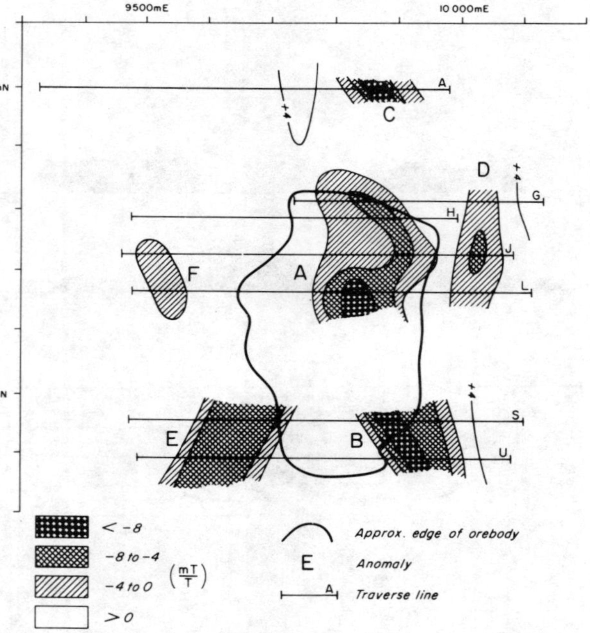

Figure 5. Chargeability contours

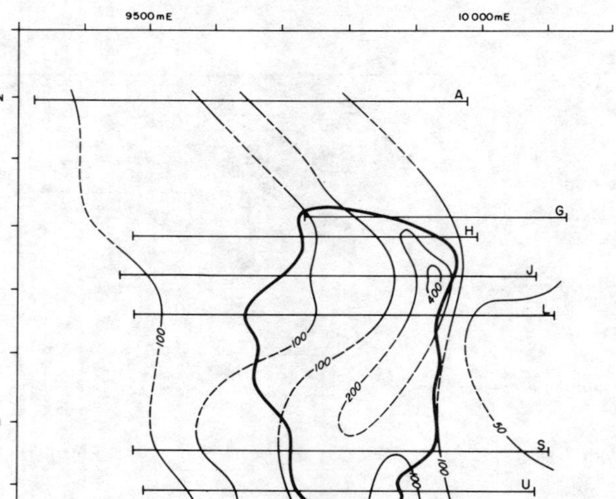

Figure 6. Normalized horizontal magnetic field contours.

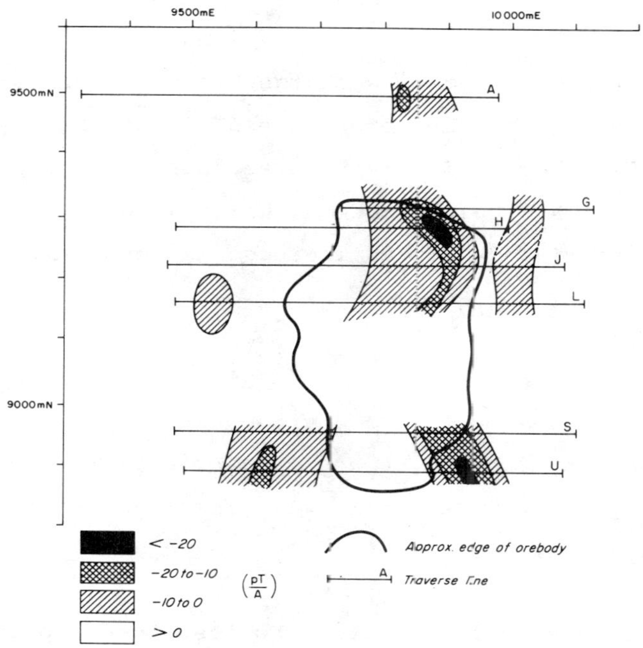

Figure 7. Secondary horizontal magnetic field contours.

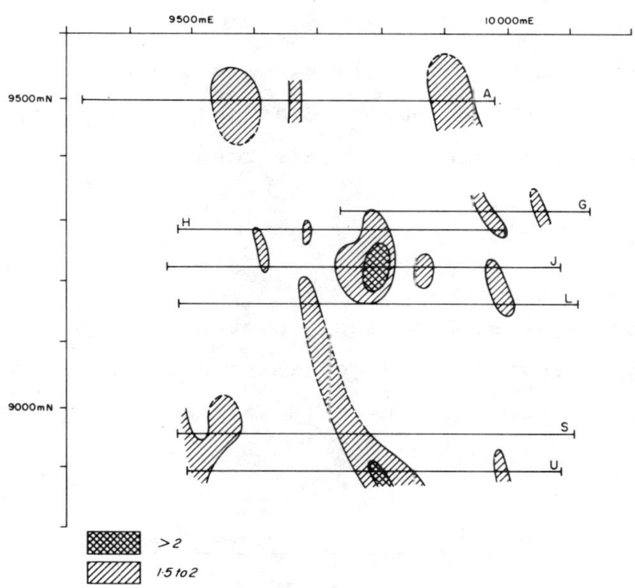

Figure 8. Chargeability decay ratio contours.

Zone B

A zone of strong negative chargeabilities occurs over mineralized volcanics at the southern end of the orebody. Coincident with the negative chargeabilities are secondary magnetic fields in excess of -40pT/A. Unlike zone A, which is over the main body, zone B is not associated with an increase in the normalized magnetic field.

Zones C, D and E

Chargeability zones C, D and E have similar magnitudes of approximately -4mT/T, and show only minor variations in the normalized magnetic field.

Zone C occurs over dolerite; however the response could be caused by a buried rubbish pit at 9815mE, line A. The secondary magnetic field results for zone C are similar to the chargeability results.

Zone D lies mainly over black shale and may be caused by disseminated pyrite in the shale. Small negative secondary magnetic fields are recorded over this zone.

Zone F

This zone is located directly over a small gossan west of the main orebody. Negative chargeabilities of -3mT/T occur with secondary fields of -2pT/A. There is no change in the normalized magnetic field.

Results from the Black Shale

Fig. 9 shows the MIP results and geological cross section across the pyritic black shale. A peak chargeable response of

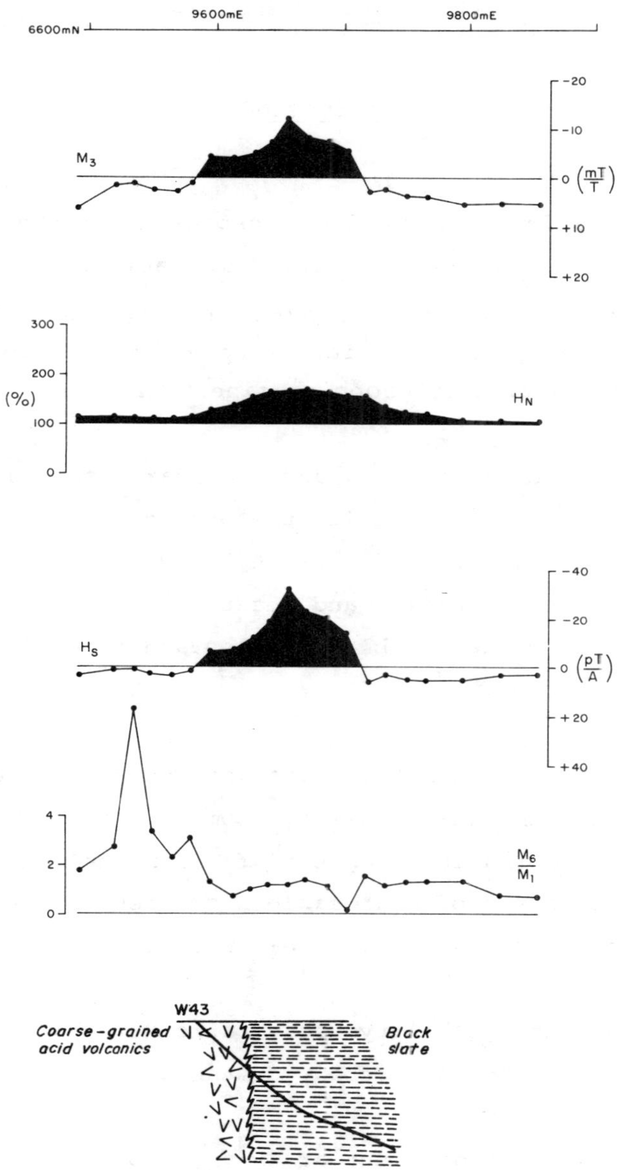

Figure 9. MIP results, Black Shale.

-12mT/T was observed within a broad chargeable zone of about -5mT/T. Broad normalized magnetic fields in excess of 170 per cent are associated with the chargeable response.

Discussion

Interpretation

The MIP results show a strong response over the orebody. However, only the normalized horizontal magnetic field results (Fig. 6) clearly discriminate the known orebody from other sources. The strong normalized magnetic field response is probably caused by the highly conductive subcropping mineralization.

Although the chargeability and secondary magnetic field parameters do not uniquely identify the orebody, a large negative chargeability zone was recorded over the massive ore. Smaller zones of negative chargeability and negative secondary magnetic fields appear to be caused by black shales and mineralized volcanics and dolerites.

A comparison of the MIP results and the geological cross sections shows that the chargeability anomalies over the orebody are downdip of the normalized magnetic field anomalies, which suggests that the source of chargeable anomalies is deeper than the source of normalized magnetic field anomalies.

Contours of chargeability decay ratio M_6/M_1 (Fig. 8) provide no useful information on the orebody or the geology of the area.

Results of an EIP gradient array and dipole dipole array surveys (Quick and Cifali, this volume, p 333), have been replotted in simplified form in Figs. 10, 11 and 12 to allow comparison of MIP and EIP results. Only qualitative comparison should be expected since electrode array geometry is fundamentally different for both systems.

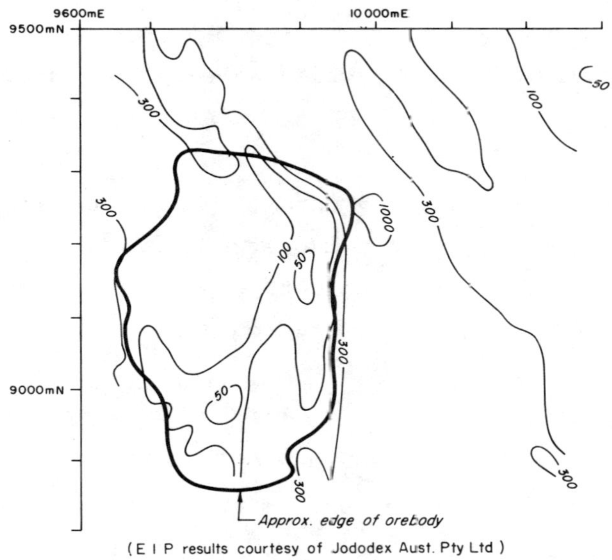

Figure 10. EIP gradient array apparent resistivity contours.

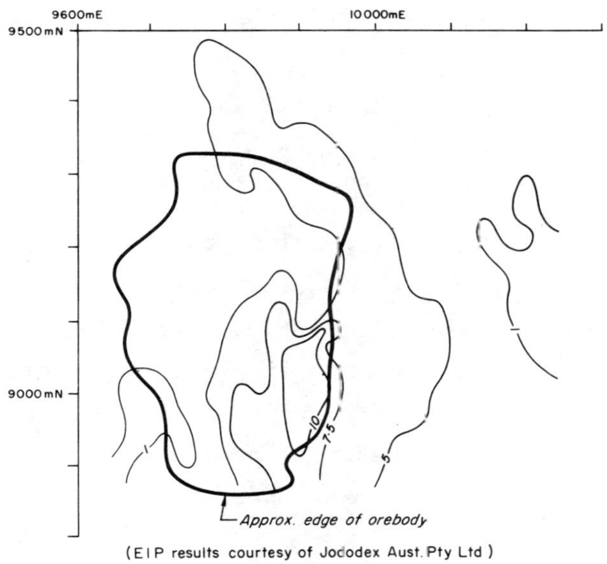

Figure 11. EIP gradient array frequency effect contours.

The results of the EIP gradient array survey centred on the Woodlawn orebody are shown in Figs. 10 and 11. This array used current electrodes 1830m apart oriented east to west and a 30m potential dipole.

The apparent resistivity data (Fig. 10) outline the subcropping mineralization in a manner similar to the MIP Hn results. However, they also indicate a second low resistivity zone at 9000mN, 9770mE (line Q, station Q16).

EIP frequency effects (Fig. 11) are greater than 10% over the mineralized volcanics in the southern part of the orebody. Surrounding this anomalous feature is a broader zone of greater than 5%, which clearly outlines the pyritic footwall rocks east of the orebody. The orebody occurs along the flank of this anomalous zone. An MIP chargeability anomaly (Fig. 5) occurs directly over the massive ore in the north.

The MIP and EIP dipole dipole array data and the geological cross section along line G are shown in Fig. 12. The dipole dipole array used a dipole spacing of 30m, and a dipole separation of 30m, 90m and 150m (n = 1, 3 and 5).

The dipole dipole array results for n = 1 show a strong, broad anomalous IP zone indicating that both the orebody and pyritic footwall rocks are polarizable. The apparent resistivities (n = 1) clearly outline the narrow very conductive massive sulphide body. The dipole dipole results for n = 3 and n = 5 are included for interest only as the dipole separations are too large to clearly resolve the narrow conductive ore zone.

The MIP current dipole is oriented so that it preferentially energizes the ore zone (Table 1). Therefore the MIP negative

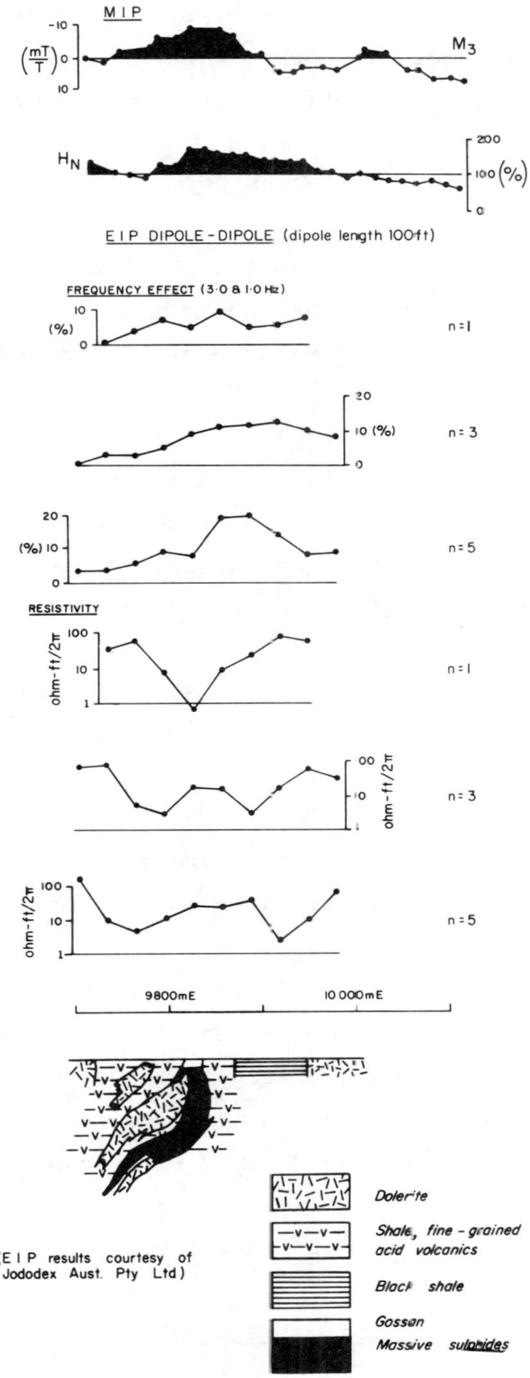

Figure 12. EIP dipole dipole array and MIP profiles, line G.

chargeabilities broadly outline the orebody in preference to the adjacent polarizable footwall rocks. The MIP normalized magnetic field indicates a broad conductive high over the ore zone.

Conclusions

MIP successfully delineates the massive orebody at Woodlawn. The high conductivity of the subcropping mineralization was particularly obvious in the normalized primary magnetic field (Hn) data.

The orebody also produces a definite MIP chargeability response, although it is difficult to discriminate the chargeability anomaly over the orebody from anomalies over black shales and mineralized dolerites and volcanics.

Secondary horizontal magnetic fields (Hs) are similar to the chargeability responses, and do not provide any additional information.

The chargeability decay ratio parameter is not characteristic of any geological units in the Woodlawn area.

Comparison of MIP and EIP data over Woodlawn indicates that MIP chargeability anomaly outlines the main massive sulphide orebody better than EIP chargeability anomaly. This is considered to be largely because of the fundamentally different current electrode arrangements used for MIP and EIP surveying, the along strike MIP current electrode arrangement giving a sharper definition of the massive sulphide orebody.

Acknowledgement

This paper is published with the permission of the Director, Bureau of Mineral Resources, Canberra.

References

Dolan, W.M. & McLaughlin, G.H., 1967. Considerations concerning measurement standards and design of IP equipment. Proceedings of the Symposium on Induced Electrical Polarization, Berkeley, University of California, p. 2-31.

Howland-Rose, A.W., 1976. The magnetic induced polarization method - simple method of interpretation of anomaly forms. 25th International Geological Congress, Sydney, Australia, Abstracts, Volume 2, Section 9B, p. 392.

Seigel, H.O., 1974. The magnetic induced polarization (MIP) method. Geophysics, 39, p. 321-339.

CHAPTER 9

GROUND GEOPHYSICAL SURVEYS: MAGNETIC AND GRAVITY METHODS

9.1. Ground Magnetic Surveys at Woodlawn

A. AGOSTINI, T. C. WOOTTON AND G. CIFALI

Contents

Abstract

Both vertical component and total field magnetic surveys at Woodlawn indicate that the orebody is not a magnetic target.

Magnetic anomalies in this area have been used to delineate dolerite intrusions which occur around the orebody. However, as these intrusions post date the mineralized sequence and appear to be unrelated to the mineralizing process, they cannot be used as magnetic marker units in the search for similar stratiform volcanic deposits.

Introduction

Vertical component ground magnetic surveys over the Woodlawn orebody and adjoining areas were completed in 1970, by Jododex Australia Proprietary Limited, as part of an integrated programme of geophysical exploration. During July, 1974 and January, 1975, the Geological Survey of New South Wales completed detailed ground magnetometer surveys in the area, to accurately define local magnetic variations associated with the Woodlawn orebody and surrounding rocks. Total magnetic field intensities were measured over the area of the orebody and to the north of it.

Data Collection

A Scintrex Mark 2 fluxgate magnetometer was used to measure vertical magnetic intensity. In the orebody area, measurements were taken at stations 30.5m apart, along east to west lines 61m apart.

A Geometrics proton precession magnetometer model G806 with a repeatability of ± 1 nanoteslas (nT) was used to measure the variations in total magnetic intensity. Two or more readings were taken at each station to eliminate any spurious values. The sensor was held 2.5m above ground level and the sensor orientation was kept constant throughout the survey. The diurnal variation was continuously monitored with a Geometrics model G801 proton precession magnetometer with an accuracy set at 0.5 nT and a repetition rate of six seconds. This magnetometer was installed 3km to the north of the survey area, at a place free from strong magnetic relief and away from human or vehicular interference.

The total field magnetic survey was completed along the east to west lines of the Woodlawn Test Grid. This survey was extended to the north to include additional east to west lines 30.5m apart referred to as Z, Y, X, W, V, Un, Tn and Sn where Sn is the northermost line.

Data Presentation

The vertical magnetic map over the Woodlawn orebody and its immediate surrounds is shown in Fig. 1. The total magnetic field data

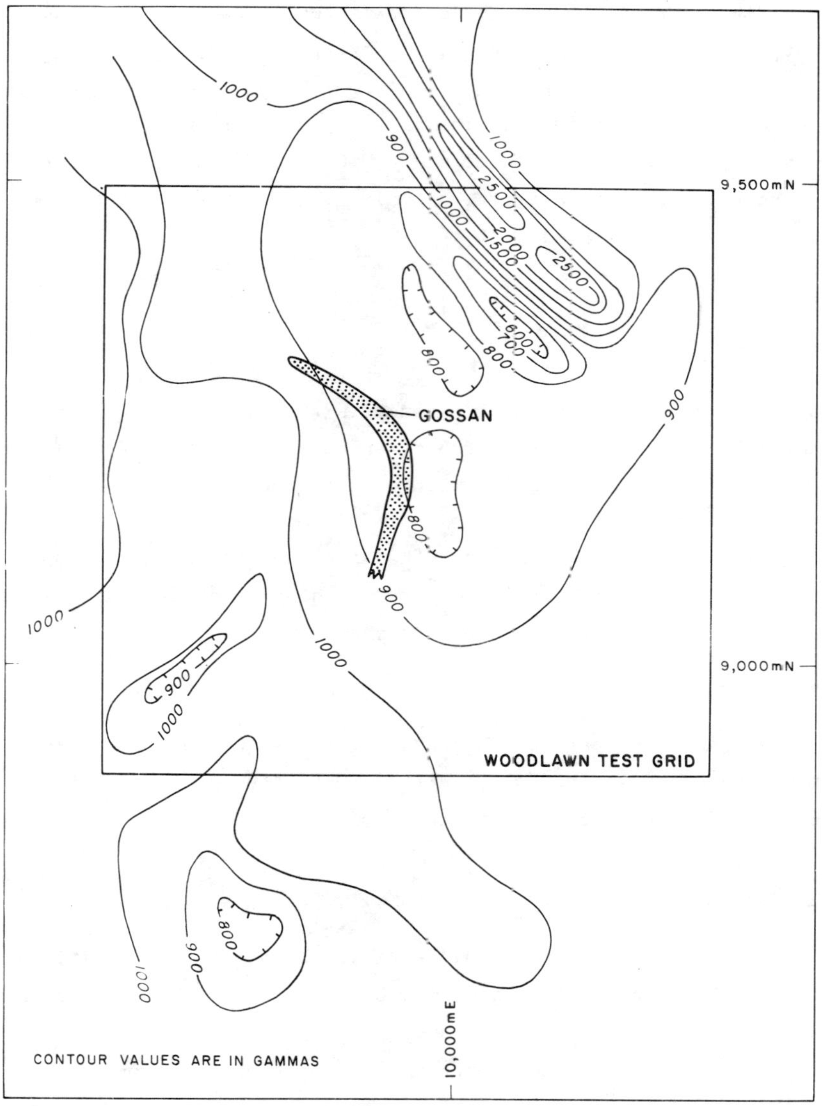

Figure 1 Surface magnetic intensity (vertical component)

are plotted in stacked profile form on Fig. 2a,* from a base level of 58,700 nT. The northern anomalies with amplitudes of 1900 nT dominate this figure. Fig. 2b shows the magnetic results on lines F to U of the Woodlawn Test Grid in the immediate vicinity of the orebody.

These data are also presented in Fig. 3 as contours of total magnetic intensity, using a base level of 58,000 nT and a contour interval of 100 nT. The anomalies of interest, anomaly 1, anomaly 2 and the central low zone are indicated on Fig. 3.

Interpretation

Vertical Component Data

The largest vertical component anomaly is the elongate high (2500 nT) situated some 250m northeast of the orebody and striking 320°M (Fig 1). Southwest of this high is a broad low within which the orebody is located.

A comparison with geology reveals that the magnetic high corresponds fairly closely to the outcrop outline of steeply dipping dolerite sills. Polarity effects from this high are certainly a contributing influence to the general low around the orebody.

Dolerite also forms part of the hanging wall sequence. Hence the slight increase in magnetic intensity in this region to the west of the low over the orebody, is attributed to this dolerite.

Total Field Data

The total magnetic intensity data exhibit two strong, well defined anomalies in the northern section and a low zone in the area of the orebody.

Of the two anomalies in the northern section the northernmost, anomaly 1, extends from line Y to line Sn; the other anomaly 2, extends from line F to line A. These anomalies are amenable to quantitative interpretation. Theoretical profiles were calculated for two dimensional (2D) dyke like bodies, (Grant and West, 1965, p 320, equation 11 to 21), and compared with the field profiles. The 2D profiles were calculated for east

*The folded Figs. 2a, 2b and 3 can be found in the envelope accompanying the book.

to west traverses to allow direct comparison of the theoretical and field profiles.

Anomaly 1

This anomaly has a total amplitude of 1,920 nT and trends at 155°M, with smooth, well defined profiles on four lines 30.5m apart. The profile on line W, through the centre of the anomaly, is presented in Fig. 4. The base level was selected at 58,660 nT by comparison with local base levels close to the profile. Fig. 4 also shows the best fit calculated profile obtained from a 2D model with the following parameters:

Susceptibility	121×10^{-3} (SI)
Strike	155° M
Dip	70° to 245°M
True width	43m
Depth to top	12m
Depth extent	500m

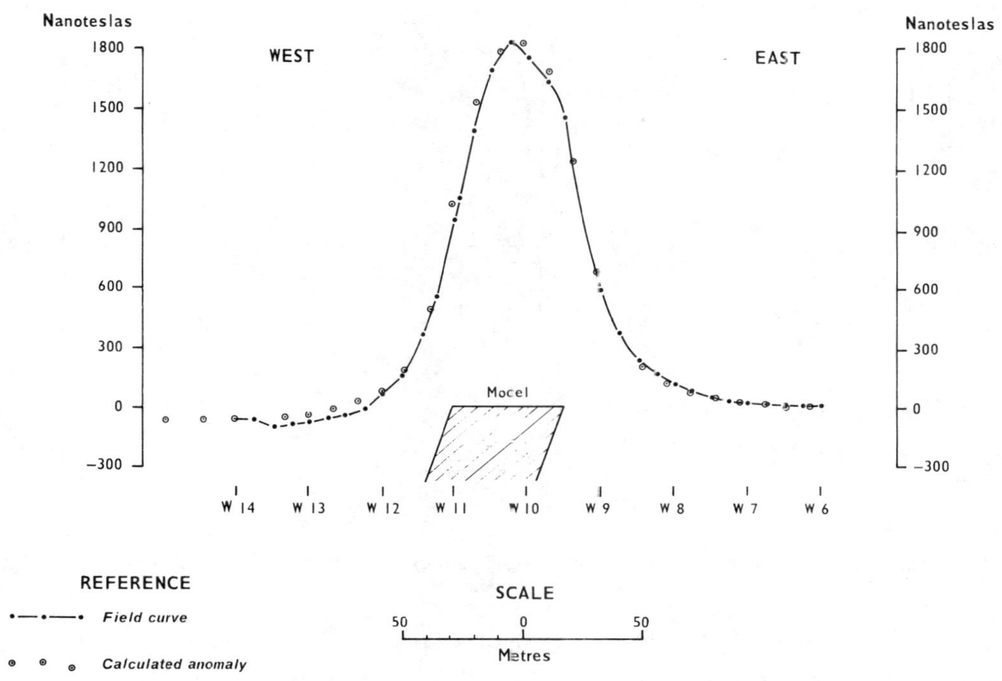

Figure 4 Line W, two dimensional interpretation of anomaly No.1

Anomaly 2

This anomaly has a total amplitude of 1,800 nT, trends at 136°M. and is well defined on five profiles 30.5m apart. The profile on line C, through the centre of the body, is quite smooth, and is shown in Fig. 5. The zero level has been selected at 58,660 nT. Fig. 5 also shows the best fit calculated profile for a 2D model with the following parameters:

Susceptibility	107.5×10^{-3} (SI)
Strike	136°M
Dip	vertical
True width	33m
Depth to top	12m
Depth extent	250m

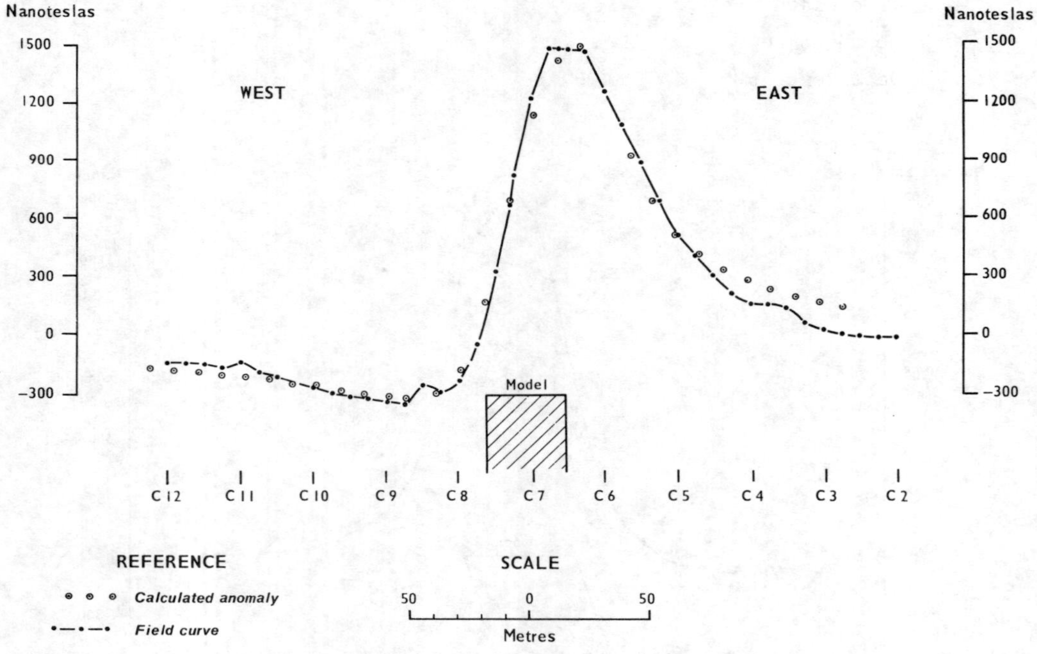

Figure 5 Line C, two dimensional interpretation of anomaly No.2

The parameters of the two models used to match the anomalies are very similar. The pattern of magnetic intensity contours in Fig. 3 suggests that a single body, displaced at line Z, could cause both anomalies 1 and 2.

It is thus inferred that anomalies 1 and 2 could be caused by a single dolerite intrusion, truncated by a fault in the vicinity of line Z.

However, this level of magnetic anomaly is not associated with the many other dolerite intrusions in the area. Some unknown factors have produced the large magnetic anomaly associated with this extensive dolerite intrusion.

Central low zone

This zone extends from line I to line Q of the Woodlawn Test Grid and has a maximum width of 150m and a maximum amplitude of 200 nT on lines J, K and L. The shape of the anomaly is affected both by the low zone associated with anomaly 2 which has the effect of flattening and extending the central low zone east from station 5 on lines I to M and by the narrow, relatively higher amplitude anomalies associated mainly with casing abandoned in drill holes. These do not mask the low zone.

Fig. 6 shows a profile through the low zone on line L and the geological cross section on that line.

While it is possible that the low zone is produced solely by the geometrical distribution of other magnetic bodies, it is also possible that this negative anomaly is produced mainly by a reversely magnetized source, either the orebody or the dolerite intrusion intersected at depth below station L 12.5.

The magnetic susceptibility of the complex ore lens (Tyne and Idnurm, this volume, p 99) ranges from 250 to 375 x 10^{-6}(SI). If we assume that this weakly magnetic zone represents a two dimensional dyke like body along strike of the ore zone and at a depth of 5m and with a width of about 30m, then the maximum total field anomaly which would be induced by the present day geomagnetic field would be no greater than 4 nT. It therefore seems unlikely that an ancient magnetic field would have been strong enough to have produced a remanent magnetism in the orebody sufficiently strong to give the 200 nT anomaly observed.

If the magnetic low over the ore zone is produced by a reversed remanent magnetism, it is therefore much more likely that the cause would be a reversely polarized dolerite dyke below the orebody which could more realistically produce the observed anomaly.

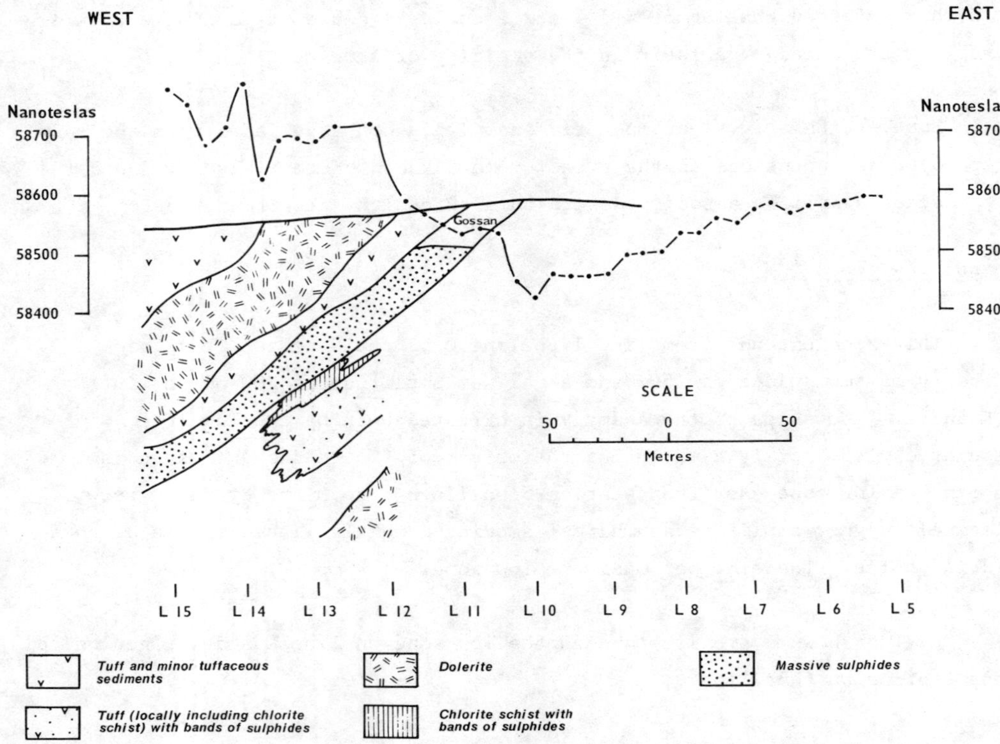

Figure 6 Line L, magnetics and geology

Fig. 6 also indicates that the hanging wall dolerites have a positive magnetic anomaly associated with them. A small positive anomaly over the gossan is probably caused by the concentration of iron oxide minerals near surface.

Conclusions

The magnetic method, in this particular geological setting, was not effective in terms of discovery as the Woodlawn orebody is essentially non magnetic.

In general, the magnetometer survey results in this area have been used to delineate some dolerite intrusions which occur in the vicinity of the orebody. However, since the intrusion of these dolerite sills post dates this mineralized sequence, neither the mapping of the sills nor detailed magnetic surveys would appear useful in the search for similar stratiform volcanogenic deposits.

Acknowledgements

This paper is published with the permission of the Under Secretary of the New South Wales Department of Mineral Resources and Development.

Reference

Grant, F.S., and West, G.F., 1965. Interpretation Theory in Applied Geophysics. McGraw Hill Book Company, New York.

9.2. Gravity Surveys at Woodlawn

R. J. TEMPLETON

Contents

Abstract

A residual gravity anomaly of 0.8 milligals was obtained over the Woodlawn orebody after removal of the regional field using a biased third order trend surface. The shape and areal extent of the anomaly closely reflects the massive sulphide deposit outlined by drilling.

Simple interpretation suggested a tabular shape and clearly indicated that the massive sulphides were the major cause of the anomaly.

Excess mass was estimated at 1.5 million tonnes. The actual mass of the causative body of massive sulphides was estimated with reasonable precision using the average density of the massive sulphides as a whole.

Although a number of smaller gravity anomalies were also discovered, no other major massive sulphides are likely to be present near the surface in the vicinity of the Woodlawn orebody.

Introduction

After the existence of a major orebody was established at Woodlawn, it was suggested that the gravity method would provide a possible means of estimating ore reserves. As a result, a gravity survey (Survey 1) was carried out from the 26th to 30th October, 1970. An extended gravity survey (Survey 2) was conducted at Woodlawn to test possible extensions of the known orebody to the northwest. This survey was conducted from the 23rd to 31st October, 1972.

Survey Methods and Data Reduction

Equipment

A Canadian Sharpe gravity meter (serial No.130) was used in both surveys. At the time of the first gravity survey the gravity meter had a calibration constant of 0.11664 milligals per scale division while the calibration constant for the second survey was 0.11663 milligals per scale division.

Woodlawn Survey

Both survey grids were constructed and levelled prior to the gravity surveys. In the first survey (Fig.1)* 129 gravity stations were read on lines spaced 61m (200ft) apart in the north to south direction and 30.5m (100 ft) apart in the east to west direction. 502 stations were read in the second survey. Most were 25m apart except for a number of stations to the east and west (Fig.1). Three stations of the first survey were repeated in the second survey to accurately tie both the surveys.

Drift corrections

In the first survey, drifts were calculated from repeated readings every half hour. In the second survey drift control readings were taken at a base station (located at 9575mE, 9500mN) every half hour. Drift corrections were calculated by assuming a linear change between repeated readings.

Latitude corrections

Latitude corrections were applied to both surveys using the formula adopted by the International Union of Geodesy and Geophysics (Grant and West, 1965, p 236).

Elevation corrections

In both surveys, free air and Bouguer corrections using a Bouguer density of $2650 kg/m^3$, were applied to the observed gravity values to reduce them to a datum at 792.5m (2600ft) above mean sea level.

---

*The folded Figs. 1, 2 and 3 can be found in the envelope accompanying the book

Terrain corrections

The observed gravity values of the first survey were corrected for terrain effects using Hammer's (1939) methods. Terrain corrections for the second survey were insignificant.

Data analysis

Initial analysis of the Bouguer gravity data involved the separation of broad smoothly varying, large scale gravity anomalies (regional anomalies) from smaller local anomalies (residual anomalies) that are related to shallow and often economically important subsurface heterogeneities.

In general, gravity readings are not spaced closely enough to delineate every small detail of the gravity field, and hence the observations appear to have a random component which is termed geological noise. This noise is rarely, if ever, purely random but is spatial in character and is caused by the gravity effects of density inhomogeneities (such as Bouguer density variations and terrain effects). These are commonly not of interest in a survey such as that at Woodlawn, and further interpretation of the data is facilitated by their removal in data analysis.

A strong regional gradient increasing to the northwest is mostly produced by a major circular gravity high to the northwest of the Woodlawn area (Malone and Whiteley, this volume, p 23) (Fig. 1). A least squares trend surface approach was used for the removal of this regional.

Since least squares trend surface fitting produces residual anomalies that are about equally positive and negative a bias, which consists of reducing the trend surface level by 0.20 milligals in this case, was applied to produce residual anomalies that are dominantly positive. Trend surface analysis calculations were carried out using a computer program developed by Templeton (1973).

Both a biased first order trend surface (Fig. 2) and a biased third order trend surface (Fig. 3) were calculated. The third order surface was adopted for further analysis as it minimized geological noise and was consistent with the regional anomalies.

Results

Description of anomalies

The area of the first gravity survey is dominated by a well defined positive anomaly, located at 9880mE, 9220mN (Station J12.3, Line J), which occupies a position over the Woodlawn orebody. This anomaly is greater than 0.80 milligals in magnitude, strikes roughly north to south and is curved concave to the west. About 400m north of this anomaly at 9850mE, 9550mN is an anomaly which has not been completely defined but is at least 0.60 milligals in magnitude. This anomaly appears to be similar in character to the anomaly over the Woodlawn orebody. It is likely, however, that the source of this anomaly is a near surface dolerite sill which has been defined by a magnetic survey (Agostini et al, this volume, p475).

Model Studies

The source of the gravity anomaly over the Woodlawn orebody can be approximately modelled as a tabular mass 20m thick, dipping 70° to the west, with a strike length of 300m and extending from a depth of 10m to a depth of 250m, using densities of $4300kg/m^3$ for the massive sulphides and $2900kg/m^3$ for the host rocks. (The density of $4300kg/m^3$ is appropriate for the higher grade and particularly lead rich massive sulphides; a density of $3900kg/m^3$ is a better average value for the massive sulphides as a whole - Ed.)

The known shape of the orebody obtained by drilling (Malone, this volume, p 49) was used in further modelling. The orebody was approximated by a set of vertical prisms of uniform density but variable depth and thickness. A theoretical gravity anomaly was calculated at a regularly spaced array of stations at the surface by summing the gravitational attraction of each prism using a computer program by Templeton (1974).

This anomaly was then contoured and is shown in Fig. 4. Comparison of this anomaly with the residual anomaly in Fig.3 shows reasonable agreement in magnitude shape and areal extent. This result indicates that the massive sulphides of the known Woodlawn orebody contribute almost all of the measured gravity anomaly.

Mass determination

The excess mass causing the gravity anomaly over the Woodlawn orebody was determined using the method described by Grant and West (1965, p 269). This determination is independant of the shape and density of the causative body. The method was applied by numerically evaluating the volume enclosed by the residual gravity surface using a nine point Newton-Cotes quadrature closed type integration formula. This formula was first applied along east to west profiles and then along north to south profiles. The calculations yielded an excess mass for the Woodlawn orebody of about 1.5 million tonnes. (It is of interest to note that the simplified tabular model gave an excess mass of about 1.4 million tonnes - Ed.)

To calculate total mass from excess mass it is necessary to know the densities. The relation between total mass and excess mass is given in equation 1 below:

$$M = \left(\frac{\rho 1}{\rho 1 - \rho 2} \right) m \qquad (1)$$

where M = total mass

m = excess mass

$\rho 1$ = density of causative body

$\rho 2$ = density of host rocks.

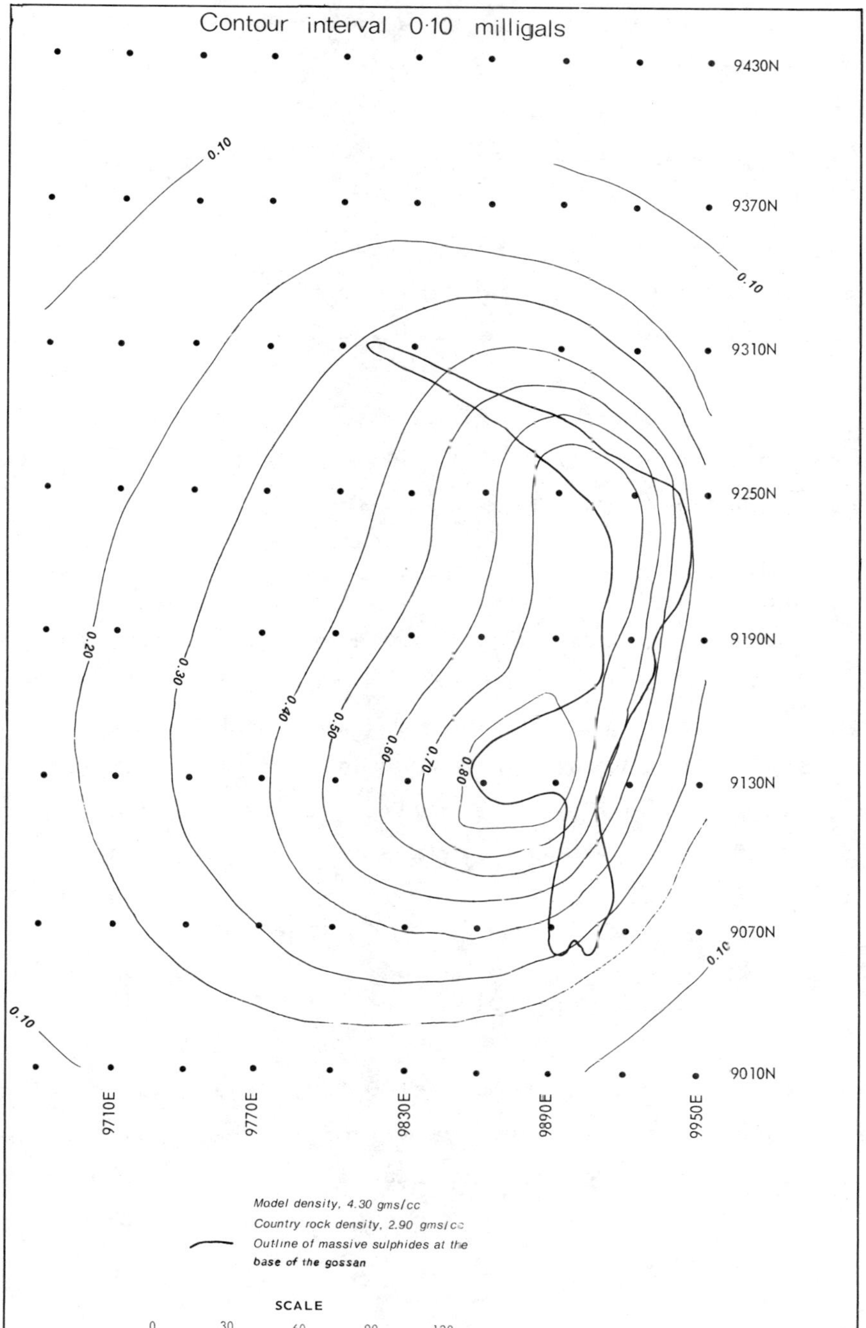

Figure 4 Calculated Gravity anomaly from known orebody

Using $4300 kg/m^3$, for the density of the causative body and $2900 kg/m^3$ for the host rocks equation 1 yielded an estimate of about 4.6 million tonnes. This value is only about 73% of the proven 6.3 million tonnes of massive sulphides ore.

An average density of the massive sulphides ore of about $3900 kg/m^3$ (Tyne, Idnurm & Malone, this volume, p 99) was used in the calculation of ore reserves from drilling and results of mining to date have confirmed that it is a satisfactory average. Insertion of this density value into equation 1 gave an estimate of the massive sulphides ore of about 5.9 million tonnes which is within 7% of the known reserves.

In addition to the massive sulphides ore, 3.7 million tonnes of footwall copper ore are present giving total reserves in the Woodlawn orebody of 10 million tonnes. The density of the footwall ore is $2900 kg/m^3$, which is the average density of the host rocks (Tyne, Idnurm & Malone, this volume, p 99). The lack of density contrast means the footwall copper ore is unlikely to be contributing much to the measured gravity anomaly. This is also substantiated by the earlier modelling.

Conclusions

A residual gravity anomaly of 0.8 milligals was obtained over the Woodlawn orebody after removal of the regional field using a biased third order trend surface. The shape and areal extent of the anomaly closely reflects the massive sulphide deposit outlined by drilling.

Model studies have shown that, to a first approximation, the orebody is tabular in shape.

Calculation of a theoretical anomaly from the known shape of the massive sulphides further suggests that they are sufficient to explain the observed anomaly. The footwall copper ore does not markedly influence the gravity anomaly.

Numerical integration of the residual gravity anomaly gave an excess mass of 1.5 million tonnes.

The actual mass of the massive sulphides causing the anomaly is considerably underestimated using the density 4300kg/m^3 in the initial modelling. An estimate much closer to the known reserves is obtained by using a density of 3900kg/m^3, which is the average density of the massive sulphides as a whole.

An extended gravity survey to the north of the known orebody defined a number of weak anomalies. Although the cause of these is uncertain they are largely attributed to dolerite sills which occur in the vicinity. The extended survey indicates that no major massive sulphide deposit is likely to be present at shallow depth along strike to the north of the known orebody.

Acknowledgement

This paper is published with the permission of the Under Secretary, New South Wales Department of Mineral Resources and Development.

References

Grant, F.S. and West, G.F., 1965. Interpretation Theory in Applied Geophysics. McGraw-Hill Book Company, New York.

Hammer, S., 1939. Terrain corrections for gravimeter stations. Geophysics, 4, No. 3, p 184.

Templeton, R.J., 1973. A Fortran V Program to Perform First to Third Order Polynomial Trend Surface Analysis on Irregularly Spaced Data. Rep. Geol. Surv. N.S.W., GS 1973/297 (Unpubl.).

Templeton, R.J., 1974. A Fortran V Program to Calculate the Gravity Anomaly Caused by an Arbitrary Three-Dimensional Mass. Rep. Geol. Surv. N.S.W., GS 1974/004 (unpubl.).

CHAPTER 10

GROUND GEOPHYSICAL SURVEYS: SEISMIC REFRACTION

10.1. Shallow Seismic Refraction Survey of the Woodlawn Orebody

L. V. HAWKINS AND R. J. WHITELEY

Contents

Abstract

A seismic refraction survey over the Woodlawn orebody clearly determined the subcrop extent of the massive sulphides as a zone of low seismic velocity set within the much higher velocity host rocks. The seismic method is a useful anciliary method for delineating shallow, massive lead-zinc sulphide deposits. The travel-time curves have unusual and distinctive features generally showing negative or infinite apparent velocities on the opposite side of the orebody to the location of the shotpoints. These are more marked for shots at larger offset distances from the recording geophone spread, and there is some migration of the commencement of the negative or infinite apparent velocity segments in the direction of the shotpoints for the more distant shots. This effect is produced by the recording of arrivals of waves which have travelled through the wall rocks of the orebody after being diffracted from points below the top of the orebody. These diffracted waves are then refracted from the wall rocks through the orebody back to the surface. Neither the very low *in situ* seismic velocities observed within the massive sulphides of the Woodlawn orebody, nor these unusual features of the travel time curves across the orebody were observed over the pyritic and electrically conductive Black Shale. The seismic refraction method would, therefore, also seem to be a useful, additional geophysical screening method in distinguishing between shallow, massive sulphide orebodies such as Woodlawn, and uneconomic, graphitic conductive zones.

Introduction

Eleven seismic refraction spreads on five traverses were recorded over the Woodlawn orebody during 1975. A further three seismic traverses were recorded across the electrically conductive Black Shale about 2.5km southwest of the orebody. The aims of the surveys were to investigate the application of the seismic refraction method to the delineation of the subcrop extent of shallow, massive sulphide orebodies, and the use of the method as an additional screening tool in the distinguishing of these deposits from shallow, uneconomic, pyritic conductive zones such as the Black Shale. The method also provided relevant engineering information on overburden thickness and host rock strength for the planned open cut mining of the orebody.

A 12 channel SIE RS4 refraction seismograph was used for the survey with charges of AN60 gelignite in shallow augered holes as seismic sources. Amplifier gains were calibrated to examine whether significant differences in attenuation of seismic energy could be observed on geophone spreads crossing the orebody. A geophone spacing of 3m was used with a two geophone overlap on adjacent spreads. Each seismic spread recorded from a centre shotpoint and up to four different shotpoint offset distances from each end of the geophone spread, i.e., up to nine shotpoints per spread. In addition, a number of seismic spreads across the orebody also recorded from broadside shotpoints offset orthogonally to the traverse line. A series of shots located at different depths down water filled test shafts and in existing deep auger holes were also recorded on traverses located near these shafts.

In all some 200 seismic records were obtained during the survey although only four of the representive traverses across the orebody will be discussed here.

Results

The locations of the seismic traverses over the Woodlawn orebody are shown in Fig. 1. The travel-time curves of the four traverses across the orebody on lines H, I, K and M are shown in Figs. 2, 3, 4 and 5 respectively. The location of the gossan and the cross-section obtained from drilling of the upper part of the orebody beneath the traverses are also shown on Figs. 2, 3 and 5.

Particular attention was paid, as is always necessary in seismic refraction surveys, to obtaining good record quality. As may be seen from the number of shotpoint offset distances and travel time curves for the traverses shown, good travel-time curve overlap was also obtained. In this case, because of the experimental nature of the survey, the overlap may appear to give unnecessary duplication. However, a certain amount of what may appear to be redundancy is necessary in seismic refraction data, particularly in complex situations, in order to resolve interpretational ambiguities and ensure that data used in analyses are relevant to the problem.

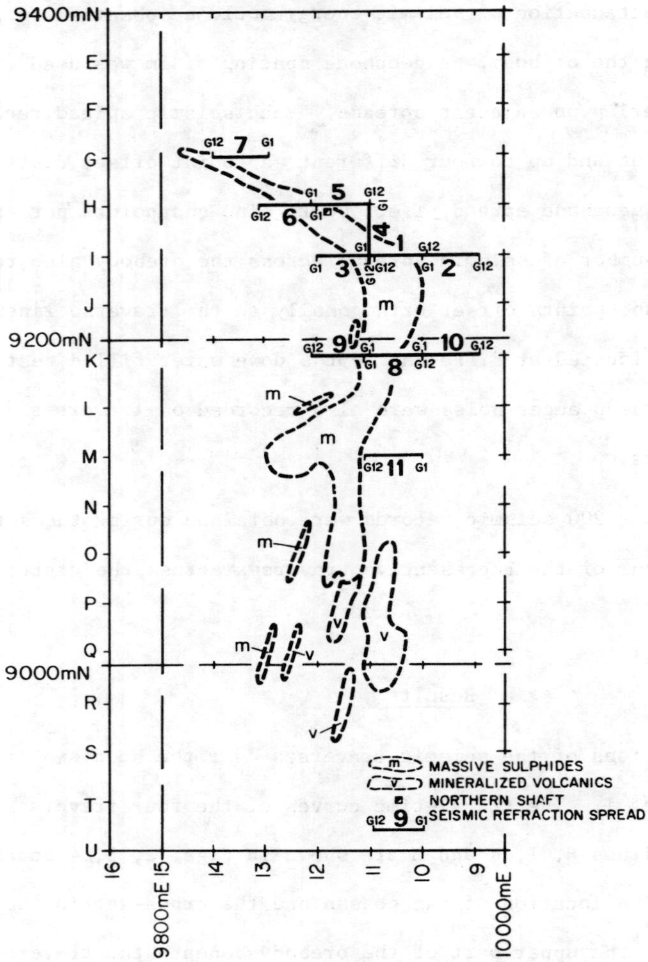

Fig. 1 Location of Seismic refraction traverses over the Woodlawn orebody.

Unusual features of the travel time graphs

In the travel time graphs shown in Figs. 2, 3, 4 and 5, the plotted first arrival travel time points are joined for each record or shotpoint.

For easier identification and to avoid confusion interpretation line segments for all refractors are not drawn. However, one set of interpretational segments relating to the bedrock profile are shown for each direction

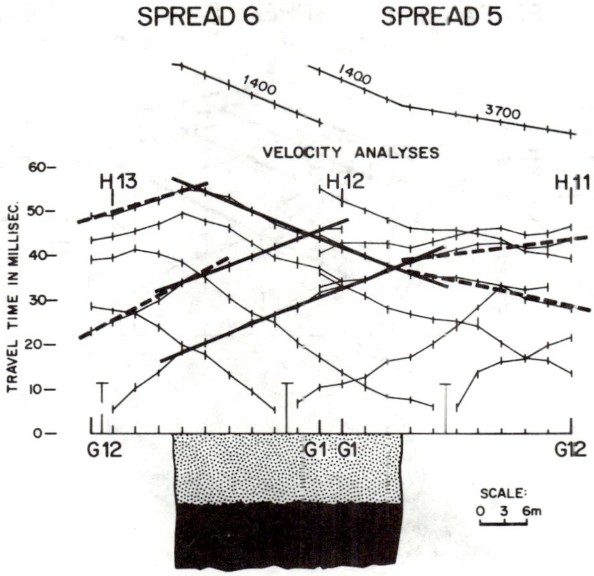

Fig. 2 Travel time curves across the Woodlawn orebody along line H.

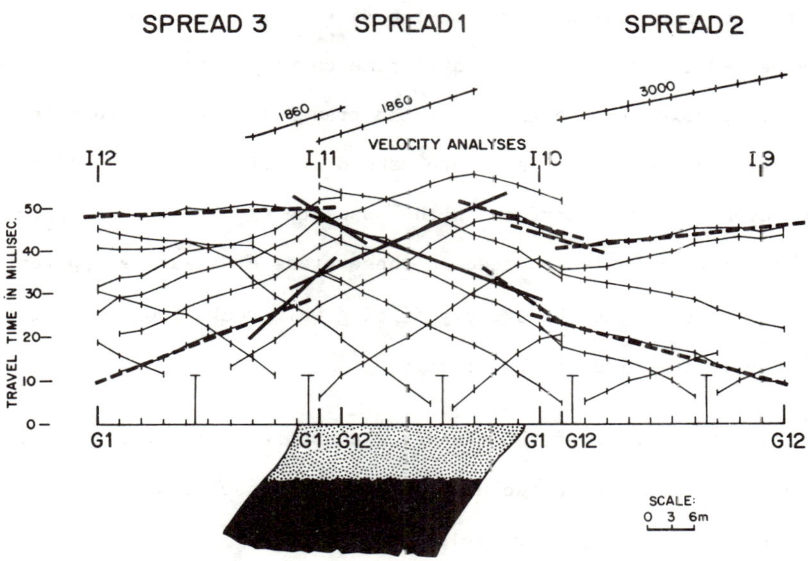

Fig. 3 Travel time curves across the Woodlawn orebody along line I.

of shooting. The interpretation may be verified by considering the full set of overlapping and reversed travel time curves.

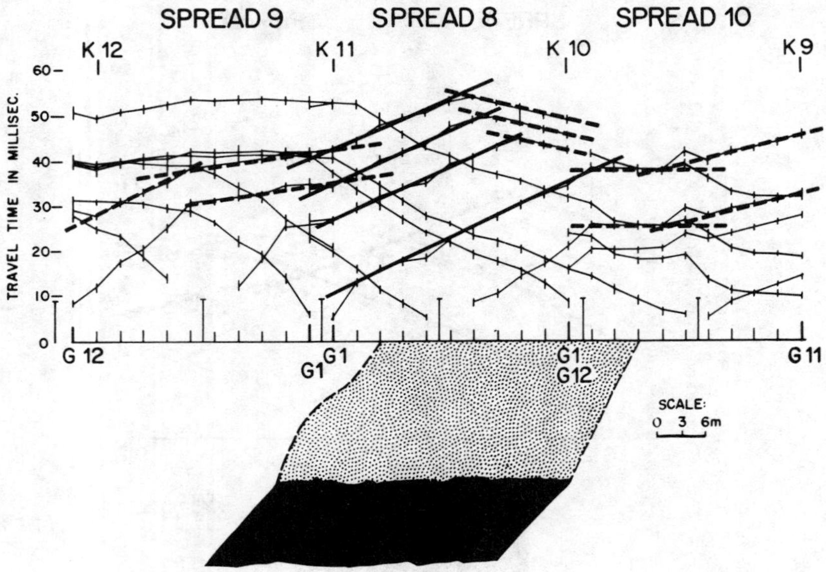

Fig. 4 Travel time curves across the Woodlawn orebody along line K.

At the edges of the orebody, on the side opposite to the shotpoints, there is a marked tendency of the travel time curves towards negative apparent velocities i.e., downturned segments of the travel-time graphs, particularly from the distant shotpoints. These are caused by first arrivals generated at the orebody host rock interface on the opposite side to the shotpoint location. Seismic arrivals recorded in these areas of negative apparent velocity have travelled up the steeply dipping wall rocks from below the top of or possibly to the side of the orebody.

On the downdip side of the orebody the apparent velocities are infinite rather than negative. From a technical viewpoint negative apparent velocities are higher than infinite apparent velocities.

Also, in general, there is a marked migration of the negative or infinite apparent velocity segments of the travel time curves in the direction of the shotpoint with increasing offset distances. This has been indicated by presenting interpretational line segments for line K on Fig.5 for the bedrock

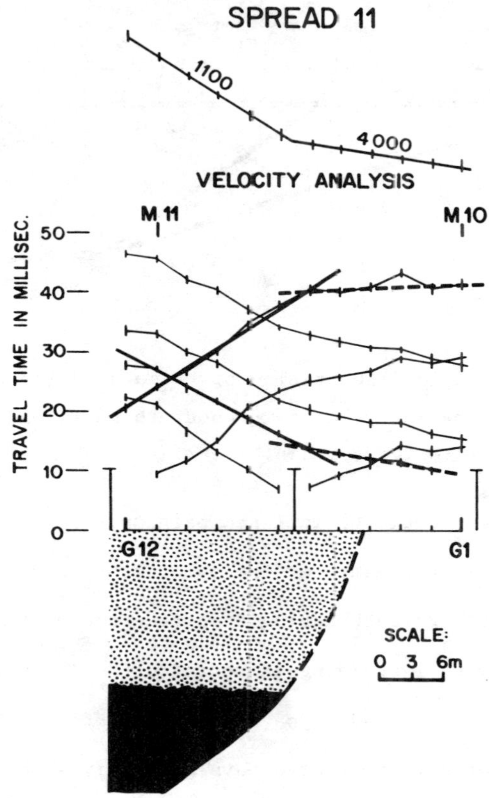

Fig. 5 Travel-time curves across the Woodlawn orebody along line M.

refractor defined by shots offset positions to the west of the recording spreads. Usually, such migration of apparent velocity segments indicates that a deeper high velocity refractor is being recorded, however, this is not the case at Woodlawn.

In the vicinity of the Woodlawn orebody seismic waves are travelling below or possibly to the side of the top of the orebody or through a deeper, thinner part of the orebody. They are being diffracted around the low velocity top part of the body orebody. These diffracted waves are then refracted back from the interface between the orebody and high velocity wall rocks through the low velocity zone of the orebody to the surface. This is shown diagramatically in Fig.6 with seismic rays rather than wavefronts for simplicity. As the

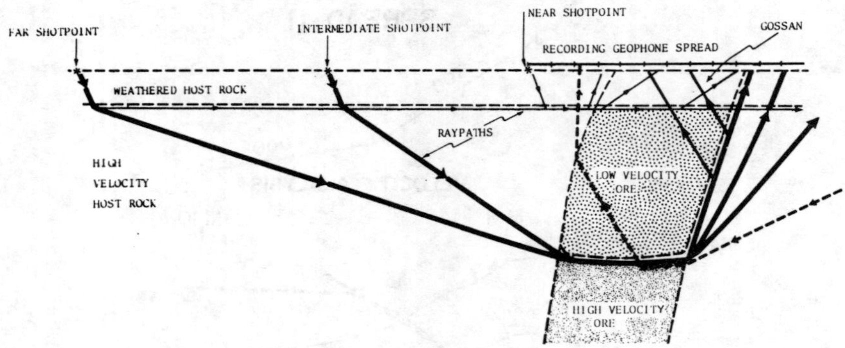

Fig. 6 Diagramatic illustration of raypaths diffracted from below the low velocity orebody to be refracted through the orebody to the surface.

shotpoint offset is increased, the inclined path approaches more closely the path of the rays normally refracted at the base of the weathered layers. This results in the diffracted raypath having a shorter travel time relative to the normal refracted wave with increasing shotpoint offset. In the case of the Woodlawn orebody, this diffracted raypath for large offsets has a smaller travel time than the normally refracted rays and arrives before them.

It should be noted that the path of the diffracted ray, when being refracted back through the orebody to the surface, is inclined backwards relative to the general direction of ray travel. This is necessary for a negative apparent velocity to be observed from the travel time curve. Also, a limit to the migration of the onset of the negative or infinite apparent velocity segments will be approached with very large offset distances. Similar although smaller migration effects of this type have been observed by the authors in association with diffraction below the downthrown fault block in the vicinity of the Mt. Bulga deposit (Malone and Whiteley, this volume p. 23).

From these results other subcropping, massive sulphide deposits with a low velocity upper zone of limited length and depth extent would be expected to show similar features on seismic travel time curves.

An understanding of the cause of this observed migration is very important, since in carrying out normal depth analyses of a bedrock profile, and velocity analyses of the seismic wave velocities through the host rocks and orebody, the travel times used must result from normally refracted and not the diffracted waves. If the sequence of diffracted wave travel times are used in the normal interpretational techniques (Hawkins, 1961; Hagedoorn, 1959; Dampney and Whiteley, 1980) significant errors in the calculated velocities and depths would result in those sections of the profile where the diffracted arrivals were used.

Seismic Wave Attenuation

It was expected that some observable difference in seismic wave attenuation between waves which travelled only in the host rocks and those which had travelled through the orebody would be apparent. However, no clear and consistent differences in recorded wave amplitudes and characteristics were observed which would allow mapping of the subcrop of the orebody by this means.

Velocity Analysis

Relevant portions of seismic travel time data were subjected to velocity analysis using Hagedoorn's (1959) minus method. These analyses are also shown in Figs. 2, 3 and 4 above the travel time data. A marked difference in the seismic compressional wave velocities within the orebody relative to the host rocks was obtained.

The fresh host rock velocities varied from about 3,000 to 4,000 m/s.

Analysis of the travel time data from above the orebody gave velocities from about 1,100 to 1,850 m/s. These are similar to velocities in the near surface gossan.

Apparent velocities measured from travel time data in the vicinity of the orebody also allow the angles at which first arrival rays intercept the surface to be calculated. These calculations indicate that the first arrival rays have travelled through the gossan overburden after diffraction at an interface beneath the gossan and altered or weathered ore at a depth below the general level of the unweathered host rock. This low velocity zone has a limited extent is believed to cause the negative apparent velocity segments on the time-distance graphs. Geologically this zone may be related to a somewhat thicker gossan and the observed vugular and cavernous zone at the base of the gossan near the top of the supergene ore and fresh ore beneath.

Comparison with results over the Black Shales

The three seismic traverses across the Black Shale at Woodlawn showed bedrock seismic velocities to be about 4000m/s in the shale and about 5000m/s in the adjacent volcanics. There was no low velocity zone within the shale as was present over the massive Woodlawn orebody. Also, none of the special features of the travel time curves over the Woodlawn orebody discussed earlier were present.

Conclusions

The shallow seismic refraction method at Woodlawn clearly distinguished between the very low velocity zone of limited dimensions, consistent with the location of the upper part of the shallow massive sulphide orebody. This was absent over the Black Shale. The method would therefore appear to be a very useful additional geophysical method in exploration for massive sulphides in both confirming the presence and delineating the subcrop extent of shallow, massive sulphide orebodies such as the Woodlawn deposit, and to be a useful additional screening technique in distinguishing these from uneconomic, electrically conductive shales.

REFERENCES

Dampney, C.N.G. and Whiteley, R.J., 1980. Velocity Determination and Error Analysis for the Seismic Refraction Method Geophysical Prospecting v.28 pp. 1-17.

Hagedoorn, J.G., 1959. The Plus-Minus Method of Interpreting Seismic Refraction Sections: Geophysical Prospecting, v.7. n. 2, p. 158-182.

Hawkins, L.V., 1961. The Reciprocal Method of Routine Shallow Seismic Refraction Investigations: Geophysics, v. 26, pp. 806-819.

CHAPTER 11

DOWNHOLE GEOPHYSICAL SURVEYS

11.1. Surface and Downhole Applied Potential (Mise-a-la-Masse) Surveys at Woodlawn

R. J. TEMPLETON, E. D. TYNE AND K. P. QUICK

Contents

Abstract

Surface and downhole applied potential surveys at Woodlawn effectively outlined the dipping massive sulphide orebody and clearly illustrate that the very conductive massive sulphides are electrically continuous between drill hole intersections and the surface. Results from the downhole survey clearly indicate that in this environment, once a massive sulphide body is intersected by drilling and electrically energized, the subsurface location of the body can be inferred by downhole applied potential measurements even in drill holes which pass up to 60m away from the body.

Introduction

The applied potential (mise-a-la-masse) method has been described by Parasnis (1967). In this method the orebody is directly energized by means of a current electrode placed in contact with it usually in a drill hole. The return current electrode is placed at some distance from the energizing electrode. Voltages are then measured either at the surface or in adjacent drill holes.

The initial applied potential survey of the Woodlawn deposit was carried out by Jododex Australia Proprietary Limited in July, 1974, after completion of the early drilling programmes. This survey was later extended and five drill holes intersecting the orebody were surveyed with the downhole applied potential method by the Geological Survey of New South Wales in 1975.

The aims of the surveys were to test the efficacy of the applied potential method for outlining shallow massive sulphide lenses in this geological environment, to determine if the ore intersections in the drill holes were electrically connected and to obtain a three dimensional electrical picture of the orebody.

Geological Model

The orebody approximates a curved dipping tabular body which dips about 40° to the west on average, is between 10 and 20m thick, and extends for at least 250m down dip. (Malone, this volume, p 49) Fig. 1*shows the outline of the orebody at the base of the gossan. Dolerite sills, acid tuff, and pelitic metasediments are the dominant host rocks.

Typical resistivities obtained from laboratory measurements (Tyne, Idnurm and Malone, this volume, p 99) were 100 ohm m to 1000 ohm m for the acid tuffs and pelitic metasediments, about 6500 ohm m for the dolerite and about 0.06 ohm m for the massive sulphides.

The resistivity contrast between host rocks and massive sulphides is thus of the order of 10,000 to 1 and the massive sulphides are particularly conductive. This ensures favourable conditions for the application of the applied potential method.

*The folded fig.1 can be found in the envelope accompanying the book.

Field Procedures and Equipment

An energizing current electrode was emplaced in the massive sulphides in drill hole WE5 which intersected 36m of massive sulphides towards the northern end of the orebody. The current electrode was positioned in the central part of the massive sulphide intersection in WE5 at a vertical depth of 46m and consisted of a copper tube (weighted with lead shot), 50mm in diameter and 3m in length. A remote current electrode was placed approximately 1670m southwest of WE5 (Fig.1).

In the initial survey the orebody was energized with a low frequency alternating current at 0.3Hz using a Heinrichs Geoexploration induced polarization (IP) transmitter. A Geoscience IP transmitter was used for the later surface and the downhole surveys.

The remote potential electrode was placed 1280m northwest of WE5 (Fig.1). Electric potential measurements were made at station intervals of 15.25m and 30.5m along the east to west Woodlawn Test Grid lines, using a Heinrichs IP receiver and Fluke high impedance voltmeter.

Discussion of Surface Applied Potential Results

The results of both the initial and extended surface surveys are shown in Fig. 1.

The results for the two surveys generally agree well. The discrepancies on lines L to U may possibly be explained by assuming different soil moisture conditions for the two surveys. Such differences will change the near surface resistivities and hence the distribution of surface potentials measured in the applied potential method.

Equipotential contours shown in Fig. 1 outline the near surface shape of the orebody extremely well and faithfully indicate its changes in strike and dip. The 200mV/A contour encloses the top of the massive sulphides at the base of the gossan. The highest potentials coincide with the top part of the body, outlining even the subordinate ore lenses on either side of the main lens near its central region.

Lower potential gradients on the western side of the orebody clearly indicate the westerly dip of the body. The plunge of the massive sulphide lens to the northwest is also suggested from the contour pattern.

The lack of steep gradients around the central and southern regions of the body is most likely an indication of the shallower dips observed in this portion of the orebody and also of the increase in overburden thickness from 10m to 20m.

Down Hole Data Collection

Applied potential measurements were made at 10m intervals down hole using a shielded logging cable (Tyne, this volume, p 533). Inductive coupling was minimized by using a low frequency current (0.3Hz) to energize the orebody and by maintaining the remote potential connecting wire at right angles to the remote current wire.

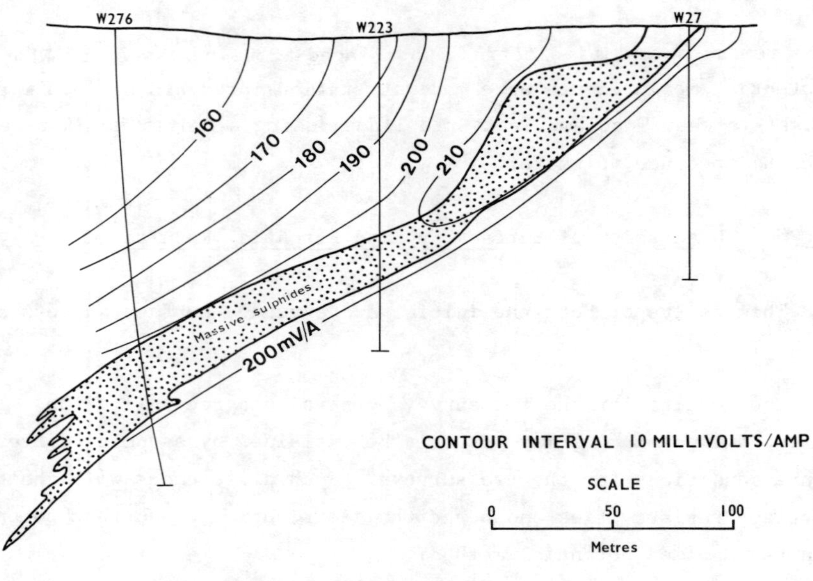

Figure 2 Drill Hole Applied Potential Section through W276, W223 and W27.

Interpretation

In Fig. 2 the hole to hole and hole to surface results have been contoured along a section through drill holes W276, W223 and W27. These contours very effectively outline the dipping orebody, and as with the surface data, the orebody is enclosed by the 200mV/A contour.

Figs. 3 to 7 show the observed applied potentials for each of the drill holes surveyed.

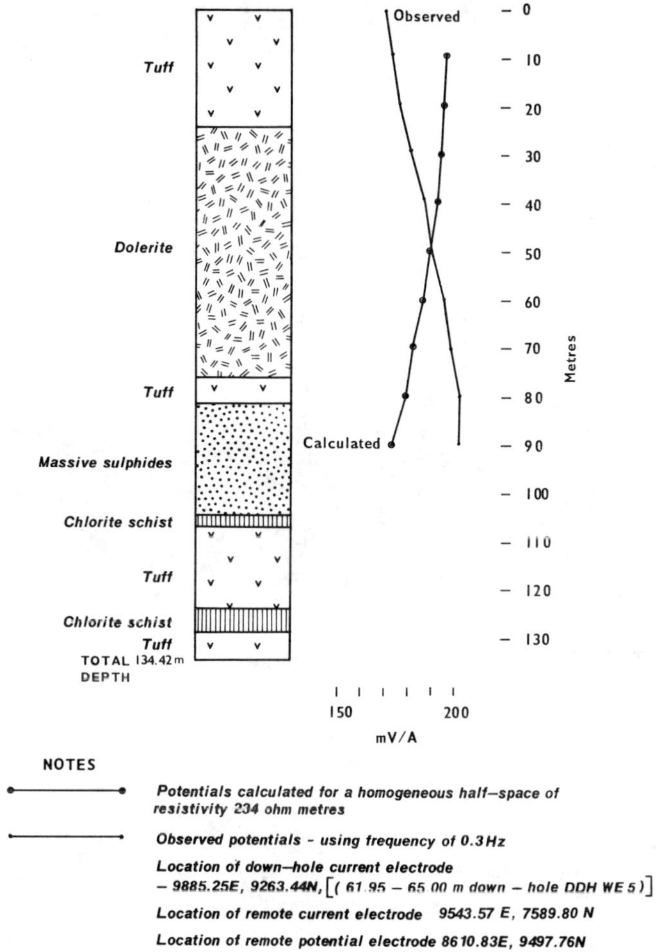

Figure 3 Drill Hole Applied Potential Survey, W223

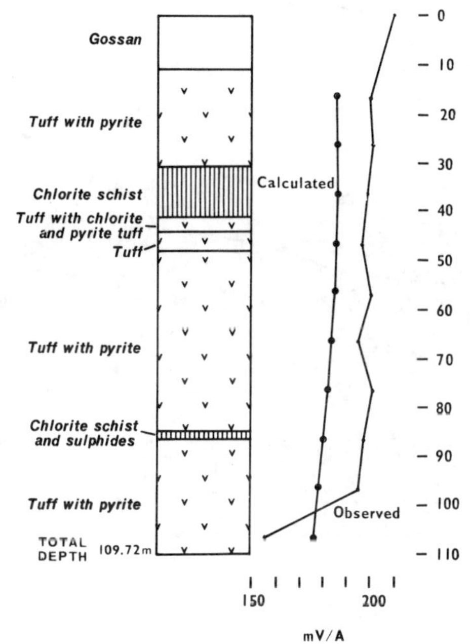

Figure 4 Drill Hole Applied Potential Survey, W27

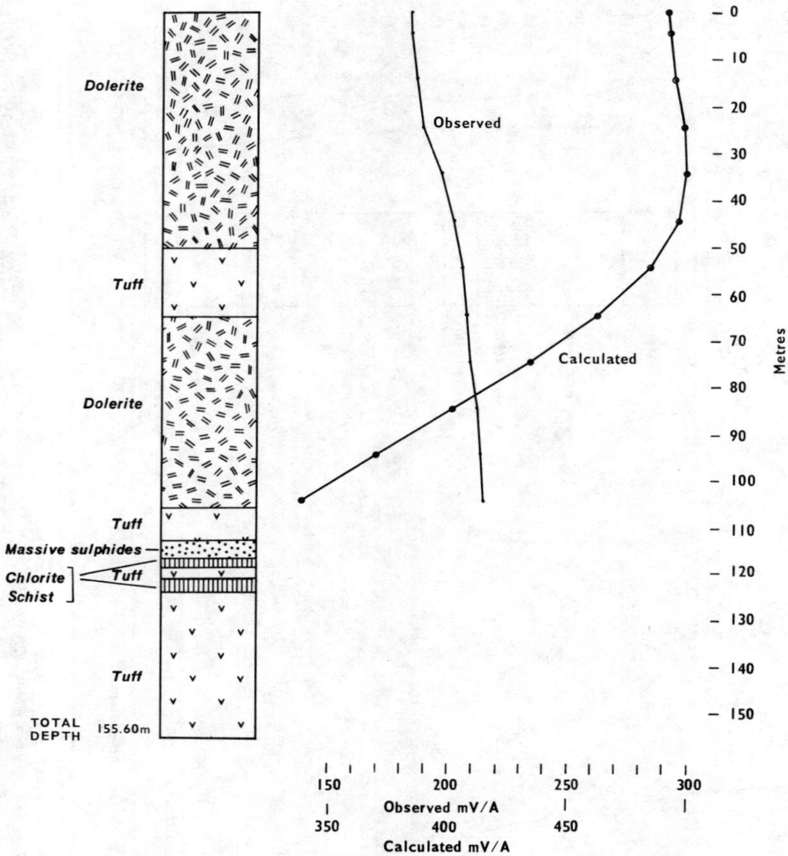

Figure 5 Drill Hole Applied Potential Survey, W212

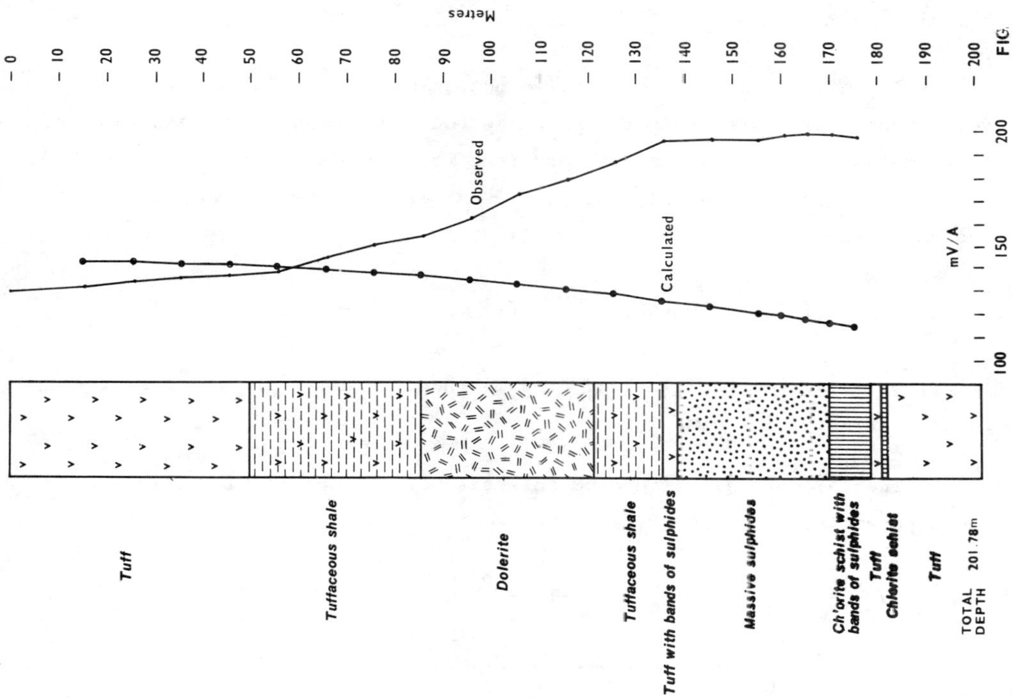

Figure 7 Drill Hole Applied Potential Survey, W276

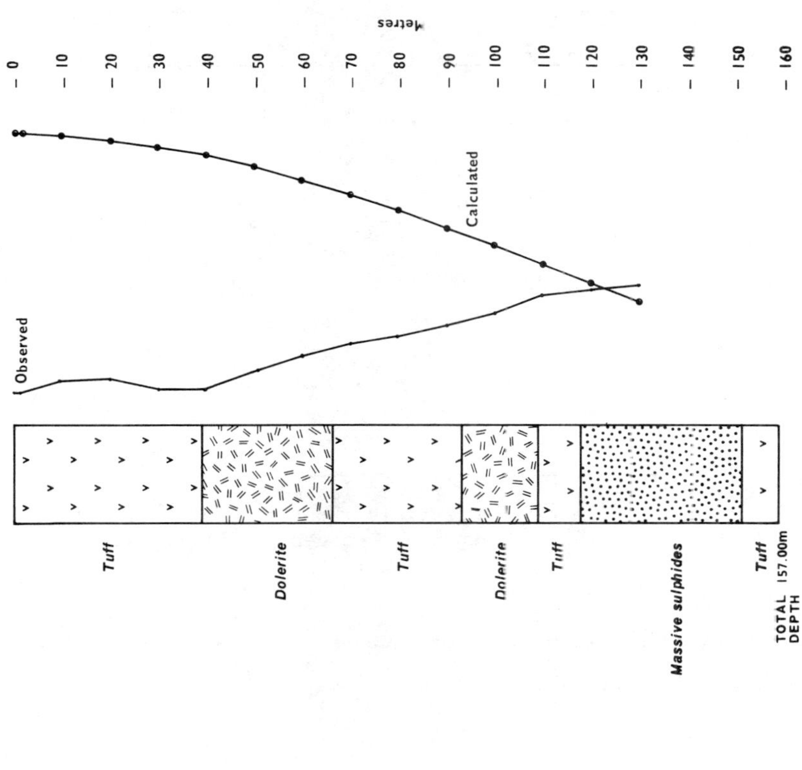

Figure 6 Drill Hole Applied Potential Survey, W254

Theoretical applied potentials

To better evaluate the applied potential results in each hole surveyed, theoretical potentials for a homogeneous medium have been calculated using a formula derived from image theory (see for example Van Nostrand and Cook 1966) for the electric potential V (x,y,z) at a point (x,y,z) in an isotropic, homogeneous medium:

$$V(x,y,z) = \rho \frac{I}{4\pi} \{ (x^2 + y^2 + (z-d)^2)^{-\frac{1}{2}} + (x^2 + y^2 + (z+d)^2)^{-\frac{1}{2}} \} \quad (1)$$

where x, y, z, d, are defined in Fig. 8, I is the current at the point source I, ρ is the resistivity of the medium.

The theoretical calculated potential difference (Δ V) between the drill hole potential electrode P_1 and the remote potential electrode P_2 is given by:

$$\Delta V = (V_{P_1 I_1} - V_{P_1 I_2}) - (V_{P_2 I_1} - V_{P_2 I_2}) \quad (2)$$

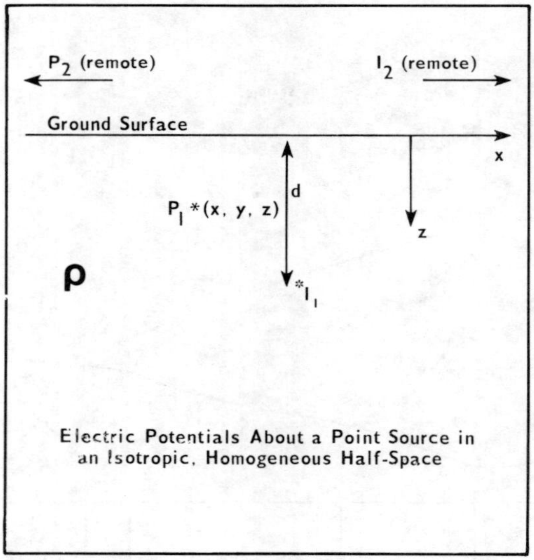

Figure 8 Electric Potentials about a Point Source in an Isotropic, Homogeneous Half Space

In this calculation, the coordinates of the measurement points were derived from Pajari surveys of the drill holes.

An estimate of ρ can be obtained using surface data at points remote from the orebody. At such points the surface applied potentials can be assumed to closely approximate the potentials in homogeneous medium.

Equation (1) can then be used to calculate the estimate of ρ . A value of 234 ohm m was obtained, using the observed potentials at stations H20, I20, J20 and K20.

The theoretical potentials calculated for each of the five holes surveyed are shown in Figs. 3 to 7.

Discussion of Down Hole Results

Observed potentials increase with depth in all drill holes except W27. In contrast to this the potentials for a homogeneous subsurface decrease with depth, and only in W27 is there any similarity between the observed potentials and the potentials for a homogeneous subsurface. W27 is the only hole of those surveyed that did not penetrate the orebody at depth. This hole intersected gossan at the surface and continued away from the orebody into footwall rocks (Fig.2).

Clearly, the presence of the orebody causes a considerable disturbance in the observed potentials. The trend in W276 suggests that this disturbing influence can be detected at least 60m up the drill hole above the orebody. In W212 the proximity of the orebody is suggested by the observed potentials below 80m down hole.

The observed potentials in drill holes W223, W254 and W276 remain constant within the massive sulphide intersections, and reach a level of about 200mV/A in each of the three holes. It is thus inferred that the orebody is electrically continuous between those holes and that the orebody is physically continuous between these holes and the surface.

Conclusions

The applied potential surveys at Woodlawn show the value of both surface and down hole surveys in allowing a meaningful three dimensional interpretation of the conductive target geometry to be made. Results illustrate the ability of the applied potential method in this environment to yield accurate information on shape, attitude and near surface extent of a massive sulphide body and to confirm the continuity of massive sulphide intersections between drill holes and the surface.

Comparison of the observed and theoretical potentials in the holes surveyed indicates that the presence of massive sulphides in this type of environment can be inferred from applied potential measurements even if the drill hole passes up to 60m away from the massive sulphides. Drill hole potential measurements conducted concurrently with a drilling programme for massive sulphides can indicate whether a hole in progress has penetrated the ore zone or needs to be deepened to reach the ore zone. Applied potential surveys of completed holes can assist in the effective siting of further test holes.

Acknowledgements

This paper is published with the permission of the Under Secretary, New South Wales Department of Mineral Resources and Development.

References

Parasnis, D.S., 1967. Three Dimensional Mise-a-la-Masse Surveys of an Irregular Pb-Zn-Cu Deposit in Central Sweden. G.P. V15, no.3, 407 to 437.

Van Nostrand, R.G., and Cook, K.L., 1966. Interpretation of resistivity data. U.S. Geol. Surv. Prof. Pap. 499.

11.2. Drill Hole Electrical and Gamma Radiation Surveys at Woodlawn

I. G. HONE AND G. A. YOUNG

Contents

Abstract

Resistivity, induced polarization (IP), self potential, single point resistance and gamma radiation measurements were made in four exploration drill holes at Woodlawn.

Drill hole resistivity, resistance and gamma radiation methods proved useful for logging lithologies and rock boundaries. Self potential, resistivity and IP measurements suggest that two holes passed close to sulphide mineralization.

Large contrasts in resistivity and IP effect were observed between the orebody and the surrounding rocks. In situ resistivities of the massive sulphides approached 0.1 ohm m, whereas those of the country rocks were about 10^3 to 10^4 ohm m. Where resistivities around the drill holes were very low, IP readings were impossible to take, however the logs show that the massive sulphides have an IP frequency effect of over 30 per cent.

Introduction

In February 1973, the Bureau of Mineral Resources (BMR) made resistivity, induced polarization (IP), self potential (SP), single point resistance and gamma radiation measurements in four exploration drill holes at Woodlawn. Two of these holes W256 and W276 intersected the massive sulphides of the Woodlawn orebody, W16 came close to but did not intersect the orebody and W55 was remote from the orebody and did not intersect any significant mineralization (Fig.1).

The drill hole geophysical surveys were made to measure the in situ physical properties of the ore and host rocks, and to investigate the use in logging, in exploring for adjacent sulphide mineralization not intersected by the drill hole.

Geophysical Techniques

SP, single point resistance and gamma radiation measurements were made with a Widco Porta logger. These measurements record changes in the properties of the rocks close to the hole and are more useful for observing lithological changes than detecting adjacent mineralization. The gamma radiation logs record the natural gamma ray intensity in drill hole. Changes in lithology are indicated by intensity variations attributed to changes in the concentration of radioactive minerals in the rocks close to

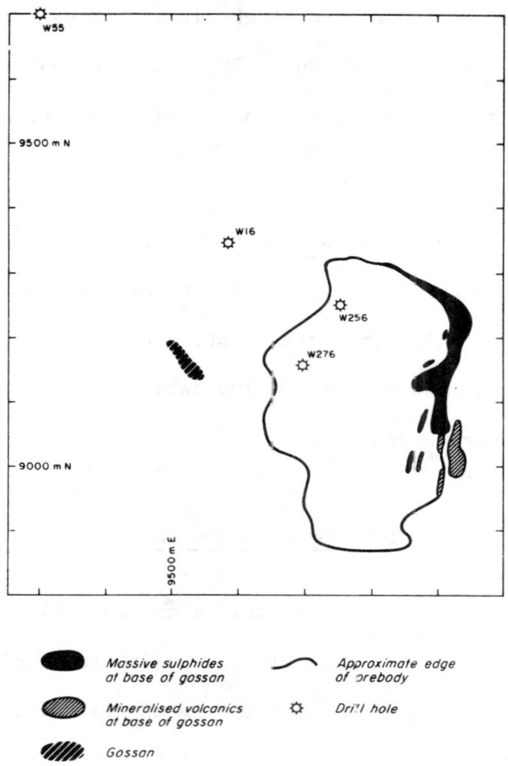

Figure 1 Location of drill holes.

the probe. The resistance and SP logs record the resistance and potential between a moving downhole electrode and a fixed electrode located in a mud pit at the surface. Resistance and SP logs are particularly influenced by changes in the electrical properties of rocks near the probe.

Resistivity and IP measurements were made using downhole probes with a dipole dipole array and a McPhar frequency domain IP system. The probes had equally spaced electrodes (i.e., n=1 geometry) 10ft (3.1m) or 30ft (9.1m) apart. The upper, transmitting electrodes were of lead and the lower, receiving electrodes were porous, ceramic pot, non-polarizing, copper in copper sulphate electrodes. Resistivity and IP measurements were made with the probe stationary, and the reading at a station was assigned to the position of the centre of the array. Frequency effect measurements were made between 0.3Hz and 5.0Hz, and apparent resistivities were calculated assuming that the probe was buried in an infinite homogeneous earth.

The resistivity and IP measurements record the average resistivity and IP effect of a volume of rock in the vicinity of the probe and are therefore useful for exploring for mineralization adjacent to, but not intersected by, the drill hole as well as indicating the in situ resistivity and IP of the ore and surrounding rocks.

Results and Discussion

Results of the geological, electric,gamma radiation and IP surveys of the four holes are shown in Figs. 2 to 5. These figures show that specific rock types, including the sulphide mineralization, have distinctive geophysical responses. Young (1976) has described correlations between the geophysical logs and the geology, and these are discussed below.

Gamma radiation logs

The gamma radiation logs are generally low in the intervals where the probe is adjacent to dolerite or sulphide mineralization. This correlation is shown in the logs of W16, W55, and W276.

The logs of W55 and W276 show that the gamma radiation associated with welded tuffs and tuffaceous shales corresponds to levels of 0.03 and 0.02 millirontgens per hour (mR/h) respectively above the radiation level of the dolerite. (For conversion to SI units 1R = 2.58×10^{-4} coulombs per kilogram), the radiation level associated with mineralized volcanic rocks is much higher than the level associated with the massive sulphides or dolerite.

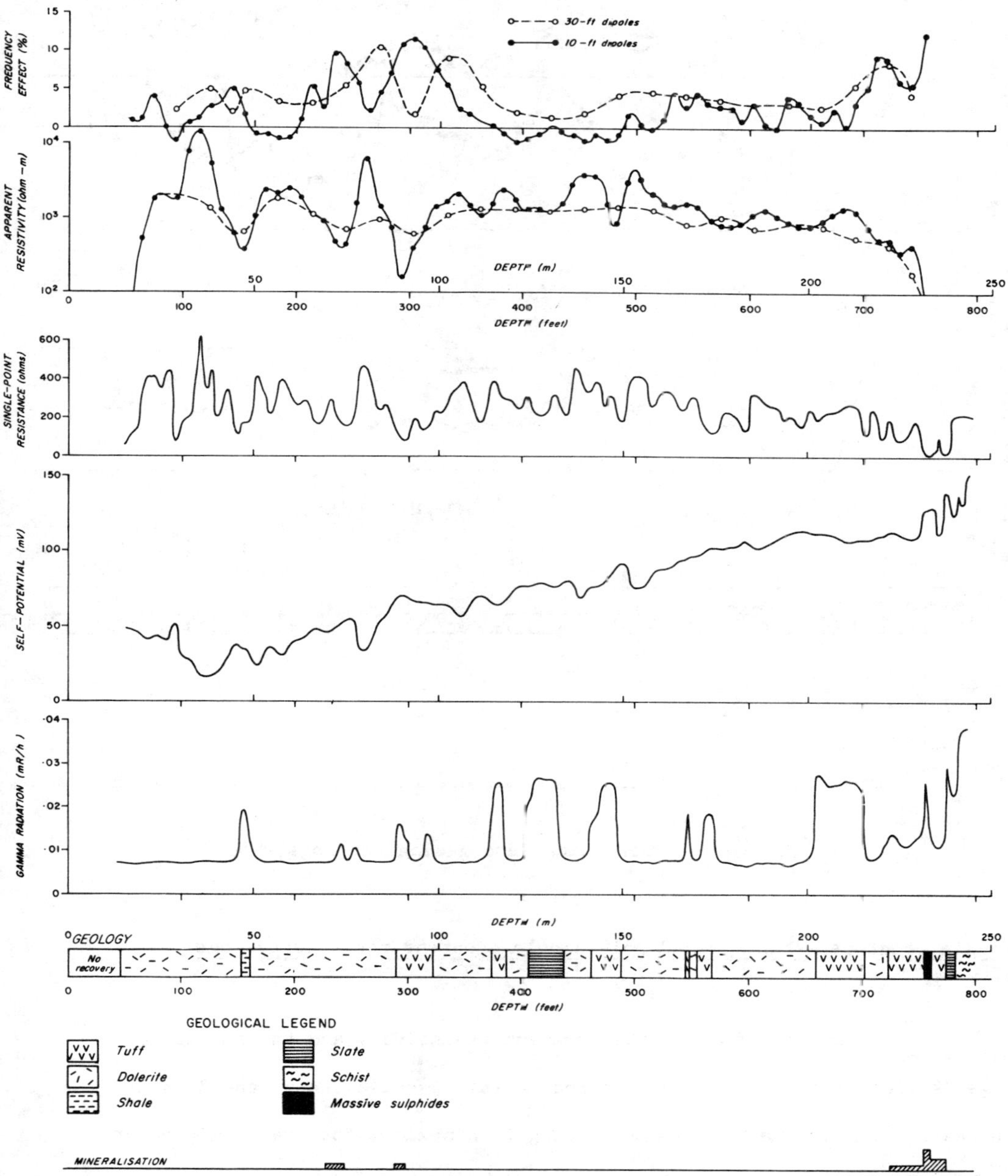

Figure 2 Geophysical and Geological logs, W16.

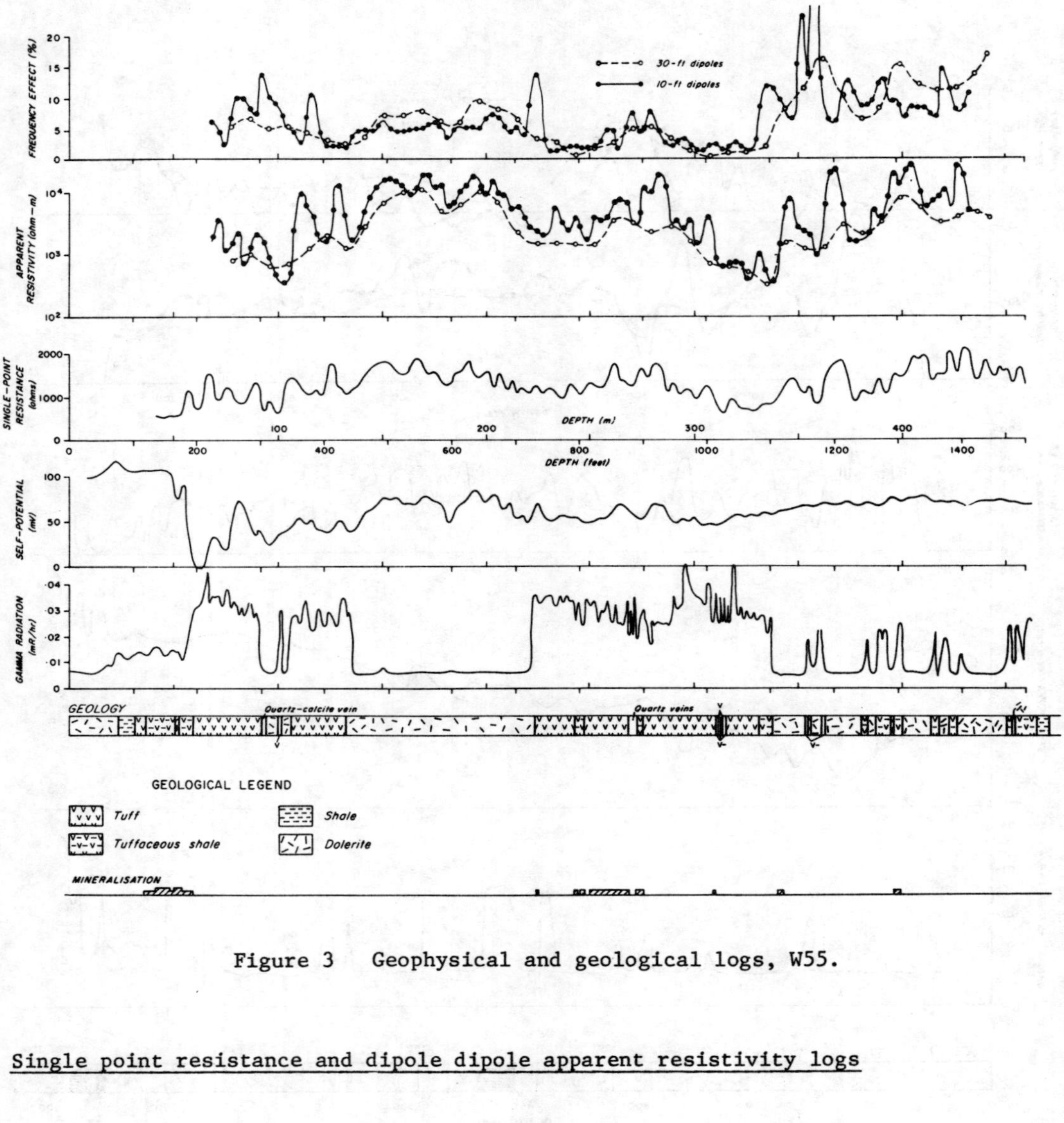

Figure 3 Geophysical and geological logs, W55.

Single point resistance and dipole dipole apparent resistivity logs

The overall form of the apparent resistivity and the resistance logs is similar for the three electrode arrays used, although the detail increases with decreasing dipole spacing to a maximum for the single point probe. Data yielded by the 10ft (3.1m) dipoles, however, are the most useful indication of rock resistivities where thicknesses of rock units intersected are large in comparison with the dipole size.

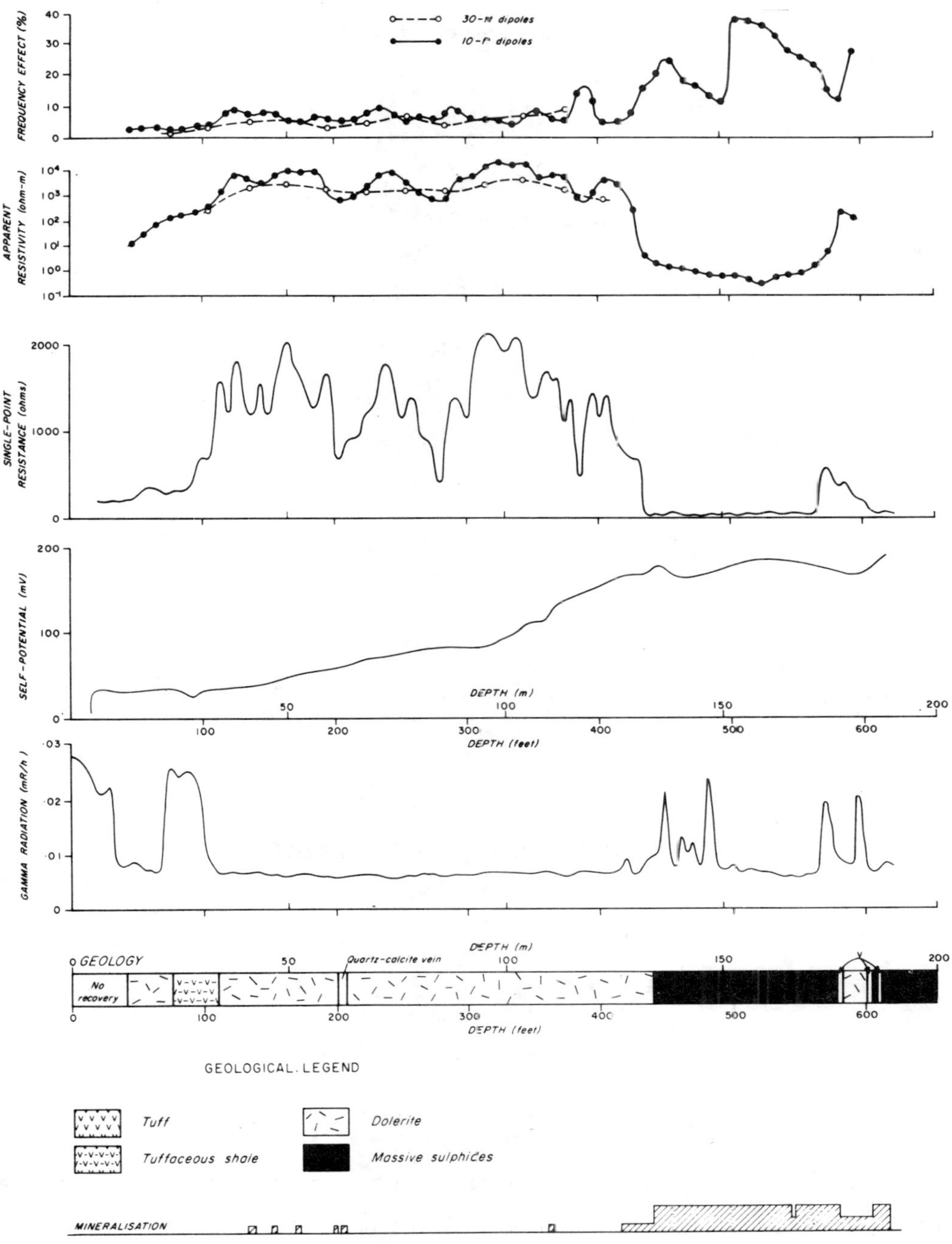

Figure 4 Geophysical and geological logs, W256.

Unmineralized rocks including dolerite and acid tuff have resistivities from 10^3 to 10^4 ohm m, with dolerite being somewhat more

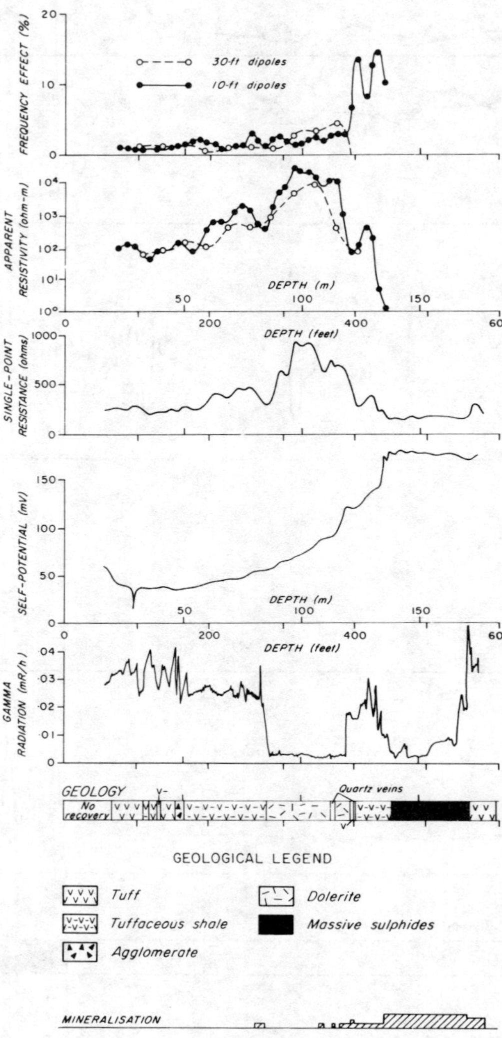

Figure 5 Geophysical and geological logs, W276.

resistive than the acid rocks. By comparison major intervals of massive sulphide mineralization have resistivities approaching 0.1 ohm m.

The presence of minor intervals of sulphide mineralization is reflected in the 10ft (3.1m) dipole dipole apparent resistivity log of W16, which shows resistivity lows coincident with sulphides associated with dolerite and tuff in the interval 230ft (70m) to 300ft (91m) (Fig.2). A similar low occurs at 155ft (47m) in W16, suggesting either that a similar minor sulphide zone may be present within a short distance from the drill hole, or that the shale intersected has a low resistivity.

SP Logs

The general character of the SP logs for W256 and W276 differs from those for W16 and W55. Logs for the latter holes show variations in potential over intervals of 20ft (6.1m) to 50ft (13.2m). These commonly reflect changes in resistivity recorded by the single point resistance probe. This effect is not so evident in the logs of W256 and W276, which show a strong increase in SP towards sulphide mineralization, and a positive potential plateau of 150mV over the main bulk of the sulphide interval.

The results of the SP logs in W256 and W276 plus surface SP data (Cifali and Whiteley, this volume, p 333) show a strong negative anomaly at the surface and an increasing SP potential with depth. This anomaly pattern is consistent with the electrochemical mechanism advocated by Sato and Mooney (1960) for the SP effect produced by a highly conducting body which extends upwards into the zone of oxidation. The mechanism advocated by Sato and Mooney requires an electronic conductor to form part of an electric circuit by connecting two electrochemical half cells situated above and below the level of oxidation.

One implication of their hypothesis is that if a drill hole intersects a conducting massive sulphide lode which extends into the oxidizing zone the potential distribution along the hole should be as shown in Fig. 6. This SP potential increases towards the conducting lode and is constant within it. An increase in potential without the development of a plateau should indicate that a drill hole is restricted to the interval

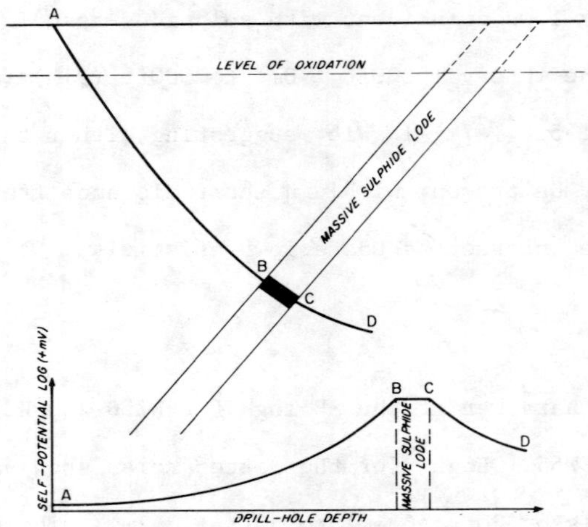

Figure 6 Expected potential distribution for drill hole which intersects sulphide lode extending into zone of oxidation.

AB, i.e. it is continuing to approach the sulphide lode. Such a response is evident in W16, which, although failing to intersect an extensive zone of massive sulphides, appears to be approaching such a zone near 800ft (244m) depth.

The absence of an increase in potential in the deeper part of W55 (Fig.3) is consistent with the drill hole failing to intersect or even closely approach massive sulphides which have continuity into the oxidized zone.

IP Logs

The IP logs show that background frequency effects are from 2 to 10%, and that the massive sulphide ore has a frequency effect in excess of 30%. Where the resistivity of the sulphides is less than 10 ohm m, highly erratic readings were observed and a reliable measure of the *in situ* frequency effect of the massive ore was not possible. The erratic readings probably arise from electromagnetic coupling between the logging system and the conductive sulphides. The IP logs of holes W55, W256 and W276 indicate that the dolerite has a slightly higher background IP effect than the more acid rocks.

They also show several increases in IP effect, which might indicate mineralization adjacent to but not intersected by the hole since the increases do not appear to be caused by rock types or sulphides identified in the geological logs. In W55 near 1160 ft (354m) both the 10 ft (3.1m) and 30 ft (9.1m) IP logs show a strong IP effect of over 20 per cent associated with a drop in resistivity from 10^4 to 10^3 ohm m. Only a small IP effect was recorded at 155 ft (47m) in W16 downgrading the significance of the resistivity low recorded at this point. A small IP anomaly coincident with a resistivity anomaly occurs in hole W256 at 390 ft (119m).

Conclusions

The logging results show the extremely large electrical contrasts between the Woodlawn orebody and the surrounding rocks. The massive sulphides have frequency effects of over 30% and resistivities as low as 0.1 ohm m. The country rocks have background frequency effects from 2 to 10%, and resistivities of the order of 10^3 to 10^4 ohm m. Dolerite is more resistive, and has a higher background IP effect than the acid rocks.

Drill hole resistivity, resistance and gamma radiation logs show characteristic responses for different rock types and were useful for logging lithologies and rock boundaries.

SP measurements show a strong potential rise towards massive sulphide mineralization below the oxidizing zone. This effect could be used for detecting mineralization remote from the drill hole.

Resistivity and IP measurements indicate possible mineralization adjacent to holes W16 and W55.

Acknowledgement

This paper is published with the permission of the Director, Bureau of Mineral Resources, Canberra.

References

Sato, M., & Mooney, H.M., 1960 - The electrochemical mechanism of sulphide self potentials. Geophysics, 25(1) 226-49

Young, G.A., 1976 - Drill hole logging and transient electromagnetic test surveys, Woodlawn deposit, New South Wales, 1973. Bureau of Mineral Resources, Australia, Record 1976/52 (unpublished).

11.3. Drill Hole Resistivity and Induced Polarization Logging at Woodlawn

E. D. TYNE

Contents

Abstract

Multiple electrode spacing, resistivity and induced polarization (IP) logging was completed at Woodlawn in holes W276 and W57 with the aim of determining in situ electrical rock properties and the persistence of conductive intersections away from the holes.

The in situ resistivity of the massive sulphides in W276 was found to be less than 0.1 ohm m while that of the host rocks was as much as several thousand ohm m. Resistivity contrast varied between 1 to 1500 and 1 to 100,000 and an IP contrast of at least 5 to 1 was evident between the orebody and the hanging wall rocks. These contrasts were consistent with those obtained in laboratory measurements on drill core.

Resistivity measurements in W57 at multiple electrode spacings confirm the efficacy of the technique and suggest that the 4.5m width of massive sulphides intersected at about 185m in the hole extends to well in excess of 50m from the hole.

Introduction

Large amounts of surface electrical and electromagnetic geophysical data have been collected over the Woodlawn deposit. Refined interpretation of these data especially by quantitative computer modelling techniques cannot proceed without realistic estimates of electrical contrast between ore and host rocks and between various rock types within the host sequence.

Limited resistivity and induced polarization (IP) measurements on Woodlawn drill core (Tyne, Idnurm and Malone, this volume, p 99), have provided estimates of the range in resistivity and IP for the orebody and some of the host rocks. Emerson (1969) has discussed the limitations for laboratory electrical resistivity testing of core samples and asserts that laboratory measurements are not a substitute for in situ experimental investigations on the rare occasions when sufficient geological control exists for these to be made. At Woodlawn excellent geological control was available and so in situ electrical rock property measurements by drill hole logging were considered essential to the interpretation of many of the geophysical surveys. The logging primarily aimed at quantifying the in situ resistivity and chargeability of the massive sulphide orebody and the various host rocks.

A secondary aim was to determine if logging using both the two and three array with multiple electrode spacing provides useful information on the geometry and persistence of the sulphide intersections away from the drill holes.

Only holes W276 and W57 were open to sufficient depths at the time of the project. W276 which intersects the main massive orebody, had collapsed on the immediate footwall side of the ore and so only the hanging wall rocks and massive ore were electrically logged. W57 intersected the northwestern extension of the orebody about 600m northwest of W276. Electrical logging extended to maximum logging cable length of 345m in W57 and included only the first of the three massive sulphide intersections in that hole.

Logging Technique

The Drill Hole Cable

A multiple electrode spacing logging system developed by the Geological Survey of N.S.W. was employed in W276 and W57 using similar techniques to those of Wagg and Seigel (1963). The logging system and the electrode arrays and spacings are shown in Fig. 1.

A special, seven conductor, shielded cable was used to carry current and potential electrodes into each drill hole. The entire cable system was designed to prevent intrusion of drill hole fluids. Current was transmitted down the cable to an end electrode. An effectively infinite current sink was placed on the surface some distance from the hole collar (at 9543.57mE, 7589.80mN). Potentials in the hole were measured using six lead and lead oxide electrodes located on the cable at various spacings. These electrodes were covered with cellulose sponge to protect them from damage in the hole and to ensure positive contact with the borehole walls in either wet or dry holes. The smallest potential to current spacing (near hole spacing) used was 1.5m. This spacing increased logarithmically to the maximum of 48m (distant spacing). The near hole spacing was used to obtain an electrical properties log of the rocks in the immediate surrounds of the drill hole. The larger spacings were used to investigate progressively farther away from the drill hole.

The cable system employed allows use of the two or normal electrode array at six spacings or the equispaced three electrode array at five spacings. A surface potential electrode for use with the two array measurements, was

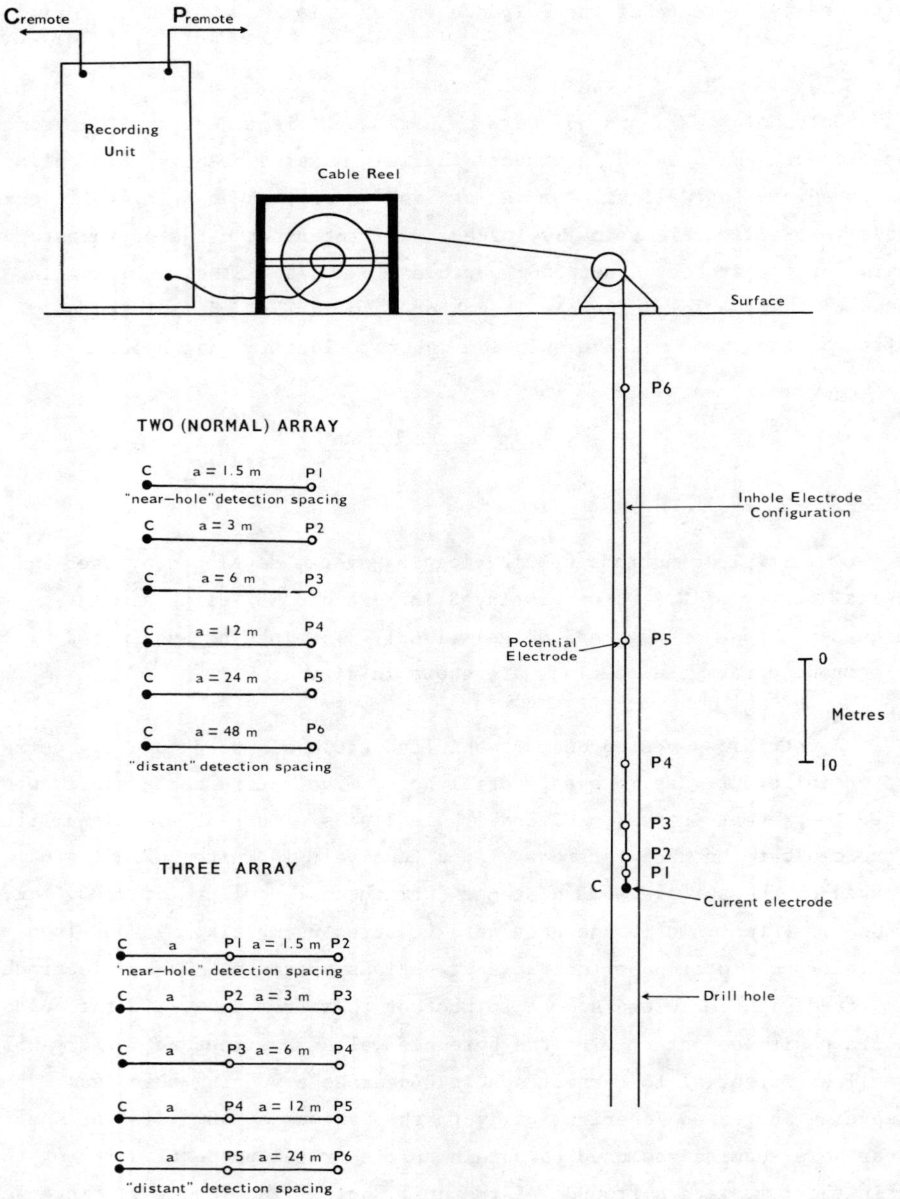

Fig. 1 Diagrammatic Representation of Detection Logging System

placed at 8610.83mE, 9497.76mN, effectively at an infinite distance away from the holes.

Electrode Arrays

The three array is commonly used for logging because it is more sensitive to lateral rock boundaries traversed by the drill hole than the two electrode array (Whiteley,1973, p.9) and provides more reliable estimates of true resistivity for conductive units. Accurate measurement of very low resistivities with the two array is also dependent on the positioning of the remote potential electrode. The three array is less susceptible to capacitive and inductive electromagnetic (EM) coupling in the drill hole and produces much larger resistivity and IP anomalies for a given source (Seigel, 1959). However, the two array is more advantageous because it produces simple symmetrical anomalies which are sometimes easier to interpret in logs of a complex geological sequence. This is in comparison with the asymmetrical double peaked anomalies, characteristic of the three array (Seigel, 1959).

The potential electrode spacings on the logging cable used are such that if only normal logs are measured in the drill hole then three array logs may be computed from two array logs in the following manner:

$$\rho_3 = 2.\rho_a - \rho_{2a} \qquad (1)$$

where a = electrode spacing, 1.5, 3, 6, 12 and 24m.

ρ_3 = three array apparent resistivity at spacing a.

ρ_a = two array apparent resistivity at spacing a.

ρ_{2a} = two array apparent resistivity at spacing 2a.

If both two and three array logs are obtained then this constitutes a check on measurement errors and EM coupling. The above formula was routinely used to check for measurement errors in the two and three array logs from W276 and W57.

Recording and Instrumentation

Unlike conventional well logging practice, the multiple spacing logs are not recorded in a continuous fashion down the hole. The logging cable was lowered (or raised) in equal station intervals of 3 to 6m, or less where higher resolution was required. Resistivity and chargeability were measured at each station for each of the electrode spacings.

Both two and three array resistivity logs were run in W276 and W57 at all the electrode spacings available. Current was supplied by a Geoscience Model 5170 square wave transmitter at a frequency of 0.1Hz and resistivity readings were taken using a frequency domain Heinrichs Mark III receiver.

IP logs were subsequently obtained from W276 using a 6m and a 24m three array. An 8 second time domain current wave form was supplied by a 2.5kW Huntec transmitter and a Huntec Mark III receiver was used to record resistivity and IP measurements. The IP parameter, chargeability, was recorded over the time interval 0.48s to 1.53s after current switch off. The long delay time of 0.48s was chosen to reduce capacitive and inductive interference to a minimum.

Geological and Petrophysical Features Affecting in situ Electrical Measurements

Rhyolitic volcanics, tuff, interbedded shales and metadolerite sills were intersected by W276 and occur extensively in the hanging wall sequence. Similar rock types were intersected by W57 to the northwest. The east to west geological section through W276 shows the extensive tabular nature of the dolerite sills and the orebody (Fig.2).

The massive sulphides tested in both holes consist of about 75% sulphides, mainly pyrite, sphalerite, galena and chalcopyrite in order of decreasing abundance (Malone, this volume, p 49). Pyrite, galena and chalcopyrite are all good conductors with a resistivity usually much less than 0.5 ohm m. However sphalerite is only an intermediate conductor which may have a resistivity varying from about 0.003 to 12,000 ohm m (Keller and Frischknecht, 1966, p.5). Because the massive sulphide orebody contains a high percentage of mixed sulphides it was expected to act as a good conductor.

The massive nature of the sulphides compared with the unmineralized hanging wall rocks and pyritic footwall rocks suggested that there would be a strong resistivity and chargeability contrast between the orebody and the host rocks. Also variations in silicification and chloritization of different rock types suggested distinctive changes in electrical rock properties with changes in rock type throughout the host sequence could be expected.

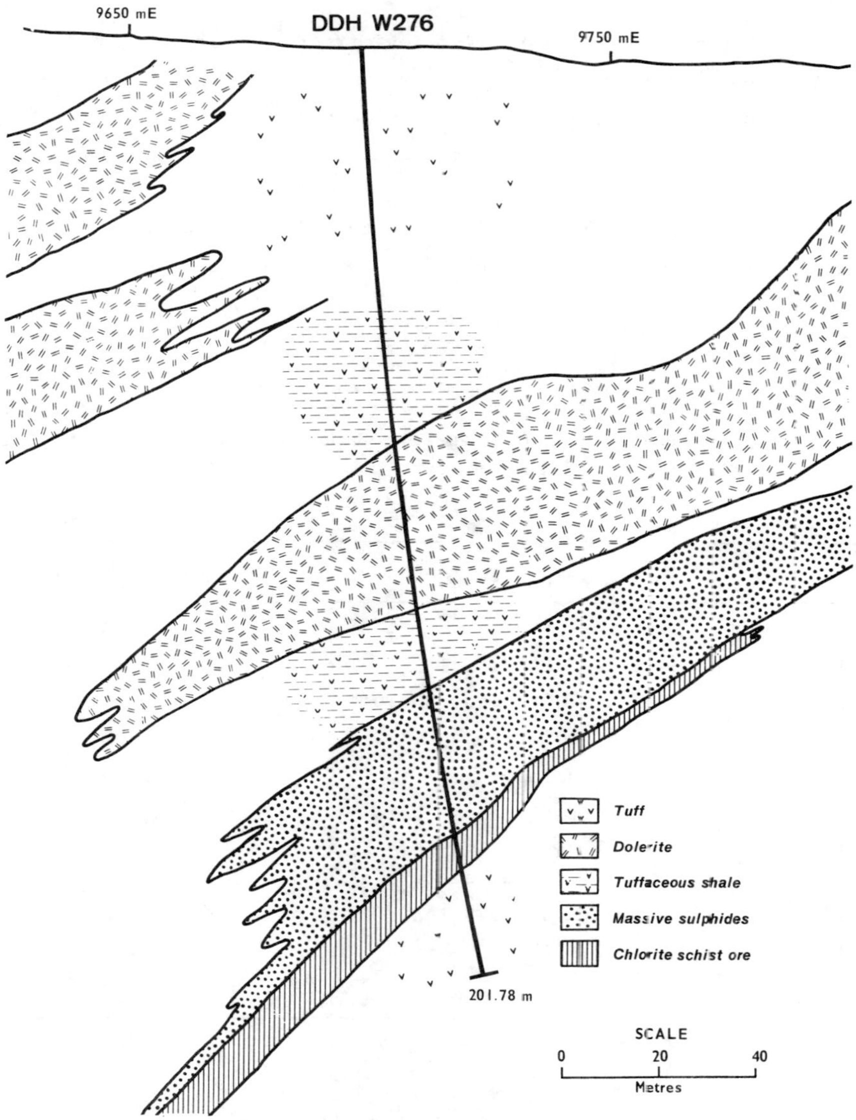

Fig. 2 Simplified East West Section at 9160 mN, showing W276

Electrical Properties from Laboratory Testing of Drill Core Samples

Table 1 summarizes the resistivities of core samples from W276 measured by the Geological Survey of New South Wales. These are reported by Tyne, Idnurm and Malone (this volume p 99) and are presented as a reference for comparison with the *in situ* resistivity and IP measurements in W276 and W57.

The data in Table 1 clearly indicate the highly conductive nature of the massive sulphides as well as the large resistivity contrast between the orebody and host rocks.

Table 1 Resistivity of unevacuated Core Samples from W276 (after Tyne, Idnurm and Malone, this volume, p 99).

Depth (m)	Lithology	Resistivity (ohm m)
27.4	Tuff	279
29.3	Tuff	178
56.7	Tuffaceous Shale	677
93.5	Dolerite	7,438
114.0	Dolerite	7,430
134.9	Tuffaceous Shale	110,000
135.9	Massive Sulphide with intercalated Tuff	0.1
145.8	Massive Sulphide	0.01
179.0	Tuff	28,779
191.0	Tuff	43,188

Resistivity and IP Logs from W276

The most striking feature of both sets of resistivity logs (Figs.3,4) is the large range of resistivities measured in the hole. This is over three decades on the two array logs and five decades on the three array logs. The resistivity contrasts between the dominant rock units intersected by the hole are very considerable.

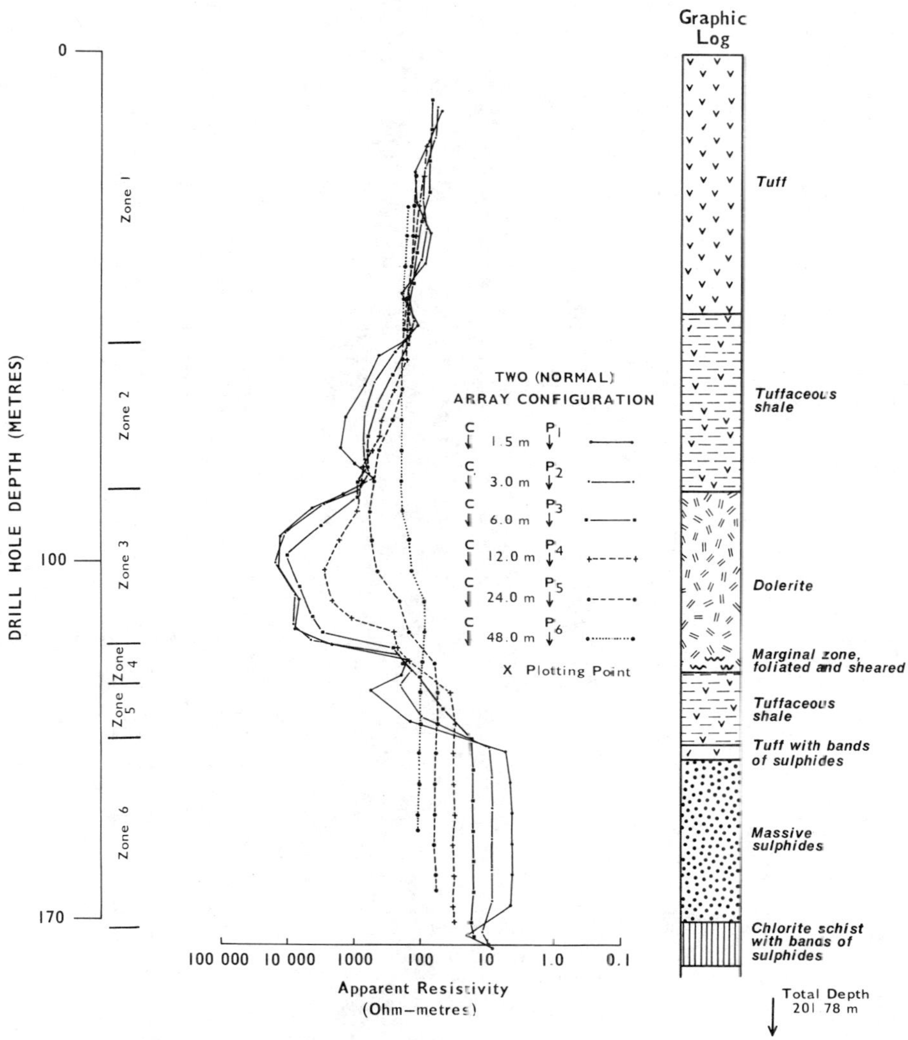

Fig. 3 Apparent Resistivity Logs , Two Array Woodlawn W276

The pronounced apparent resistivity anomalies arising from the dolerite sill and the orebody decrease in amplitude with increasing spacing. The rate of decrease for each anomaly is consistent with that expected for a tabular unit extending well away from the drill hole as illustrated by the geological section through W276 (Fig.2).

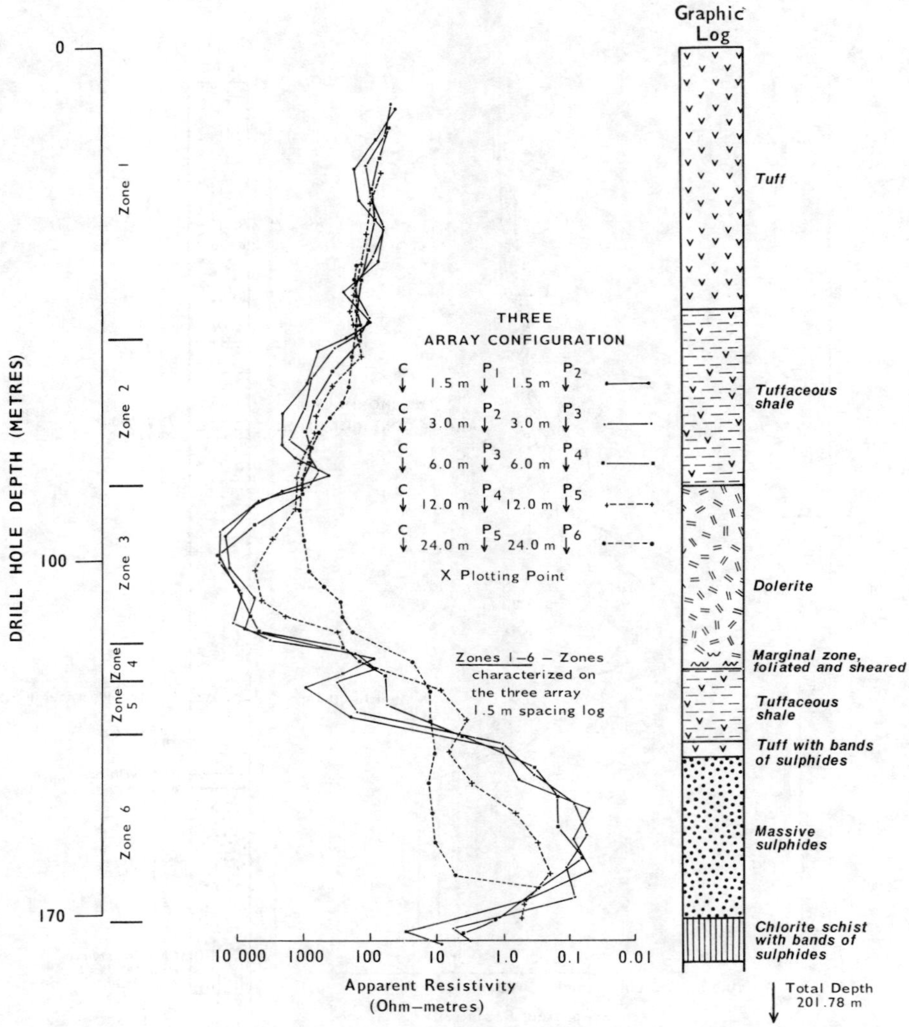

Fig. 4 Apparent Resistivity Logs , Three Array Woodlawn W276

Comparison of Two and Three Array Logs

The two and three array logs (Figs. 3 and 4) are very similar in shape for resistivities greater than 100 ohm m. However, the three array logs show larger amplitude variations than the two array logs reflecting the greater sensitivity of the three array to lateral resistivity variations.

Comparison of both sets of logs at resistivities below 100 ohm m and particularly within the massive sulphides, reveals a dramatic difference in anomaly shape and amplitude. The three array logs exhibit resistivities at least 100 times smaller than those on the two array logs. The same absolute voltage measurement was recorded for each two array spacing from 1.5m to 48m within the ore thus indicating the extremely conductive nature of the sulphide lens. The increase in apparent resistivity with increasing spacing in the ore zone as shown on Fig.3 simply reflects the fact that the geometric array factor is proportional to the spacing. It appears that the minimum absolute voltage recorded with the two array is highly dependent on the remote potential pole location.

The comparison of the two and three array logs indicates that the two array is suitable for defining the boundaries and resistivities of the major resistive rock types in W276 but is not capable of portraying the true resistivity of very conductive units such as the massive sulphides.

Physical Property Logs

The 1.5m logs for both arrays can be used to define zones with a distinctive resistivity level. Six zones have been defined and these are shown with the 1.5m logs on Fig. 5.

Zone 1 is characterized by resistivities of the order of 100 ohm m and correlates with mainly acid tuffs. A gradual increase in apparent resistivity with depth probably reflects the decrease in weathering, and alteration with depth. The top of the tuffaceous shale has an apparent resistivity comparable to the tuff because it is very porous at this point.

Zone 2 has an average apparent resistivity about ten times that of zone 1 and corresponds to a tuffaceous shale. The sharp drop in apparent resistivity just above the lower boundary of zone 2, appears to be related to minor pyrite mineralization.

The coarse grained dolerite sill, which is the most resistive host rock, correlates with zone 3. Resistivities of the order of 10,000 ohm m characterize this zone.

The resistivity zones 4 and 5 are narrow and correspond to the basal contact of the dolerite sill and to tuffaceous shale respectively. A

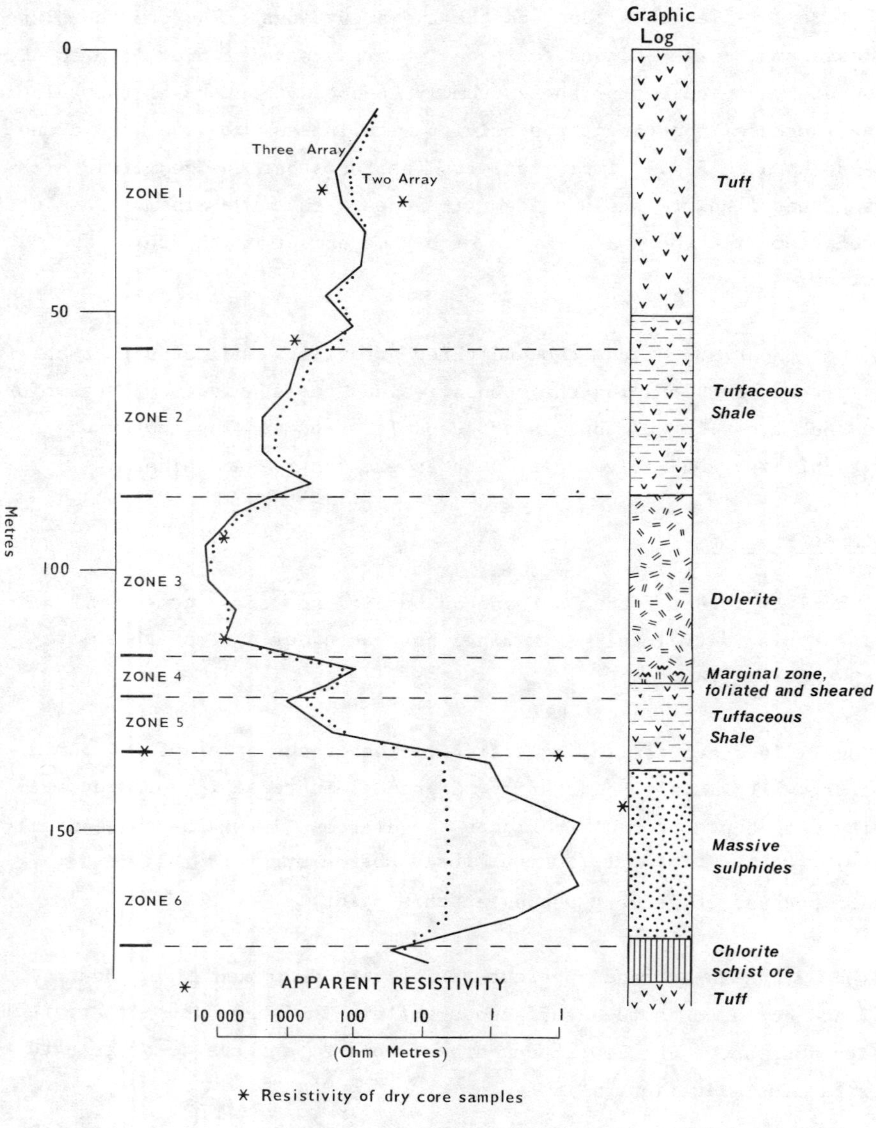

Fig. 5 Physical Property Logs W276

decrease in resistivity in zone 4 is probably caused by foliation and shearing along this contact. Zone 5 shows similar levels of resistivity recorded in the tuffaceous shale of zone 2 above.

Resistivities as low as 0.1 ohm m correlate with the 30m wide massive orebody and define zone 6.

It is in the widest zones (1, 2, 3 and 6) that true resistivities may be calculated from the logs.

Resistivities of the unevacuated core samples are plotted on Fig. 5. There is a satisfactory correlation between the 1.5m resistivity log and the laboratory determined resistivities, except at the lithological boundary zones, for example at 135m. Although this correlation cannot be taken as quantitative, it suggests that well controlled core measurements may satisfactorily represent the true *in situ* electrical properties of the rocks.

Calculation of True Resistivities

Lynch (1962, p.121) has shown that the apparent resistivity indicated by a two array system reaches 90% of the true resistivity of a resistive bed if the two electrode spacing is about 1/10 that of the bed thickness. The electrode spacing to bed thickness ratio required for determining true resistivities is greater for conductive beds.

The four major units shown on the graphic log all have thicknesses in excess of 30m. The smallest two and three array spacing were 1.5m which is less than 1/20 of the minimum bed thickness. Data obtained at these spacings should accurately determine the true resistivity of these four units after correction for extraneous influences of the drill hole and drill hole fluids.

To obtain the true *in situ* resistivity of the rocks surrounding a drill hole the masking effect due to the resistivity contrasts between the drill hole fluid and the wall rocks must be evaluated. Resistivity departure curves for this purpose, such as those presented by Lynch (1962, p.123) and Brant *et al* (1966) for the two array, have been in common use in oil well logging for some time. Departure curves for the equispaced three array do not appear in the literature, however, curves for the normal (two array) device may be used to compute those for the three array.

The resistivity log obtained from an equispaced three array, with an electrode spacing a, can be computed from two normal logs at spacings a and 2a, using equation (1). This equation was used to compute the three array drill hole corrections at Woodlawn from the normal departure curves. For accurate evaluation of the masking effect it is necessary to know the resistivity of the drill hole fluid and the effective drill hole diameter. A resistivity for this fluid of 6.45 ohm m was determined from laboratoratory measurements.

The drill hole diameter of W276 was 0.1m. The effective drill hole diameter or the diameter of the wall rock invaded by drilling fluid required for correction was then determined by analysing a section of constant response on the two array logs, using the procedure described Brant et al (1966). There appeared to be almost no invasion of the wall rock in W276 and therefore the effective hole diameter was approximately 0.1m.

The two array departure curves show that for a hole of 0.1m and spacing of 1.5m, measured apparent resistivities which are less than that of the hole fluid (6.45 ohm m) are equal to the true resistivity of the wall rock. For apparent resistivities greater than that of the hole fluid, a correction must be made to the logs to obtain true resistivity.

The true resistivity of the massive sulphides was obtained directly from the three array log. The two and three array logs were used in conjunction with the normal departure curves to determine true resistivity of the resistive units, tuff, tuffaceous shale and dolerite.

Table 2 lists the corrected resistivities of each of the dominant rock types in W276. Corrected resistivities for both arrays agree to within 10%. The extremely low resistivity of the sulphides is representative of the very massive material in the centre of the orebody and therefore may not be indicative of the average resistivity of the total orebody. The resistivity contrast between sulphides and host rocks varies from 1 to 1500 to about 1 to 100,000. The massive sulphides are effectively an infinitely conductive medium in comparison with the host rocks. These contrasts are consistent with those obtained in laboratory core measurements (Table 1).

IP Logs

Fig. 6 shows IP logs at the 6m and 24m three array spacings. IP measurements were difficult to make in the vicinity of the very conductive massive sulphides because the very small absolute primary and secondary voltage levels are swamped by large inductive coupling voltages. Although

Table 2 True Resistivity of Dominant Rock Types in W276

Dominant Rock Type (Zone)	Average Peak Resistivity (ohm m)		True (Corrected) Resistivity (ohm m)	
	Two Array	Three Array	Two Array	Three Array
Acid Tuff (Zone 1)	95	115	90	100
Tuffaceous Shale (Zone 2)	1,550	2,100	975	1,000
Dolerite (Zone 3)	10,415	10,495	6,125	6,770
Massive Sulphides (Zone 6)	-	0.06	-	0.06

Drill hole diameter = 0.1m Drill hole fluid resistivity = 6.45 ohm m

coupling has probably decayed away to a low level at the start of the chargeability recording interval (0.48s), the large absolute amplitude of the coupling voltage at current switch off causes synchronization failure between the transmitter and recording instruments.

The 6m spacingwas too large for determination of the true chargeability of each rock type but the 6m log does provide some measure of the relative difference in chargeability for each unit.

Analysis of the 6m log shows that chargeability is constant at about 10ms within the tuff. Chargeability increases gradually through the dolerite to a maximum of 30ms at which point it increases sharply to about 60ms at the upper polarizable boundary of the orebody.

The logs suggest that there is a significant chargeability contrast of at least 5 to 1 between the massive sulphide boundary zone and the unmineralized hanging wall rocks, therefore rendering the massive sulphide body a good target for surface IP, as well as resistivity profiling.

Resistivity Logs from W57

These logs also show as much as five orders of magnitude variation in measured resistivities.

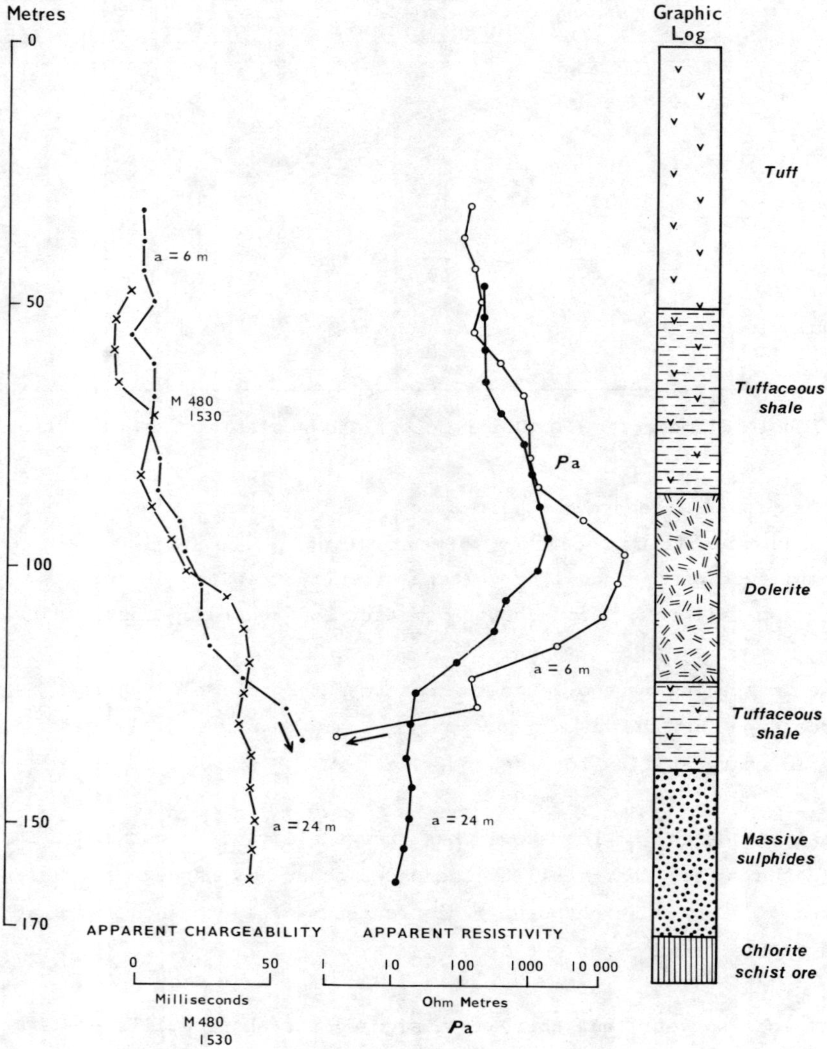

Fig. 6 Induced Polarization and Resistivity Logs Woodlawn W276

Comparison of the two and three array logs on Figs. 7 and 8 shows close similarity in amplitude and shape for resistivities above 100 ohm m. At lower resistivity levels the three array logs show the larger deflections, as observed in W276.

Physical Property Logs

The graphic log of W57 (Fig. 7) shows rock types similar to those present in W276, but in W57 they occur in thinner units which are repeated throughout the hole. The 1.5m logs for the two and three array (Fig. 9) are used to define three major zones with distinctive levels of apparent resistivity.

Zone 1 is characterized by a relatively constant resistivity base level at about 1,000 ohm m. The only major departure from this average occurs opposite a rhyolite unit which is almost 10 times more resistive than the surrounding rocks. The rock types in zone 1 are predominantly tuff and tuffaceous shale.

Zone 2 is defined as a very conductive region. The logs in Fig. 9 show two anomalous peaks within this zone. The larger deflection with a minimum resistivity of 0.25 ohm m correlates with a 4.5m intersection of massive sulphides at 185m hole depth. The smaller peak in zone 2 correlates with some very narrow massive sulphide intersections at the base of the thickest sulphide unit. The electrode spacing of 1.5m is not small enough to determine true resistivity of the sulphides. The tuffaceous shale unit immediately underlying the massive sulphides contains various amounts of sulphides.

The underlying rock types, mainly dolerite, coincide with zone 3. The zone is characterized by apparent resistivities which fluctuate between 1,000 and 10,000 ohm m, with an average apparent resistivity of about 5,000 to 6,000 ohm m. Variations in apparent resistivity within the dolerites in zone 3 indicate that they are not electrically homogeneous, and therefore, probably not petrologically homogeneous.

The true resistivity for each rock type, including the massive sulphides, cannot be calculated because the occurrences of each rock type are too thin. However, the relative resistivity contrast between each of the major rock types may still be obtained from the logs.

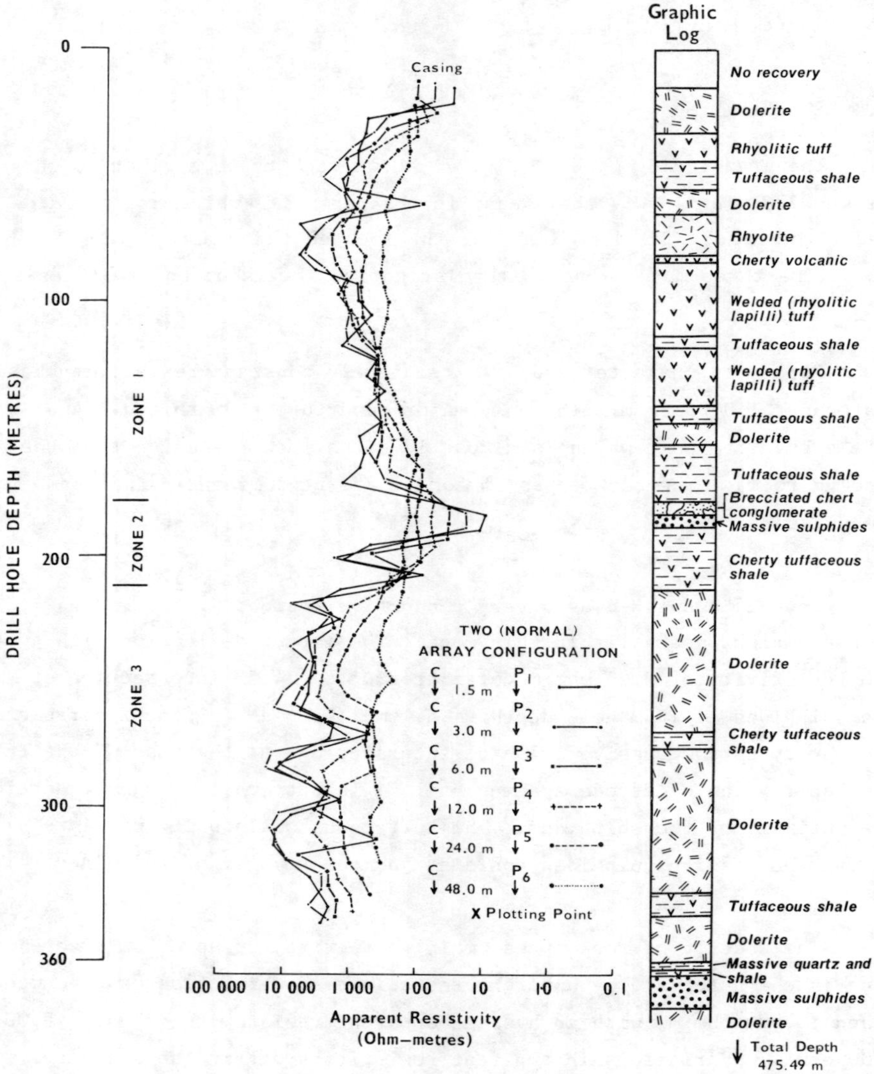

Fig. 7 Apparent Resistivity Logs, Two Array Woodlawn W57

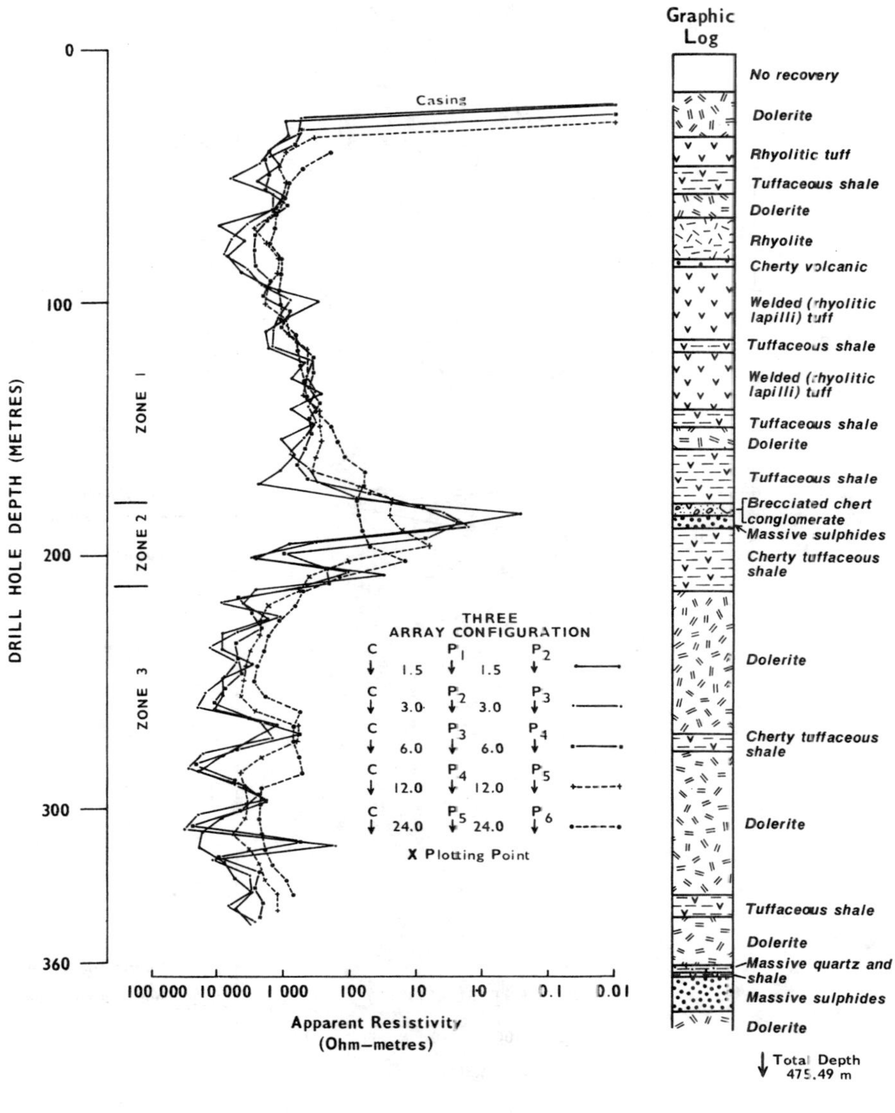

Fig. 8 Apparent Resistivity Logs, Three Array Woodlawn W57

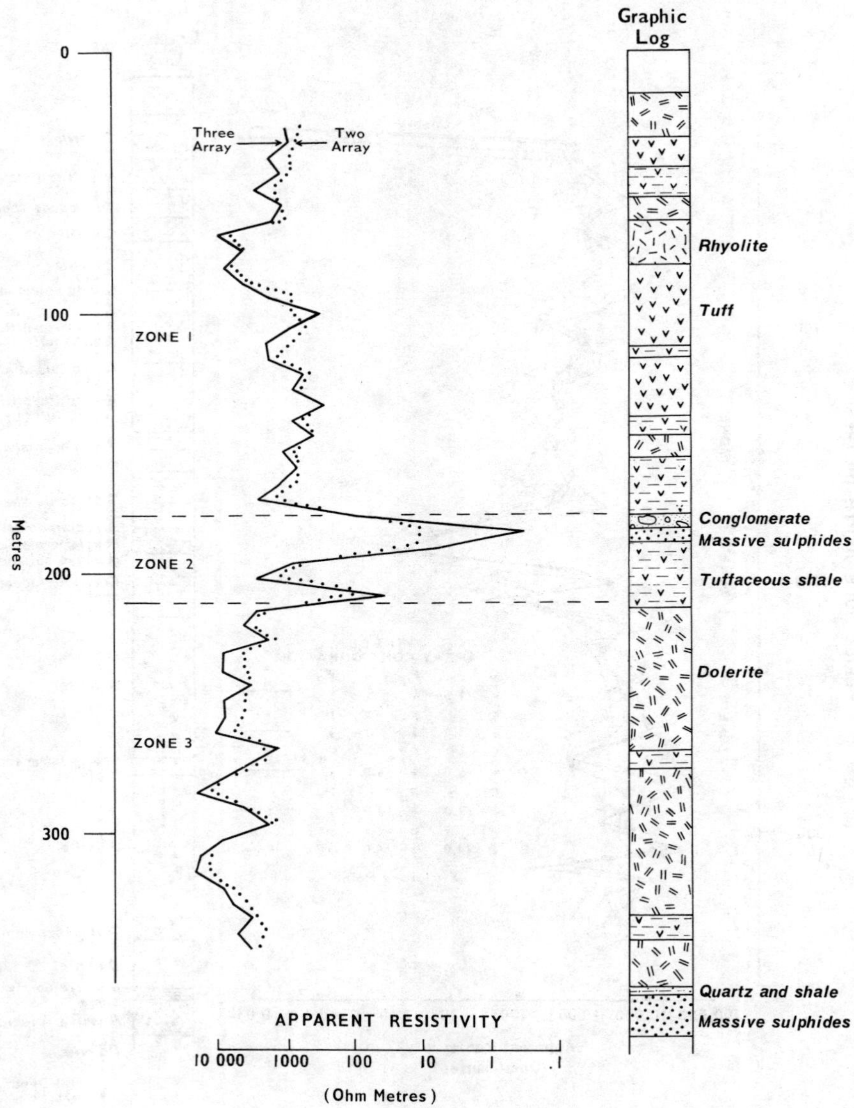

Fig. 9 Physical Property Logs W57

Table 3 lists the average observed resistivities and the average resistivities corrected for the drill hole effect. These values must be considered as bulk resistivities for each zone rather than true resistivities for the individual rock types which dominate each zone.

Table 3 Average Resistivity of Dominant Rock Types in W57

Dominant Rock Type (Zone)	Average (Observed) Resistivity (ohm m)		Average (Corrected) Resistivity (ohm m)	
	Two Array	Three Array	Two Array	Three Array
Acid Tuff and Tuffaceous Shale (Zone 1)	750	1,100	540	590
Massive Sulphide (Zone 2)	-	0.25	-	0.25
Dolerite (Zone 3)	4,920	6,770	2,800	3,600

Drill hole fluid resistivity = 8.13 ohm m. Drill hole diameter = 0.1m.

For this reason the bulk resistivities listed in Table 3 do not compare in absolute amplitude with the resistivities obtained in W276 (Table 2). However, it is noteworthy that the resistivity contrast between each of the major rock types is consistent for both drill holes. The resistivity contrast between tuff, tuffaceous shale and dolerite is about 1 to 6 in each hole. The minimum resistivity contrast between the massive sulphide intersection and the surrounding rocks in W57 is 1 to 2000 which is consistent with the contrast measured in W276.

Analysis of the logs from W57 indicate that the rocks intersected by this hole are very similar in electrical properties to some types of rocks intersected by W276 in the main ore zone.

Persistence of a Massive Sulphide Intersection Away from W57

A salient feature of the two and three array logs (Fig. 7 and 8) is the resistivity low anomaly arising from the 4.5m massive sulphide intersection at 185m hole depth. An anomaly is evident on each electrode spacing log. The symmetrical shape of the two array anomalies and the rate of increase in anomaly width with electrode spacing suggest that the massive sulphides intersect the drill hole approximately at right angles and that the sulphide lens does not thicken rapidly away from the drill hole.

The multiple electrode spacing logs in W57 allows an estimate to be made of the persistence of the sulphide intersection away from the hole. A scheme for determining orebody size by drill hole measurements has been proposed by Seigel (1952) using an oblate spheroid to represent an orebody intersected by a drill hole. Seigel presents resistivity departure curves using a large resistivity contrast of 1 to 200 for different eccentricities of the spheroid and electrode array spacings. Fig. 10 shows these curves for the normal array centred at a drill hole intersection of thickness d.

The 4.5m sulphide intersection in W57 may be considered to reasonably represent a conductive disc which is normal to the drill hole. The two array apparent resistivity for the 6, 12 and 24m electrode spacing logs recorded at 185m hole depth opposite the massive sulphide intersection has been normalized with a background resistivity of 400 ohm m, and plotted on Fig. 10. The log data closely match the model curves for an eccentricity of E = 100, yielding a 'b' value of less than 450m. The match implies that the oblate spheroid is a reasonably valid model for the lens intersected in W57. Seigel indicates that estimates using these departure curves may be in error by a factor of 2. However, the logging results clearly indicate that the 4.5m intersection of massive sulphides extends away from the hole, to a distance probably in excess of the largest electrode spacing of 48m.

Summary and Conclusions

Resistivity and IP logging in the drill holes at Woodlawn with a multiple electrode system enabled the electrical properties for the rock units intersected by the hole to be determined. A knowledge of these properties allows a realistic and more complete interpretation of surface electrical data. Secondly, the logging enabled the persistence of conductive rock units away from the drill hole to be estimated from the change in anomalous response on logs with increasing electrode spacings.

At Woodlawn, logging results show that the three electrode array is better suited to physical property logging than the two array because it produces a larger amplitude response and allows more realistic estimates to

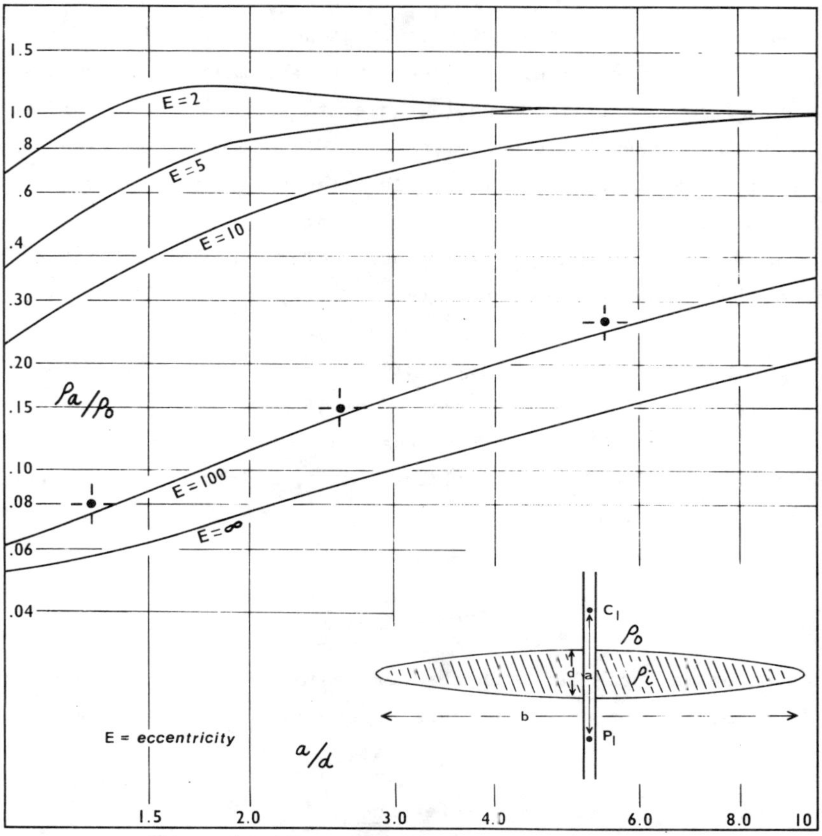

Oblate spheroid resistivity departure curves, normal array

(after Seigel 1952, p. 913)

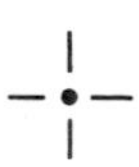

Fig. 10 Normal Array Departure Curves for a Conductive Body Intersected by a Drill Hole.

be made of true resistivity of conductive units. The two array fails in this last instance partly because of its insensitivity to very low resistivities and because it is affected by the positioning of the remote electrodes.

Zones of distinct resistivity contrast were defined by the logging in W276 and W57. These zones correlate closely with the major rock types intersected i.e., tuff, tuffaceous shale, dolerite and massive sulphides. In situ true resistivities range from less than 0.1 ohm m for massive sulphides to several thousand ohm m for dolerite sills. There is a minimum resistivity contrast of 1 to 1500 and an IP chargeability contrast of at least 5 to 1 between the orebody and the hanging wall rocks. These contrasts confirm those obtained from laboratory core measurements.

Analysis of multiple spacing resistivity logs from W57 suggests that the narrow massive sulphide lens intersected at 185m extends beyond about 50m from the hole. These results illustrate that this logging technique can be used effectively in a single hole to estimate the persistence of very conductive sulphide lenses away from a drill hole.

Acknowledgements

This paper is published with the permission of the Under Secretary, New South Wales Department of Mineral Resources and Development, Sydney.

References

Brant A.A. and the Newmont Exploration Staff. Examples of Induced Polarization Field Results in the Time Domain, Mining Geophysics, Vol.1 Society of Exploration Geophysicists, 1966.

Keller, G.V. and Frischknecht, F.C., 1966, Electrical Methods in Geophysical Prospecting. 519pp. 1st. edn. Pergammon Press Inc., Oxford.

Lynch, E.J., 1962, Formation Evaluation. 422pp. 1st edn. Harper and Row, Publishers Inc., New York

Malone, E.J. et al., 1975, Woodlawn Copper-Lead-Zinc Orebody. Economic Geology of Australia and Papua New Guinea, 1. Metals, AIMM Monograph Series, No.5.

Seigel, H.O., 1952. Orebody Size Determination in Electrical Prospecting. Geophysics, V.17, 907 to 914.

Seigel, H.O., 1959. Mathematical Formulation and Type Curves for Induced Polarization. Geophysics, V.24, 547 to 565.

Wagg, D.M. and Seigel, H.O., 1963, Induced Polarization in Drill Holes. Can. Min. J., V. 84, 54 to 59

Whiteley, R.J., 1973, Electrode Arrays in Resistivity and IP Prospecting: A Review. Bull. Aust. Soc. Explor. Geophys., V. 4, 1 to 29.

11.4. *Downhole Electromagnetic Surveys at Woodlawn*

R. D. OGILVY

Contents

Abstract

Field tests were made early in 1975 over the Woodlawn orebody using a Scintrex DHP4, fixed source, electromagnetic (EM) borehole prospecting system. The tests were designed to evaluate the detection capabilities of the DHP4 equipment and to develop guidelines for its use in mineral exploration.

The field trials indicate that detectability depends upon the mutual EM coupling relationships between the transmitter, conductor and downhole receiver, and varies significantly for differing geometrical situations. For optimum EM coupling the detection radius was shown to be of the order of 30m for the Woodlawn orebody. Owing to the limited nature of the trials the maximum operating depth could not be determined. However results from drill holes that intersected ore show that the operating depth may be severely restricted by the screening effect of shallower ore zones.

Theoretical considerations indicate that a free space approximation cannot be used to normalize downhole amplitude data. Free space calculations can however assist in survey design and data interpretation.

Although a response from the Woodlawn orebody was obtained, the tests showed that the DHP4 has only limited applications in its present form. However, several factors favour the continued development of fixed source, downhole methods. Inherent advantages include the insensitivity of the method to small conductance targets, and the possibility of directional information. Detectability could be improved by measuring orthogonal components of the subsurface magnetic field.

Introduction

During the period January to March 1975, the Bureau of Mineral Resources (BMR) conducted downhole electromagnetic (EM) surveys with the Scintrex DHP4 equipment in six exploration drill holes at Woodlawn. The surveys were made to investigate the ability of the DHP4 equipment to detect massive sulphide mineralization away from the drill hole, and to develop guidelines for designing and interpreting downhole EM surveys.

Geophysical Background

Suitability of Woodlawn as a test site

The Woodlawn orebody is a very conductive EM target (Hone and Young, this volume, p 519) with excellent geological and geophysical control being provided by exploration drilling and test geophysical surveys. The holes available for surveying included several which intersected the orebody, one which passed 30m below the orebody, and one remote from the orebody that did not intersect any significant mineralization. In combination these holes provided a good site at which to carry out tests with the DHP4 equipment. The barren drill hole could be used to determine the normal field and the magnitude of geologic noise. The near miss and intersection holes could be used to estimate the search radius of the DHP4 equipment and the response of the orebody.

Description of the DHP4 equipment

Principle of operation The principle of operation of the DHP4 equipment is based on the fact that if a conductor is subjected to an alternating primary electromagnetic field, eddy currents are induced in the conductor and produce a secondary electromagnetic field. The DHP4 equipment employs a large surface transmitting loop to establish the primary field and a downhole receiving coil to detect the primary and secondary fields in the hole. The amplitude and phase of the magnetic field component coaxial with the drill hole are measured with respect to the amplitude and phase of the signal in a fixed reference coil at the drill hole collar. The presence of a conductor near the drill hole is indicated by anomalous phase readings and departures of the total field from the normal primary field. With the DHP4 unit, measurements are made for a fixed frequency of 2000Hz. A schematic representation of a drill hole EM survey using the DHP4 equipment is shown in Fig.1.

Instrumentation The DHP4 equipment consists of three standard units. The transmitting unit comprises a 1kW motor generator, a current regulator, and a large single turn insulated loop. The reference and downhole coils are each constructed from 2000 turns of insulated copper wire wound on silicon-iron strips. The outside diameter of each coil is 19mm (0.75ins).

The measuring unit comprises a cable drum, 300m of cable, and an amplifier unit with appropriate gain and phase shifting networks. The signal

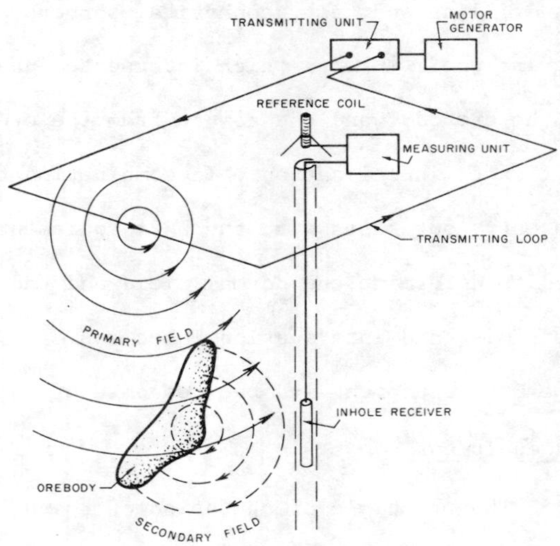

Figure 1. Schematic representation of DHP4 survey.

from the receiving coil is compared with that in the reference coil to give relative signal strength in millivolts (mv) and phase difference in degrees.

Data presentation

Ideally a system of plotting which shows the departure of the total field from the normal primary field is required. Such a method would be similar to the reduced ratio presentation used in Turam interpretation (Haren, this volume, p 219). However an attempt to normalize the downhole amplitude using the free space approximation was in most cases unsuccessful, and discrepancies between the observed and calculated primary amplitudes were noted. The failure of the free space approximation is attributed to scattering of the primary field in the earth.

However, theoretical free space calculations of the primary field do provide a means of differentiating between amplitude deviations caused by geometrical effects (e.g. drill hole deviations and field reversals), and deviations caused by subsurface zones of anomalous conductivity. Hence theoretical amplitude curves are presented with the field results to assist the interpretation of the data. In all cases, the axial component of the free space primary field has been used to permit direct comparison with the observed amplitude data. The derivation of the formula used in calculating the free space axial field due to a square current loop is based on the magnetic field formula for a finite line source of current.

The DHP4 measures amplitude in millivolts, normalized with respect to the reference signal. The amplitude measurements have been expressed as a percentage of the appropriate free space primary field strength at the top of the drill hole. Phase angle measurements are given in degrees.

Directional information

The response of a downhole EM system to a conductor depends on the coupling between the conductor, receiver and transmitter. With the DHP4 system, the coupling can easily be altered by changing the size or location of the transmitting loop. Directional information may be obtained by running surveys with the loops in each quadrant surrounding the drill hole. The quadrant in which the maximum response is recorded indicates the direction to the conductor.

Field reversals

If a drill hole passes beneath the circumference of the surface transmitting loop, the direction of the axial component of the primary field will undergo a 180° change at some point, and the amplitude curve will pass through a minimum. In interpreting the results of DHP4 surveys, it is important to separate these geometric effects from anomalies caused by conductive bodies.

Survey Details

Six drill holes were surveyed with the DHP4 equipment at Woodlawn. W55 was a barren hole remote from the orebody, W29 was a near miss hole at the northern end of the orebody and WE5, W251, W254, and W276 intersected the orebody. The location of the drill holes is shown in Fig.2. W29, W254

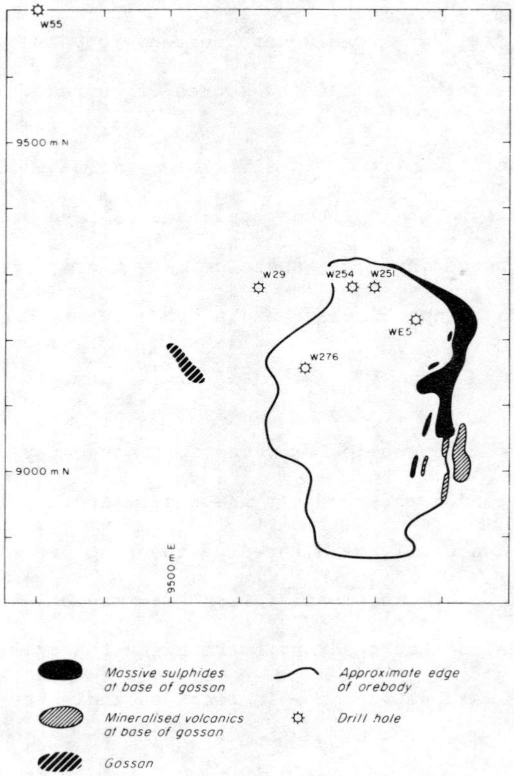

Figure 2. Location of drill holes.

and W251 were drilled along traverse 9280mN (Line H), and the location of these holes relative to the orebody is indicated in the cross section shown in Fig. 3.

Transmitting loops were square, and had side lengths of 100m to 400m. Some holes were surveyed with a variety of loop sizes and locations (Table 1) in an attempt to increase the response of the orebody by changing the coupling characteristics of the survey. The sides of the loops were oriented parallel to the Woodlawn Test Grid.

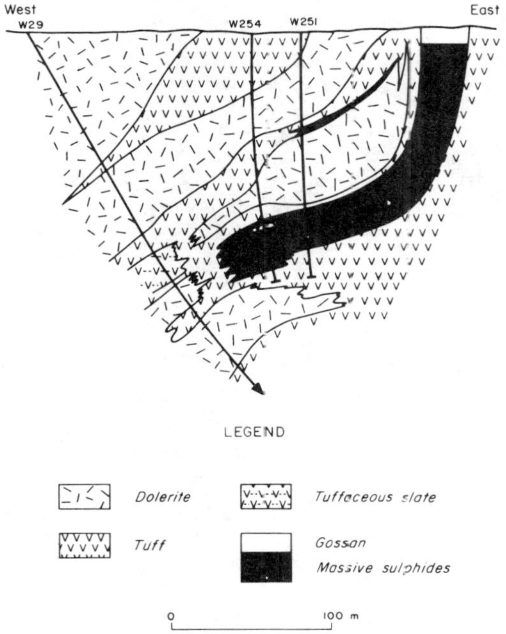

Figure 3. Cross-section, traverse 9280mN. (Line H)

Results

Drill Hole W55

W55 is remote from the Woodlawn orebody and was surveyed with a 400 by 400m loop centred on the hole collar. The results of the downhole EM survey and the geological log of the hole are shown in Fig.4.

Downhole resistivity logging of this hole (Hone and Young, this volume, p 519), showed that the dolerite and tuff have resistivities in the range 100 to 10,000 ohm m. However, both the amplitude and the phase curves are smooth and no anomalies are evident throughout this region of variable resistivity. The amplitude curve decreases monotonically down the hole and is similar to the calculated free space amplitude curve. The phase curve shows a gradual increase with depth.

The smooth amplitude and phase curves indicate that the DHP4 system is insensitive to variations in conductance associated with changes

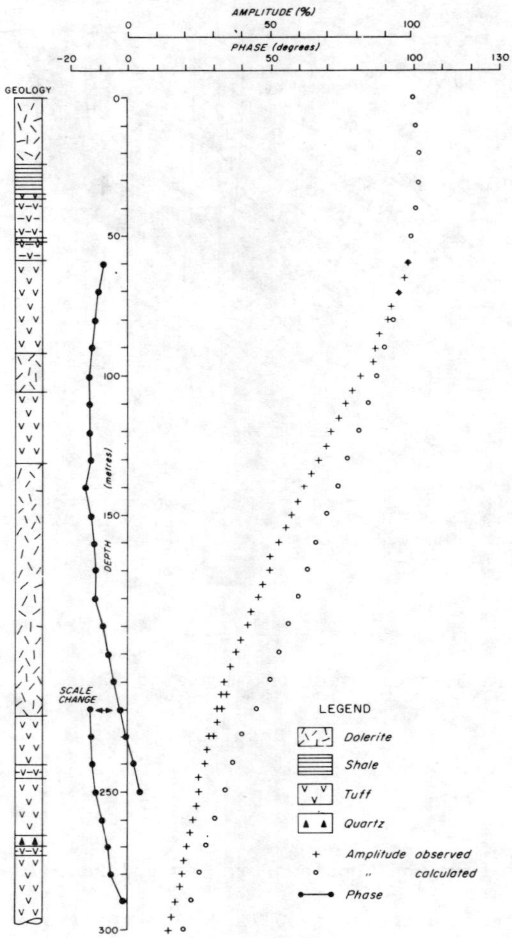

Figure 4. W55 survey.

in lithology. However, as only the coaxial field component was measured, the lack of response to changes in lithology might in part be attributed to poor coupling between the loop, probe and rocks. The gradual increase in phase with depth can be attributed to the scattering of the primary field in a conductive earth.

Drill hole W29

W29 represented an ideal test hole in that it was relatively free of mineralization and was thought to pass within 30 m of the orebody as shown in Fig. 3. This hole was surveyed with a variety of loop sizes and loop locations, as shown in Table 1.

TABLE 1 - Survey details

Drill Hole	Loop size (length of side in m)	Loop centre
W55	400	W55
W29		
Case 1	300	W29
Case 2	400	420 m E of W29 on traverse 9280mN
Case 3	200	W251
WE5	200	WE5
W251	200	W251
W254	200	W251
W276		
Case 1	300	W276
Case 2	100	W276

Case 1. Fig. 5 shows the EM results obtained for a 300 m by 300 m loop centred directly over W29. The observed amplitude curve correlates almost exactly with the theoretical amplitude curve and no anomalous response is evident. The phase curve shows a gradual increase opposite the projected

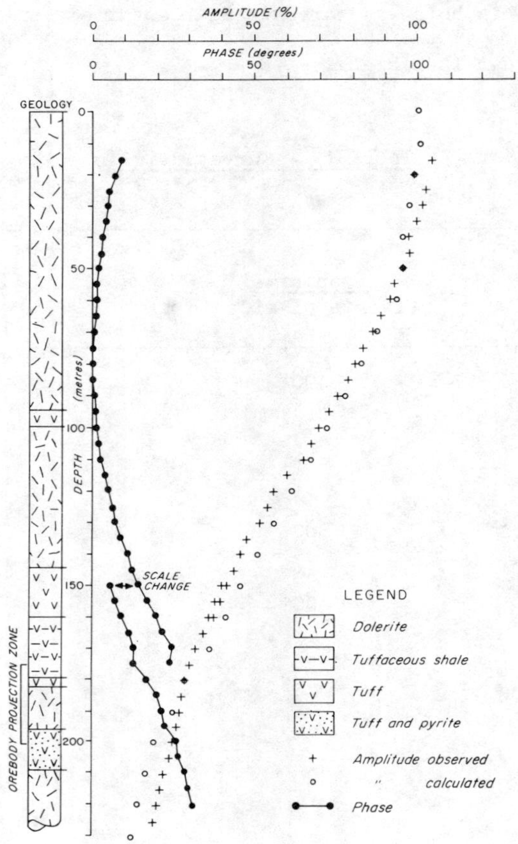

Figure 5. W29 Case 1.

position of the orebody, but the increase in phase is similar to that observed in W55, and is only marginally larger.

The poor response to the orebody in this case may be due to unfavourable EM coupling as indicated by the schematic representation of the primary and secondary field geometries shown in Fig. 6. Note that as the primary field parallels the orebody minimum induction occurs, and a weak secondary field results. Furthermore, as both secondary and primary fields have strong flux linkages with the receiving coil, the relative strength of the orebody response is small.

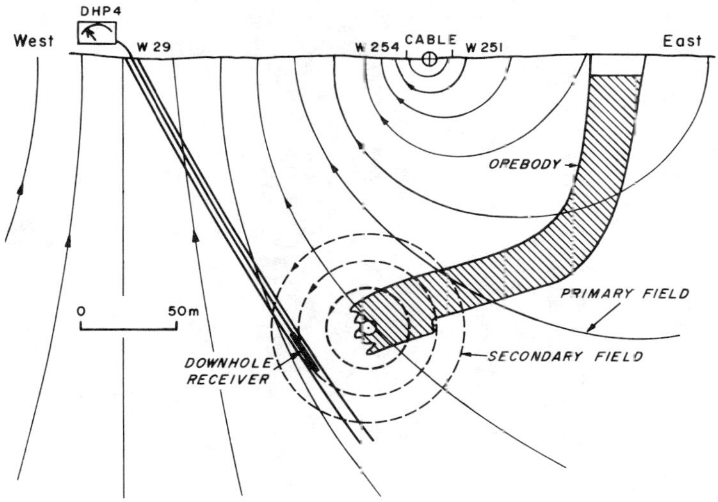

Figure 6. Field geometry, DDHW29 Case 1.

Case 2. Fig. 7 shows the EM results obtained for a 400 m by 400 m loop centred 420 m east of W29. This loop position was chosen because the primary field of a loop in this position would have greater flux linkage with the orebody.

Although the observed and theoretical amplitude curves differ in magnitude, their basic shape is similar. In particular, the small minimum observed in the amplitude curve at 160m is not far from the minimum in the theoretical curve at a depth of 170m. If the anomalous behaviour of the amplitude curve at 160m is attributed to a geometric effect, then no anomalous EM responses were observed in this case.

Case 3. Fig.8 shows the results obtained in W29 for a 200 x 200m loop centred on drill hole W251. A simplified view of the field geometries for this position as shown in Fig.9 suggests that optimum EM coupling would be obtained for the enhancement and detection of secondary magnetic fields. As the primary field is near vertical at the centre of the loop, maximum flux

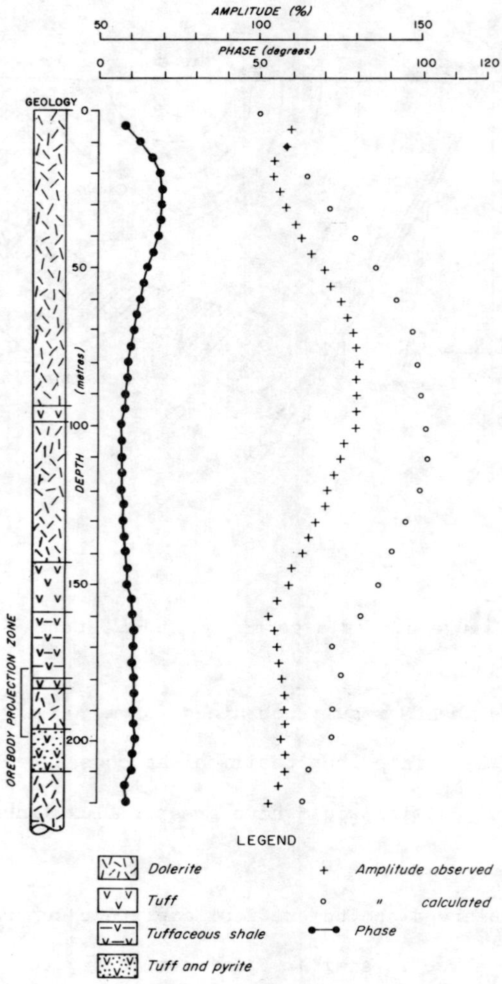

Figure 7. W29 Case 1.

linkage with the orebody and minimum linkage with the receiving coil will occur. Also secondary fields generated in the orebody will have maximum flux linkage with the receiving coil.

The theoretical amplitude curve shows a field reversal at 130m but there appears to be no corresponding reversal in the field data. This disparity may be attributed to distortions in the primary field caused by the orebody, or it may be that the reversal is not discernible at such low field intensities.

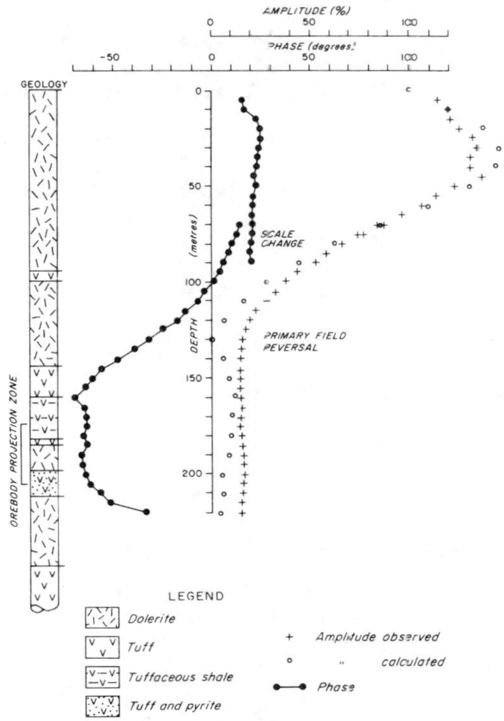

Figure 8. W29 Case 3.

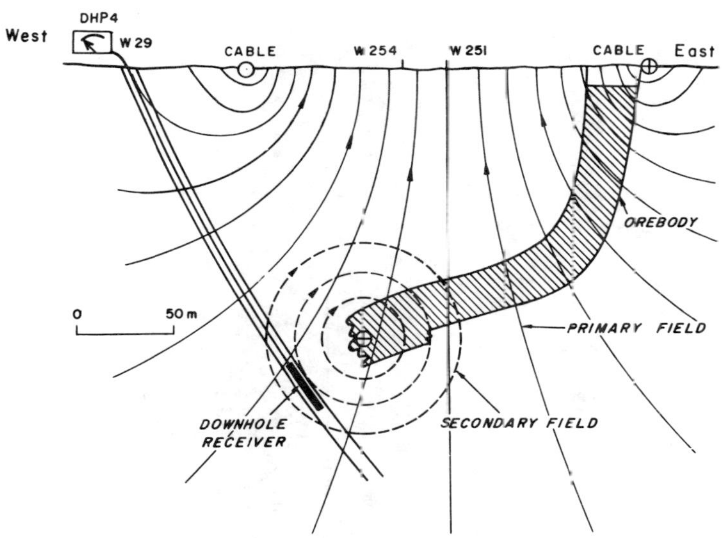

Figure 9. Field geometry, DDHW29 Case 3.

A substantial negative phase anomaly was recorded opposite the orebody projection zone accompanied by what appears to be a negligible amplitude response. The nature of this anomalous response suggests that it is produced by a poor electrical conductor and most probably reflects the presence of nearby disseminated mineralization rather than being a direct response from the massive ore intersected in drill holes W251 and W254. Nevertheless, this result illustrates the advantages of using a simplified free space approximation both as a guide for optimizing the field measuring technique and for purposes of interpretation.

Drill hole WE5

Fig.10 shows the EM response obtained with a 200m square loop centred on drill hole WE5. This drill hole intersected massive sulphides.

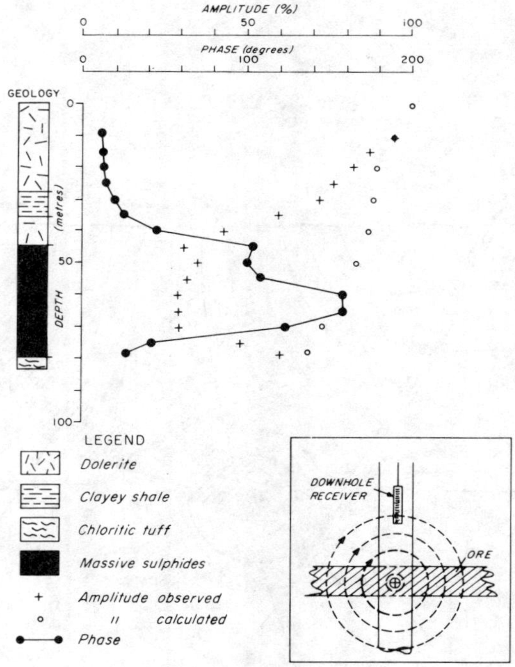

Figure 10. WE5 survey.

On passing through the sulphides, a phase anomaly of greater than 155° and a significant decrease in amplitude were recorded. The amplitude trough is attributed to attenuation of the primary field within the conductive sulphides. Assuming an average ore conductivity of 10 S/m (Spies, 1977) the sulphides are about 9 skin depths thick. However, total shielding has not occurred, owing to the steeply dipping attitude of the conductor and the straddle position of the energizing loop. The EM log does not penetrate far beyond the intersection zones, but a recovery of field strength is evident below 75m drill depth. The response illustrates an inherent disadvantage of this method of downhole measurements: with small loops intersected ore zones may attenuate the primary field, and effectively screen deeper conductors.

Anomalous amplitude and phase responses were recorded some 15 to 20m before the intersection. However, it is necessary to note that when the drill hole intersects the orebody the search coil would invariably be minimally coupled with any secondary field generated by the conductor (see insert, Fig.10). The response is not therefore truly indicative of the DHP4's optimum detection range.

Drill hole W251

Fig.11 shows the EM response obtained for a 200m square loop centred on drill hole W251. The drill hole intersected two minor shoots of massive sulphide at about 60m, and the main ore zone below 110m. The field curves show a typical amplitude trough in the major ore zone accompanied by a large phase anomaly of about 120°. No EM response was obtained on passing through the minor ore shoots, and no significant phase departure was observed further than 10 to 15m from the major ore intersection.

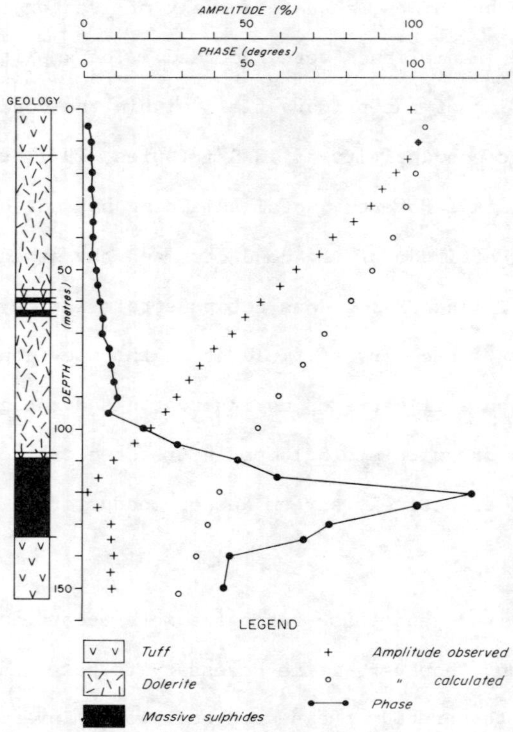

Figure 11. W251 survey.

Drill hole W254

Fig. 12 shows the EM response obtained in W254 for a 200m square loop centred on W251. This drill hole intersected massive sulphides. The observed amplitude curve is similar to the theoretical curve except near the ore. As in other intersection cases, the phase measurements respond to the orebody, but only 15 to 20m before the intersection.

Drill hole W276

Case 1 Fig. 13 shows the EM response obtained for a 300m square loop centred on W276, which intersected the orebody. The observed amplitude curve resembles the theoretical curve until the ore is intersected. Little or no

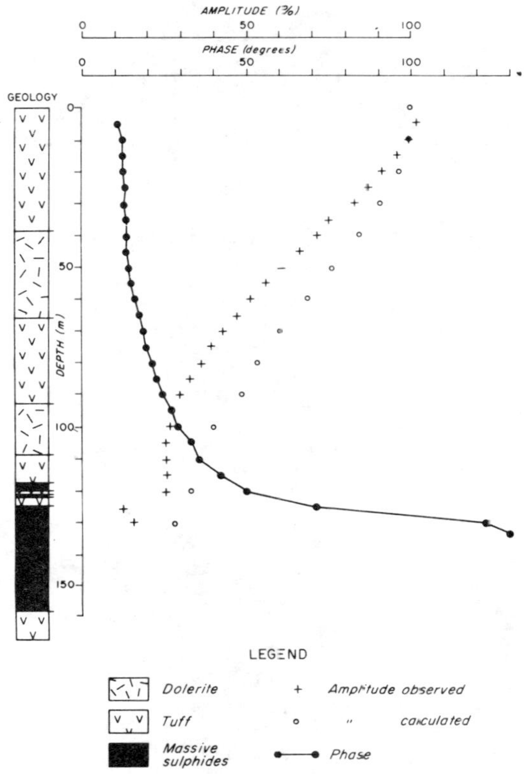

Figure 12. DDHW254 survey.

anomalous EM response to the orebody was recorded before the intersection depth. Within the ore zone, the EM response is characterised by a rapid decrease in amplitude and a large phase anomaly. Again assuming an average conductivity of 10 S/m, the ore intersection is some 9 skin depths thick. Unlike the intersection anomaly in hole WE5, there does not appear to be any recovery of the field strength below the ore zone, indicating that (in this case) total shielding has occurred. It should be noted that in such circumstances deeper conductors could go undetected with the DHP4 system.

Case 2 Fig. 14 shows the EM response obtained for a 100m square loop centred on W276. The theoretical and observed amplitude curves show good correlation and, as expected, a more rapid attenuation than was observed

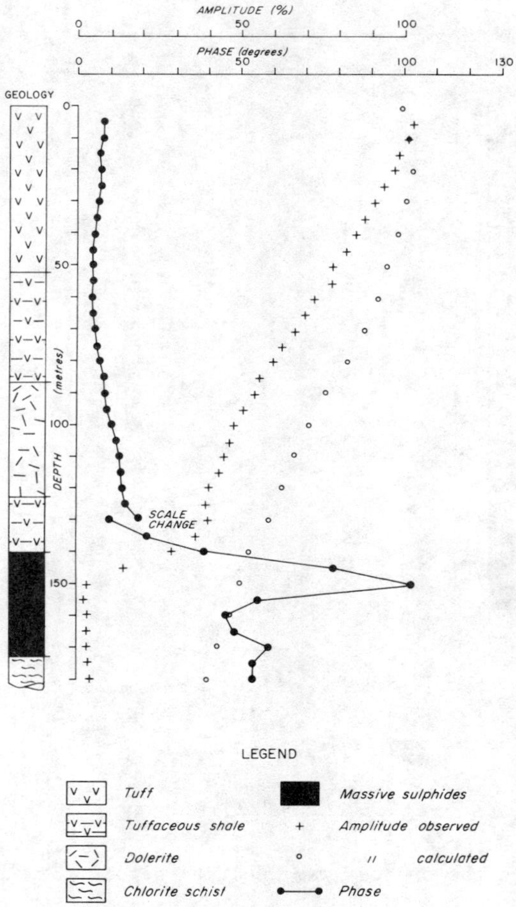

Figure 13. W276 Case 1.

with the 300m square loop. No anomalous amplitude response was recorded before the ore intersection, although the phase curve shows a significant increase some 50m from the intersection.

A significant feature, not evident using the 300m square loop, is the 180° phase reversal within the orebody at a drill depth of 150m. The theoretical curve suggests that no geometrical reversal is expected, and the reversal is therefore attributed to scattering of the primary field within the conductor.

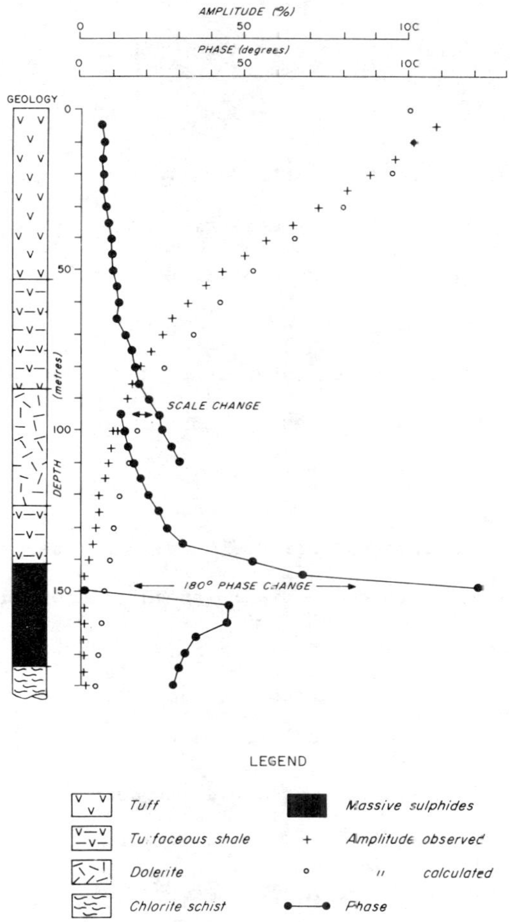

Figure 14. W276 Case 2.

Conclusions

The field tests with the DHP4 provided a guide to the effectiveness of fixed source drill hole EM surveys, and to the response characteristics of massive sulphide deposits. The tests showed that the Woodlawn orebody is an excellent EM target, but that the DHP4 is not a very effective prospecting tool.

With optimum EM coupling, the maximum range of detectability was shown to be of the order of 30m at Woodlawn. However, in less favourable geometrical situations, where the receiver to conductor coupling is minimal, or where there is minimal primary field linkage with the conductor, detectability range may be only a few metres.

The most serious limitation of the DHP4 instrument is the geometrical restrictions placed on detectability by measuring only a single component of the subsurface magnetic field. As the magnetic field to be observed is elliptically polarized, the coaxial magnetic component could be practically the same in either the presence or the absence of an orebody. A significant improvement in detectability could be expected by taking amplitude and phase measurements in three orthogonal directions.

Theoretical calculations established that some interpretational ambiguities may occur because of geometrical irregularities in the primary field. These difficulties may be aggravated by geometrical field reversals, which can occur when the drill hole extends beyond the transmitting loop circumference. Although unsitable for normalization purposes, the free space approximation was found in most downhole situations to be adequate for the identification of anomalous amplitude measurements. It was also shown that free space considerations could assist survey design, particularly in avoiding reversals and in optimising coupling. Where possible the use of large loops is recommended. Large loops ensure a uniform primary field distribution in the area of interest, and an adequate signal to noise ratio at the required depth.

The nature of the tests precludes any determination of maximum usable operating depth. Reliable measurements were obtained in W55 to a vertical depth of 290m. Subject to ambient noise levels, ground conductivities and transmitting power, there does not appear to be any reason why fixed source systems should not be used in extremely deep drill holes. Another limitation of this type of downhole EM method is the screening effect of intersected ore zones. This difficulty may be offset to some extent by using large transmitting loops.

In spite of the apparent ineffectiveness of the DHP4 system, the method does warrant further development. The fixed source systems have some intrinsic advantages over other downhole systems. These advantages include the possibility of obtaining directional information by suitable arrangement of the surface transmitting loop, the insensitivity of the method to small conductance targets, the utilization of complex surface instrumentation, and the small expense involved for lost probes.

Acknowledgement

This paper is published with the permission of the Director, Bureau of Mineral Resources, Canberra.

Reference

Spies, B.R., 1977 - Absolute electromagnetic scale modelling and its use in interpretation of TEM response. BMR J. Aust. Geol. & Geophysics, 2(2), 89-95.

AUTHOR INDEX

SUBJECT INDEX

AUTHOR, TITLE AND AFFILIATIONS

A. Agostini	Senior Geophysicist	Geological Survey of N.S.W.
G. Buselli	Research Officer	C.S.I.R.O.
Guido Cifali	Senior Geologist	Formerly Jododex Aust. Pty. Ltd.
Duncan Crone	President	Crone Geophysics Ltd., Ontario
Geoffrey O. Dickson	Chief Geophysicist	Formerly Newmont Pty. Ltd.
Douglas C. Fraser	President	Dighem Ltd., Toronto
R.J. Haren	Research Officer	C.S.I.R.O.
L.V. Hawkins	Professor	University of N.S.W.
I.G. Hone	Geophysicist	Bureau of Mineral Resources
M. Idnurm	Geophysicist	Bureau of Mineral Resources
R.J.G. Lewis	Lecturer	University of Tasmania
E.J. Malone	Exploration Manager	Jododex Aust. Pty. Ltd.
D.F. O'Driscoll	Chairman and Managing Director	Jododex Aust. Pty. Ltd.

R.D. Ogilvy	Geophysicist	Formerly Bureau of Mineral Resources
Rolf N. Pederson	Geophysicist	Formerly Geoterrex Pty. Ltd.
K.P. Quick	Geophysicist	Formerly Jododex Aust. Pty. Ltd.
D.F. Robson	Geophysicist	Formerly Bureau of Mineral Resources
R.J. Smith	Chief Geophysicist	C.R.A. Exploration Pty. Ltd.
B.R. Spies	Geophysicist	Bureau of Mineral Resources
R.J. Templeton	Geophysicist	Formerly Geological Survey of N.S.W.
E.D. Tyne	Geophysicist	Geological Survey of N.S.W.
K. Vozoff	Professor of Geophysics	Macquarie University
Robert J. Whiteley	Lecturer	University of N.S.W.
J.W. Williams	Geophysicist	Bureau of Mineral Resources
T.C. Wootton	Geophysicist	Formerly Geological Survey of N.S.W.
G.A. Young	Geophysicist	Bureau of Mineral Resources